Nuclear Decommissioning, Waste Management, and Environmental Site Remediation

Nuclear Decommissioning, Waste Management, and Environmental Site Remediation

Dr C. R. BAYLISS CEng FIEE
&
Dr K. F. LANGLEY CChem MRSC

ELSEVIER
BUTTERWORTH
HEINEMANN

Amsterdam   Boston   Heidelberg   London   New York   Oxford   Paris
San Diego   San Francisco   Singapore   Sydney   Tokyo

Butterworth-Heinemann
An Imprint of Elsevier
200 Wheeler Rd.
Burlington, Ma. 01803

A member of the Reed Elsevier plc group

First published 2003

British Library Cataloguing in Publication Data
A catalogue record for this book is available from the British Library.

Library of Congress Cataloguing in Publication Data

Bayliss, C. R. (Colin R.)
   Nuclear decommissioning, waste management, and environmental site remediation / C. R. Bayliss & K. F. Langley.
      p. cm.
   Includes index.
      ISBN-13: 978-0-7506-7744-8   ISBN-10: 0-7506-7744-9
      1. Nuclear facilities—Decommissioning. 2. Radioactive waste sites—Cleanup. 3. Hazardous waste site remediation. I. Langley, K. F. II. Title.

TK9152.2.B39 2003
621.48'3—dc21

ISBN-13: 978-0-7506-7744-8
ISBN-10: 0-7506-7744-9

2003045383

Transferred to Digital Printing 2010

Contents

Decommissioning
Chapter 3
Decommissioning — Introduction and Overview 41

Chapter 4
Typical Government Policy on Decommissioning 47

Chapter 5
The Transition from Operations to Decommissioning 53

Chapter 6
Reactor Decommissioning — The Safestore Concept 57

Chapter 7
Decommissioning PIE and Other Facilities 69

Chapter 8
Preparation of Documentation for Decommissioning 77

Chapter 9
Radiological Characterisation 83

Chapter 10
Decontamination Techniques 89

Chapter 11
Dismantling Techniques 99

Project and Program Management

Chapter 12
Site Environmental Restoration Program Management 113

Chapter 13
Project Investment Appraisal and Contract Strategy 127

Chapter 14
Hazard Reduction and Project Prioritisation   141

Chapter 15
Decommissioning Cost Estimating   153

Waste Management
Chapter 16
Waste Management — Introduction and Overview   161

Chapter 17
Waste Management Strategy   167

Chapter 18
Policy and Regulatory Aspects of Waste Management 177

Chapter 19
Management of Low Level Wastes (LLW) 193

Chapter 20
Management of Intermediate Level Wastes (ILW) 201

Chapter 21
Management of High Level Wastes (HLW) 221

Chapter 22
Transport 229

Chapter 25
Technologies for Remediating Contaminated Land   255

Site Environmental Remediation

Chapter 23
Site Remediation — Principles and Regulatory Aspects   241

Chapter 24
Characterisation of Contaminated Land   247

Appendices

Appendix 1
Country Specific Examples of Radioactive Waste Management Programs   263

Appendix 2
An Example of a Project Sanction Case — Repacking of Harwell Legacy Intermediate Level Wastes   285

Appendix 3
An Example of a Site Remediation Project — Dounreay Castle Ground Remediation   307

Appendix 4

Appendix 5
Elements and Isotopes   321

Index   327

About the authors

Colin Bayliss gained a first class honors degree in Electrical and Electronic Engineering at Nottingham University and went on to receive a PhD in Materials Science. He has worked for contractor, client, and consultancy companies on power construction projects both at home and overseas. This culminated in his work with Transmanche Link, the Channel Tunnel Main Contractors, where he rose from Principal Engineer at the early design phases to become the Fixed Equipment Engineering Director for that project. He joined United Kingdom Atomic Energy Authority (UKAEA) as Planning, Performance and Engineering Director in 1997, and has recently been appointed to the Main Board as Major Projects and Engineering Director at a time of increasing decommissioning work load.

Kevin Langley obtained his PhD in physical chemistry from Queen's University, Belfast. He has worked in universities and industry, both in the UK and Australia, before joining the UKAEA in 1978. Kevin has managed a broad range of projects on renewable energy technologies, nuclear waste retrieval, and nuclear fuel processing, including strategic studies. Since 1990, he has been assisting with the management of the decommissioning programs at the UKAEA's Harwell and other sites. He has led the Technical Services Group responsible for nuclear decommissioning, waste management, and site environmental remediation policy, and is currently Head of Southern Sites Projects Department of the Major Projects and Engineering Division.

Contributors

The preparation of a book covering such a wide range of topics within the field of nuclear decommissioning, waste management, and environmental remediation would not have been possible without contributions and advice from contractors, consultants, regulators, advisory bodies, academics, and colleagues. Indeed, encouragement for this book has come from all sectors across the nuclear industry. The preparation of a decommissioning and waste management postgraduate course at Birmingham University, England, by the authors and Professor Monty Charles gave the driver for preparation of this book. The names of the major contributors are listed below:

P. Barlow	BNFL
J. Blackmore	UKAEA
P. Booth	BNFL
E. Butcher	BNFL
M. Charles	Professor, Physics and Technology of Nuclear Reactors, Birmingham University
G. Coppins	UKAEA
F. Dennis	UKAEA
A. Dunne	BNFL
A. Eilbeck	BNFL
P. Fawcett	Springfields, BNFL
R. Francis	Nuclear Industries Directorate, Department of Trade and Industry (DTI)
A. Goldsmith	UKAEA
I. Gray	RM Consultants
E. Gunn	UKAEA
R. Guppy	UK Nirex Ltd.
G. Jessop	UKAEA
R. King	UKAEA
G. Linekar	UKAEA
J. Lloyd	UKAEA
P. Lock	UK Nirex Ltd.
R. Manning	UKAEA
D. Mathers	BNFL
J. Mathieson	UK Nirex Ltd.
P. Meddings	British Energy (BE)
S. Mobbs	National Radiological Protection Board (NRPB)
R. Nicol	UKAEA
G. Owen	British Energy (BE)
M. Pearl	UKAEA
G. Pugh	BNFL
C. Scales	BNFL
S. Tandy	Ministry of Defence (MoD)
F. Taylor	HSE Nuclear Installations Inspectorate (NII)
C. Williams	Environment Agency (EA)
M. Wise	UKAEA

Preface

This book covers the major topics likely to be encountered by nuclear decommissioning, waste management, and environmental site remediation technical engineers and managers engaged upon such international project works. Each chapter is self-contained and gives a useful practical introduction to each topic covered. The book is intended for graduate management or technician level staff, and bridges the gap between specialist university theoretical textbooks or scientific papers and detailed single topic references. It therefore provides, in a single reference text, a practical grounding in a wide range of nuclear site environmental restoration subjects. Civil nuclear decommissioning currently represents a £1bn per annum industry in the UK alone, with some 100 nuclear reactors now, or soon to be, decommissioned world-wide.

Although nuclear decommissioning is sometimes seen as the less glamorous end of a modern technology, it is this aspect of safe and secure restoration of redundant nuclear facilities which is currently the growth end of the nuclear industry sector. Therefore, the aim of this book is to assist staff in correctly approaching the huge challenge ahead to decommission redundant nuclear facilities and to restore such sites back for alternative use. Of particular interest are the chapters covering project appraisal, choosing the most appropriate decommissioning option, and the rigorous methodologies that may be adopted for seeking funding for site environmental remediation works. In addition, the book also covers modern approaches to appropriate contract strategies for nuclear decommissioning works.

Colin Bayliss & Kevin Langley

Foreword

This book is timely and should prove of value to a wide range of readers with interests in the nuclear industry.

The challenges today to those undertaking the decommissioning of redundant nuclear facilities and the associated site remediation are much greater than they will be in the future. Designers of modern nuclear equipment consider the requirements of their subsequent decommissioning and effect on the environment. Early plants were built rapidly and with little thought to their end-life state. The greatest current challenges involve early prototype reactors, fuel cycle plant and waste stores and silos, especially where malfunctions have occurred in the past.

Nuclear engineering is no longer a common course in universities and, until recently, few scientists and engineers joining the nuclear industry expected to work on, or had previous experience in, decommissioning and site restoration. Management of radioactive wastes is a more mature subject but, even so, is rarely found in a university curriculum. Hence, an educational text, based on a blend of theory and sound practical experience, is likely to prove invaluable to current practitioners in the fields covered: particularly to those who have recently joined, or are about to join, the increasing scientific and engineering effort over the next few decades.

The skills required by practitioners involve an understanding of radioactivity, proven techniques, cost estimating, safety, risk assessment, and project management; all of which are well covered in this book. However, success in dealing with the current challenges will require, in addition, considerable innovation to overcome some of the uncertainties and an ability to adapt concepts and techniques from other industries.

This book will be useful to clients, and contractors alike, as well as to others such as regulators, environmentalists, and government officials. Where significant uncertainties exist in decommissioning, site restoration and the ultimate disposal of radioactive wastes, and where the timescales involved in some tasks are relatively long, it is important that all parties involved have an understanding of the key principles, the methodologies and current best practice.

Roy Nelson OBE
January 2003

Chapter 1
Setting the Scene

1-1. Introduction

This chapter provides a background and understanding of the responsibilities of the different parties involved in policy and regulatory issues associated with decommissioning, waste management, and environmental site remediation. It briefly describes the history leading up to the current organisational arrangements within the UK and then goes on to describe the international scene. It introduces the subject of operational safety and environmental regulatory control regimes, together with those basic safety and environmental standards adopted throughout the world for the decommissioning and the safe storage and disposal of nuclear wastes.

1-2. The Evolution of the Current Organisational Arrangements in the UK

The key historical dates associated with the general development of nuclear fission are shown in Table 1-1.

Following the formation of the Atomic Energy Research Establishment in 1946, the UK Government recognised the need to coordinate the development of nuclear weapons, the potential for the development of a UK power program, and nuclear related research. The United Kingdom Atomic Energy Authority (UKAEA) was created for this task under the Atomic Energy Act in 1954, answerable to the Secretary of State for Trade and Industry. Following World War II, in 1948, the electricity generation, transmission, and distribution was nationalised by Government. Thereby, the stage was set for the development of the technology to achieve controlled nuclear power generation under the UKAEA and for station operation and power transmission and distribution by the Central Electricity Generating Board (CEGB).

What has followed has been the gradual break up of the key businesses involved into stand-alone entities and, where possible, a drive towards placement of these in the private sector. Figure 1-1 illustrates the general development of the nuclear industry in the UK since 1946.

The most significant developments have been:

1957   Academic research in nuclear physics divested and now with Government funded Research Council.

1971   British Nuclear Fuels Ltd (BNFL) created to exploit the provision of nuclear fuel cycle services. Initially, production of Magnox reactor fuel (Springfields, near Preston, Lancashire), reprocessing, and waste treatment (Sellafield/Windscale, West Cumbria).

1971   URENCO (Capenhurst) for fuel enrichment, Amersham International for medical isotope production, and later privatised in 1982, National Radiological Protection Board (NRPB) — all split off from UKAEA.

1973   Separation of nuclear weapons work into the Ministry of Defence (MoD), and formation of the Atomic Weapons Establishment (AWE).

1985   Recognition that a disposal route for nuclear waste was essential, and formation of the Nuclear Industries Radioactive Waste Executive (NIREX).

1986   Break up of the CEGB and creation of privatised power generation and rural electrification companies. Formation of Magnox for operation of older Magnox stations and British Energy (BE) for operating the Advanced Gas Reactors (AGRs) and Pressurised Water Reactor (PWR).

1995   Supply of site services divested to private industry.

1996   Divestment of consultancy and contracting services from UKAEA as AEA Technology. Later further split and nuclear elements purchased by SERCO and RWE Nukem in 2001.

2000   Work proceeding on possible privatisation of BNFL.

2001   Government White Paper — Managing the Nuclear Legacy — published proposing reorganisation of the management of all civil nuclear liabilities in the UK under a new Liabilities Management Authority.

Table 1-1. Some Key Historical Dates Associated with Development of Nuclear Fission

Year	Development
1938	Nuclear fission discovered
1941	US Government plans to develop the atomic bomb
1942	First experimental nuclear reactor
1945	Atomic bombs dropped on Japan (one uranium and one plutonium) and thereby ends WW II
1947	First UK reactor (British Experimental Pile 0 – BEP0) built at Harwell
1948	Electricity generation, transmission, and distribution nationalised
1955	First US nuclear powered submarine (*Nautilus*)
1956	Queen Elizabeth II opens first Magnox reactor (using naturally occurring U238) at Calder Hall for power generation into National Grid
1958	Campaign for Nuclear Disarmament (CND) formed
1960	First nuclear aircraft carrier (*Enterprise*)
1963	First Advanced Gas Cooled Reactor opened at Windscale (WAGR), Cumbria, for test purposes
1976	First commercial AGR reactor built at Hinkley Point, Somerset
1987	Decision reached to build first UK Pressurised Water Reactor (PWR) at Sizewell after long public enquiry

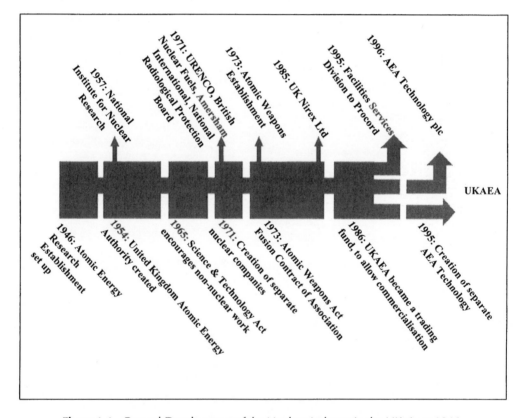

Figure 1-1. General Development of the Nuclear Industry in the UK since 1946.

The geographical location of the major civil nuclear facilities within the UK is illustrated in Figure 1-2. Table 1-2 lists the main nuclear power generating reactors (excluding research reactors and materials test reactors) built in the UK to date.

Nuclear energy currently supplies some 21% of the UK's electricity. The nuclear industry directly employs some 30,000 jobs, and twice as many such jobs indirectly, thereby contributing some £3bn to the UK Gross Domestic Product (GDP). Details of key British nuclear industry

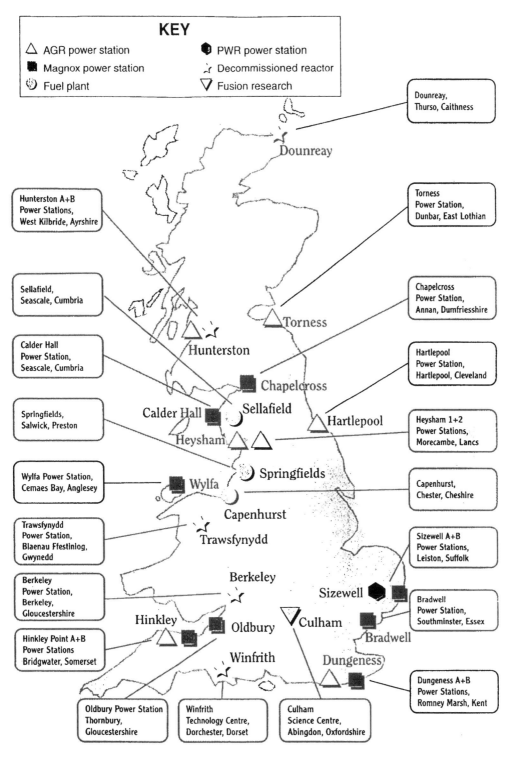

Figure 1-2. Key Nuclear Installations in the UK. Note: Currently the only fully decommissioned reactors in the UK are the Manchester/Liverpool reactor at Risley and the naval training reactor (Jason) at Greenwich in London. Other nuclear sites not shown on the map include Aldermaston, Amersham, Harwell, Imperial College London, and Risley near Manchester. There are partially decommissioned reactors at Dounreay, Windscale, Aldermaston, Harwell, Winfrith, Hunterston, and Berkeley.

Table 1-2. Main UK Nuclear Power Generating Reactors

Year	Location	Type[a]	Total capacity (MW)	Notes[b]
1956	Calder Hall	Magnox	200	Operational (2003)
1958	Chapelcross	Magnox	194	Operational (2005)
1962	Berkeley	Magnox	276	Decommissioning
1962	Bradwell	Magnox	246	Decommissioning
1964	Hunterston A	Magnox	300	Decommissioning
1965	Dungeness A	Magnox	440	Operational (2006)
1965	Hinkley Point A	Magnox	470	Decommissioning
1966	Trawsfynydd	Magnox	392	Decommissioning
1966	Sizewell A	Magnox	420	Operational (2006)
1967	Oldbury	Magnox	434	Operational (2008)
1968	Winfrith	SGHWR	100	Decommissioning
1971	Wylfa	Magnox	950	Operational (2010)
1975	Dounreay	PFR	270	Decommissioning
1976	Hinkley Point B	AGR	1200	Operational (2011)
1976	Hunterston B	AGR	1290	Operational (2011)
1984	Dungeness B	AGR	840	Operational (2008)
1984	Hartlepool	AGR	1180	Operational (2014)
1984	Heysham I	AGR	1200	Operational (2014)
1987	Torness	AGR	1564	Operational (2023)
1988	Heysham II	AGR	1344	Operational (2023)
1994	Sizewell B	PWR	1258	Operational (2035)

[a] Magnox: Natural uranium fuel contained in magnesium-based alloy; first generation gas-cooled UK nuclear reactors; PFR: Prototype Fast Reactor; SGHWR: Steam Generating Heavy Water Reactor; AGR: Advanced Gas cooled Reactor; PWR: Pressurised Water Reactor.

[b] Anticipated reactor closure dates shown in brackets.

companies are given in literature (Reports from Member Companies) obtainable from the British Nuclear Industry Forum (BNIF) [1].

1-3. A European Perspective on Nuclear Power Generation

There are currently some 150 operational reactors in Europe producing some 35% of the electricity demand. Deregulation leading to privatisation is seen as one of the greatest challenges to the economic base load supply of electricity as a whole (not only nuclear). Such large power stations require "up front" capital investment to cover the considerable construction and commissioning costs before a revenue stream from the sale of the electricity generated may start to flow. In addition, the necessarily strict regulation concerning reactor design and operations is an additional burden on profitability coupled with the relatively high end-of-life plant decommissioning costs. The current trend in nuclear power generation is, therefore, very much associated with plant lifetime extension (or clearer definition), up-rating, reduced shutdown times, etc., rather than new build. In addition, the resolution of nuclear waste management and the

development and siting of publicly acceptable waste disposal facilities, especially in densely populated Western European Countries, is considered to be key if a renaissance in nuclear power station build is to receive political support. Table 1-3 details the current nuclear power position in Europe, including Eastern Europe, Russia, and the Ukraine. Table 1-3 is intended to give the reader an insight into the scale of the nuclear power station decommissioning task ahead, and it may be noted that:

(a) France, Belgium, Bulgaria, and Lithuania have a high dependence on nuclear power (all over 50%).

(b) The political sensitivities, whilst fossil fuel prices remain relatively low, especially in Austria, Belgium, Germany, Italy, and Sweden.

1-4. An International Perspective on Radioactive Waste Management

1-4-1. Introduction

International Conventions associated with nuclear issues are detailed in Sections 1.5 and 1.6 of this Chapter, and

Table 1-3. Current Nuclear Power Station Position in Europe

Country	Power stations[a]	Total o/p (MWe)	Notes (inc. nuclear share of Country's power demand)
Armenia	1	376	26%; New plant under Government consideration
Austria	—	—	Halt in 1978 following referendum
Belgium	2 PWR in operation	5713	55%; No new build plans at present
	1 PWR in decommissioning		
Bulgaria	6	3538	60%
Czech Rep.	1 PWR in operation	1760	20%
	1 PWR under construction		
Finland	1 PWR	976	27%
	1 BWR	1680	
France	19 PWR	60045	76%
	1 FBR	235	
	1 PWR under construction		
Germany	13 PWR	14817	33%; Political pledge to phase out nuclear power
	6 BWR	6363	
Hungary	4	1840	38%
Italy	Under decommissioning	—	Moratorium in 1987 following referendum; New build terminated
Lithuania	2	2370	77%; One unit to be closed in 2005 and second by ~2009
The Netherlands	1 PWR	452	4%; No new build plans at present
Romania	1	655	8%; Second unit under construction but requires external funding
Russia	29	21242	13%; Intention to construct six new plants. Russia has three basic reactor designs: RBMK — graphite moderated, VVER 440, and VVER 1000 (similar to PWR)
Slovakia	5	2020	44%; Three new units under construction to offset older plant closures in 2008 and 2010
Slovenia	1	632	39%
Spain	7 PWR	6146	30%; No new build plans at present
	2 BWR	1491	
Sweden	1 BWR	1340	46%; 1980 decision by Parliament to phase out nuclear power by 2010. This is proving to be impracticable
	2 PWR	2710	
Switzerland	2 BWR	1435	40%; No new build plans at present
	2 PWR	1692	
UK	8 Magnox	3342	21% (note older Magnox stations produce 7% of UK electricity demand). Linkage between Kyoto Protocol and possible new nuclear plant recognised at a political level
	7 AGR	8592	
	1 PWR	1188	
Ukraine	14	12880	43%

[a] AGR: Advanced Gas cooled Reactor; BWR: Boiling Water Reactor; FBR: Fast Breeder Reactor; PWR: Pressurised Water Reactor.

Country specific examples associated with waste management given in Annex 1. Further details associated with decommissioning are given in Chapter 4, and with Waste Management in Chapter 18.

Radioactive waste is defined by the International Atomic Energy Authority (IAEA) [2] as "any material that contains or is contaminated by radionuclides or radioactivity levels greater than the exempted quantities established by the competent authorities and for which no use is foreseen." This definition may be open to slightly different interpretations, but in all cases the lack of future use means that the material may be treated as a waste and not as a resource. This is particularly important in the context of nuclear fuels. Irradiated used fuel or scraps and unirradiated residues may be treated, if economically sensible (by reprocessing, processing, or other recovery operations), as is the case in the UK and France for extraction of fissile material and potential reuse. Finland, USA, and

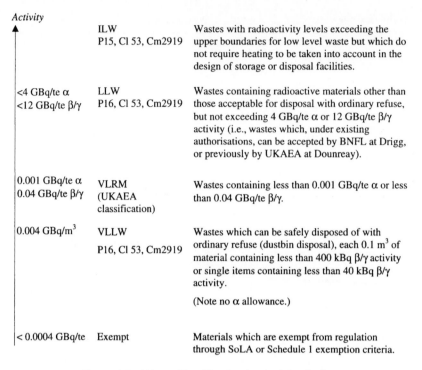

Figure 1-3. Waste Classification by Activity Scale.
Note   1 GBq/te $\equiv 10^9$ Bq/10^6 g $\equiv 10^3$ Bq/g $\therefore$ 0.0004 GBq/te $\equiv$ 0.4 Bq/g

Sweden, on the other hand, would regard such materials as waste.

If considered to be a waste then certainly interim storage and perhaps eventual disposal has to be considered. In the context of nuclear waste, disposal has the IAEA [2] definition of "the emplacement of waste in an approved specified facility ... without the intention of retrieval...." Again this is open to political and Government interpretation, with some countries requiring retrievability to be a postdisposal option. In the case of spent fuel or recovered fissile material, although perhaps without a current use, it might be considered folly to dispose irretrievably of what might be considered by future generations as a valuable energy resource. Put more simply, if security and regulatory measures could be adequately provided then such materials might be considered as part of a responsible Government's policy as "energy in the bank" and, therefore, stored for future use.

1-4-2. *General Nuclear Waste Classifications*

Nuclear waste arisings stem from:

• Nuclear power generation both for electricity and propulsion, with implications for the whole nuclear fuel cycle.
• Accidental arisings of waste from incidents such as Chernobyl.

• The military defence programmes of a number of countries.
• The application of radioactivity in medicine and industry.
• The enhancement of naturally occurring radionuclides (known by the acronym NORM) due to human activity.

There are no International standard nuclear waste definitions, although the IAEA [3] has proposed five categories, with each nation having its own classification system. In addition, the European Commission has proposed a classification system for application in Member States (see www.europa.eu.int /comm/environment /nuclear).

Figure 1-3 illustrates waste categorisations in terms of activity, and a general categorisation of waste types is given below:

(i) *Exempt Waste*: Radioactive waste that can be safely disposed of with ordinary waste.
(ii) *Transition Radioactive Waste*: Type of radioactive waste (mainly from medical uses) which will decay within the period of temporary storage and may be suitable for management outside of the regulatory controls subject to compliance with clearance levels.

In addition, large quantities of Very Low Radioactive Material (VLRM) may arise from decommissioning operations associated with the environmental restoration of contaminated ground.

In the UK, Transition Radioactive Waste is classified as Very Low Level Waste (VLLW) with the definition:

Wastes which can be safely disposed of with ordinary refuse (dust-bin disposal), each 0.1 m^3 of material containing less than 400 kBq beta/gamma activity or single items containing less than 40 kBq beta/gamma activity.

(iii) *Low Level Waste* (LLW): Consisting of trash and debris from routine operations and decommissioning. It is primarily low concentration beta/gamma contamination, but may include alpha contaminated material. It does not usually require particularly special handling, unless contaminated with alpha emitters.

In the UK, Low Level Waste (LLW) is defined as:

Wastes containing radioactive materials other than those acceptable for disposal with ordinary refuse, but not exceeding 4 GBq/te alpha or 12 GBq/te beta/gamma activity (i.e., wastes which for example, under existing authorisations, can be accepted by British Nuclear Fuels Ltd (BNFL) Drigg).

(iv) *Intermediate Level Waste* (ILW): Waste containing higher concentrations of beta/gamma contamination and sometimes alpha emitters. There is little heat output from this category of waste. These wastes usually require remote handling. Such waste originates from routine power station maintenance operations, for example used ion exchange resins and filter cartridges.

These examples may be further categorised as short-lived (usually meaning radionuclides with a half-life of less than 30 years).

Fuel reprocessing wastes, such as fuel canning materials, may also be classified as ILW but may contain long-lived species of radionuclides which may require eventual deep disposal. Some Countries, notably USA and Canada, do not specifically use the ILW classification category.

In addition, further ILW subdivision classification is possible based upon whether remote handling is necessary or whether the ILW is in solid or liquid form. Acronyms such as RHILW (Remote Handleable Intermediate Level Waste) or SILW (Solid Intermediate Level Waste) are, therefore, in common usage.

In the UK, Intermediate Level Waste (ILW) is classified by the definition:

Wastes with radioactivity levels exceeding the upper boundaries for low level wastes, but which do not require heating to be taken into account in the design of storage or disposal facilities. It should be noted from a historical perspective that a moratorium was placed on the disposal of LLW and short-lived ILW at sea under International law in 1983 (see Chapter 18).

(v) *High Level Waste* (HLW): Waste with such concentration of radionuclides that the generation of thermal power has to be considered during its storage and disposal. This heat generating waste mainly arises from the reprocessing of spent nuclear fuel. Although the amount of such HLW is relatively small in terms of volume, it contains the vast majority of the total activity of all radioactive wastes (over 95% in the UK).

Depending upon Government policy, spent fuel itself may also be considered as High Level Waste (or Intermediate Level Waste depending upon the degree of irradiation it has been subjected to in a reactor) if there is no further use for it or if there is no economic case for recovery of any useful fissile material from it.

Raffinates resulting from reprocessing contain high concentrations of beta/gamma emitting fission products and alpha emitting actinides. HLW is de facto a long-lived waste type and requires remote handling due to its high radiation levels. They may be immobilised in a suitable matrix such as glass or synthetic rock (synroc). Such highly active spent fuel or raffinate wastes are likely to have to be suitably stored for at least 50 years to allow the short-lived radionuclides to decay and heat generation to reduce prior to steps being taken for its eventual disposal.

Some countries choose to categorise alpha bearing waste separately. For example, in the USA, Transuranic Waste (TRU) is defined as: "... waste containing more than 100 nanocuries of alpha-emitting transuranic isotopes, with half lives greater than twenty years, per gramme of waste..." Such wastes arise from research laboratories, fuel fabrication and reprocessing plants.

In the UK, High Level Waste (HLW) is defined as:

Wastes in which the temperature may rise significantly as a result of their radioactivity, so that this factor has to be taken into account in designing storage or disposal facilities.

1-4-3. *Nuclear Waste Disposal Concepts*

International disposal practices may be summarised as shown in Table 1-4.

Near surface burial in shallow trenches or engineered structures is applicable to wastes that will decay to harmless levels over periods of 200–300 years. The

Table 1-4. Nuclear Waste Disposal Concepts

Waste classification	Short-lived	Long-lived
LLW	Shallow disposal	Deep disposal
ILW	Shallow disposal	Deep disposal
HLW	Not applicable	Deep disposal

design of the facility must be such as to provide an adequate means of isolation of the waste from, and prevent a return of radioactivity to, the environment over this sort of time frame. In addition, the design must, therefore, allow for monitoring of activity in the local area to give advance warning of any action that may need to be taken.

For solid LLW, the requirement for engineered barriers is minimal, and such wastes will undergo limited treatment, such as assay, compaction for waste volume minimisation purposes, and be packaged in drums or containers including immobilisation with a possible grout filling. Selected solid ILW may also be suitable for shallow burial if the beta/gamma emitters have short half lives (usually taken to be less than 30 years) and only very low concentrations of long lived alpha activity. Examples of such shallow burial facilities include:

- Drigg (West Cumbria, UK).
- Centre de l'Aube & Centre de la Manche (France).
- Rokkasho-Mura (Japan).
- El Cabril (Spain).

Facilities for mined disposal of LLW and short lived ILW to a depth of 100–500 m in hard rock or underground salt domes also exist in:

- Olkiluoto & Loviisa (Finland; hard rock).
- Forsmark (Sweden; hard rock).
- Morseleben (Germany; salt dome).
- Himdalen (Norway; mountain side rock).
- Wellenburg (Switzerland; proposal).

Deep disposal of long-lived wastes in stable geological formations is intended to reduce the risk of any return of radionuclides to the environment. A possible route for such migration to the biosphere is via groundwater pathways. Engineered and natural geological barriers are used to help prevent such movement. In addition, any inherent solubility of the waste is reduced by using suitable backfill material, well engineered waste packages, and by choosing a host geological formation in which water movement is extremely low. However, it is by no means an easy task to model the adequacy of the performance of such a nuclear waste deep repository over the long time scales (>1 million years) involved. Even more important than the satisfactory theoretical modeling of a deep disposal facility is the absolute need to gain public confidence in the adequacy of the design and approval processes. Site

specific examples of deep waste repositories include:

- WIPP (New Mexico, USA).
- Yucca Mountain (Nevada, USA).
- Gorleben (Germany).

Because of the complexity of making a suitable social, technical, and economic case for such deep waste disposal, Underground Research Laboratories (URLs) have been proposed or constructed so as to carry out full scale tests on the geology of either a preferred site or generic site. Such URL facilities include:

- Bure (France).
- Yucca Mountain (Nevada, USA).
- Onkalo (Finland; proposed).
- Gorleben (Germany).
- Wellenberg (Switzerland; proposed).
- Mol (Belgium; financed by European Community).
- Aspo (Sweden).
- Grimsel & Mont Terri (Switzerland).
- Whiteshell (Canada; now closed).
- Sellafield (UK; abandoned after public enquiry).
- Tono & Honorobe (Japan; existing and proposed, respectively).

The principal US nuclear decommissioning sites are described and illustrated in Appendix 1, Section A1.12.

1-4-4. *Management and Funding Arrangements*

Policy guidance on management approaches to radioactive waste management are included in IAEA documentation [4]. The State or Government, its independent Regulators for nuclear decommissioning, waste management and environmental site restoration, and the Waste Producers themselves all have responsibilities in what is sometimes referred to as the "classical triangle" principle. The arrangements should all be underpinned by clear Government policy.

International examples of the "classical triangle" (see Figure 1-4) approach are mostly found in Europe. However, at the working level, there are certainly differences in respect of

- waste treatment and conditioning,
- the speed of decommissioning and the time value of money,
- transport,
- storage, and
- site clearance activity levels.

One key funding principle is that of the "polluter pays." The polluter pays principle means that those who are responsible for pollution should face the costs of preventing the pollution or minimising the environmental

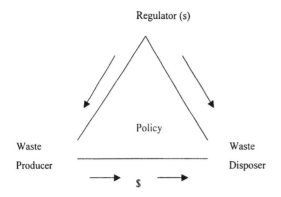

Figure 1-4. The "classical triangle."

damage. For example, there is provision under the UK Environment Act 1995 to recover clean up costs following an incident. The polluter pays principle is used to justify charging for regulatory permits. For example, in the UK, the nuclear site licence holder or power station operator pays a contribution towards the costs of running the Government Regulatory organisation.

Following this principle is intended to ensure that waste producers provide adequate financial provision and resources for the eventual safe and secure clean up programs for the wastes arising from their activities. In other words, the burden of the clean-up should best fall on those who have benefited from the activities associated with its production. Generic solutions for financing decommissioning, waste management, and site remediation issues associated with who pays and how they should pay include the following options:

- Waste producers pay directly through a tariff mechanism to a decommissioning/waste disposal/site restoration organisation.
- Electricity producers pay, through payments into a fund from levies on electricity generation, and then onto a decommissioning/waste disposal/site restoration organisation.
- Government or a third party pay, through subsidies to a decommissioning/waste disposal/site restoration organisation.

There is a general international consensus that all such decommissioning, waste management, and environmental site restoration liabilities should be identified, reported, and reviewed periodically, and that there should be mechanisms in place to meet these liabilities when they arise. In most cases, this involves the build-up of a fund to cover future costs. These funds are sometimes segregated and managed separately, either by the waste producer, the decommissioning organisation, the Government, or by independent fund managers. Cash within the funds

is retained in low risk investments such as Government bonds. More information is contained in NEA/OECD [5] and www.europa.eu.int [6] (see Table 1-5).

1-4-5. *Multinational Radioactive Waste Facilities*

International cooperation for the development of multinational facilities for the disposal of nuclear waste is a controversial subject. The IAEA have produced a document on the technical, institutional, and economic factors important for developing a multinational radioactive waste repository [7]. The issues identified include:

- Legal aspects (for example the need for harmonisation of Standards).
- Safety principles such as safety criteria to be adopted, risk assessment, intergovernmental equity, licensing, etc.
- Technical issues such as inventories and assay requirements, waste acceptance criteria, conditioning and interim storage requirements, transport, expertise availability, mixed wastes.
- Costs and liabilities.
- Institutional aspects and political continuity.
- Waste ownership.
- Ethical aspects.
- Public acceptance.
- R & D.
- Safeguards.

Note that "Safeguards" has a special meaning in this respect associated with international regulation and responsibilities for the inventory, safekeeping, and movements of fissile material. The high ratio of fixed-to-variable costs for the work required (not only to build and operate such a deep waste repository but also for the work required to receive the necessary permissions) ensures that economies of scale are applicable to a multinational nuclear waste repository. However, greater transport distances would be involved for a common multinational facility. Therefore, whilst the transport of nuclear materials is demonstrably safe there is still a huge hurdle to overcome so as to gain public acceptability that such transports will not have any significant impact on public health.

1-5. International Regulation and Collaboration

1-5-1. *The International Atomic Energy Agency (IAEA)*

The IAEA was established by the United Nations in 1957 to ensure world cooperation for the peaceful use of nuclear

Table 1-5. International Waste Management Organisations and Practices

Country	Agency	Treatment and conditioning	Transport	Storage	Disposal	Useful websites
Australia	Nat. Waste Repository Project — Dept. Science and Resources					www.ist.gov.au/
Belgium	ONDRAF/ NIRAS	ONDRAF in parallel with waste producers	ONDRAF	ONDRAF	ONDRAF	www.nirond.be. http:// hades.sckcen.be
Canada	Govt. policy being formulated	Waste producers	Waste producers	Waste producers	None, but AECL undertaking R&D on disposal and the waste producers have signed a Memorandum of Understanding to create a new agency	www.aecl.ca www.ontariopower generation.com
Czech Rep.	RAWRA					www.surao.cz/english/ indexen.html
Finland	Posiva Oy	Waste producers	N/a	Utilities	Posiva for spent fuel; utilities for ILW & LLW	www.posiva.fi www.tvo.fi www.ivo.fi
France	ANDRA	Waste producers and ANDRA for small producers	ANDRA (partially)	By industry	ANDRA	www.andra.fr
Germany	BfS (subcontracted to DBE)	Waste producers	Performed by industry after permit from BfS	By industry and/or federal centres	BfS (subcontracted to DBE)	www.bfs.de www.dbe.de
Italy	NUCLECO, ENEA and SOGIN undertake some functions	Waste producers	Commercial operators	NUCLECO	No decision on disposal taken. LLW & ILW facility being sought by ENEA	www.casaccia.enea.it/ taskforce/
Korea	None as yet	Waste producers	Industry	Waste producers	None as yet. KAERI and NETEC for R&D	www.kaeri.re.kr

Continued

Table 1-5. (continued)

Country	Agency	Treatment and conditioning	Transport	Storage	Disposal	Useful websites
Japan	New HLW organisation set up in October 2000. STA responsible for regulation	Waste producers	By industry	Waste producers	JNFL (LLW) at Rokkasho Mura. A new organisation created in October 2000 to look after HLW disposal. No website as yet but see recommended sites.	www.numo.or.jp www.infl.co.jp www.miti.go.jp www.sta.go.jp
The Netherlands	COVRA	COVRA (for low and medium level waste)	COVRA (for low and medium level waste)	COVRA for all waste types	Decision for disposal route to be taken this century	
Slovenia	Agency RAO					www.arao.si
Spain	ENRESA	Waste producers and ENRSA (in particular cases)	ENRESA	ENRESA	ENRESA	www.enresa.es
Sweden	SKB	Waste producers	SKB	SKB	SKB	www.skb.se
Switzerland	NAGRA	Waste producers		ZWILAG		www.nagra.ch
Taiwan	Fuel cycle and materials administration FCMA - regulator	Taipower	Industry	Waste producers	Taipower operates LLW facility on Lan Yu Island. FCMA for HLW disposal strategy	www.fcma.aec.gov.tw www.taipower.co.tw
UK	UK Nirex Ltd.	Waste producers	Industry	Waste producers (nuclear industry)	(i) BNFL — for LLW at Drigg (ii) UKAEA for LLW at Dounreay (iii) UK Nirex Ltd for ILW and some long lived alpha LLW (no current UK facility) (iv) No HLW facility — interim storage of vitrified HLW currently by BNFL	www.bnfl.co.uk www.ukaea.org www.nirex.co.uk

Continued

Table 1-5. (continued)

Country	Agency	Treatment and conditioning	Transport	Storage	Disposal	Useful websites
US	US DOE • OCRWM for HLW • EM State Compacts for LLW	Waste producers	Industry	Waste producers	US DOE OCRWM at Yucca Mountain US DOE EM at WIPP for TRU State compacts for LLW	www.rw.doe.gov www.em.doe.gov/dnfsbrpt/ www.wipp.carlsbad.nm.us www.envirocareurah.com/ www.ymp.gov/

energy [8]. It has some 113 member countries and is responsible for the prevention of the diversion of nuclear materials to weapons production. The IAEA has also been responsible for the development of safety guidelines associated with all stages of the nuclear lifecycle. These are set out in a series of color-coded documents. The guidance and regulations do no have a legal jurisdiction, but member countries usually endeavor to comply with IAEA recommendations.

1-5-2. *International Commission on Radiological Protection (ICRP)*

The need for adequate radiological protection dates back to the early years of the use of radiation and radioactive materials for medical purposes. The ICRP has published universal recommendations on the effects of radiation exposure on health since 1928, and these are regularly updated. Chapter 2 specifically covers this subject, as does www.icrp.org [9].

1-5-3. *The OECD Nuclear Energy Agency (OECD NEA)*

The NEA is an Agency of the OECD [10]. Membership currently consists of all European Union Member States, as well as Australia, Canada, the Czech Republic, Hungary, Iceland, Japan, the Republic of Korea, Mexico, New Zealand, Norway, Poland, Switzerland, Turkey, and the USA. The primary objective of the NEA is to promote cooperation between participating countries in the development of nuclear power as a safe, environmentally acceptable, and economic energy source. It does

this by:

• Encouraging harmonisation of national regulatory policies on the safety of nuclear installations, protection of man against ionising radiation, preservation of the environment, radioactive waste management, and nuclear third party liability and insurance;
• Assessing the contribution of nuclear power to overall energy supply by keeping under review the technical and economic aspects of nuclear power growth;
• Developing exchanges of scientific and technical information, particularly through participation in common services;
• Ensuring that the appropriate technical and economic studies on nuclear energy development and the fuel cycle are carried out; and
• Setting up international research and development programmes and joint undertakings.

The NEA works in close collaboration with the IAEA and other international nuclear organisations to help achieve these objectives.

1-5-4. *The European Commission*

Recommendations made by the ICRP, IAEA, and OECD NEA form the basis of specific Community Directives. The principles, standards, and requirements relating to nuclear and associated environmental matters in all Member States of the European Union (EU) [11] are based upon the 1957 Treaty of the European Atomic Energy Community (Euratom), the 1957 Treaty of the European Economic Community (EEC), and the single European Act of 1987.

An overview of the nuclear decommissioning and radioactive waste policy, advice, regulation, and

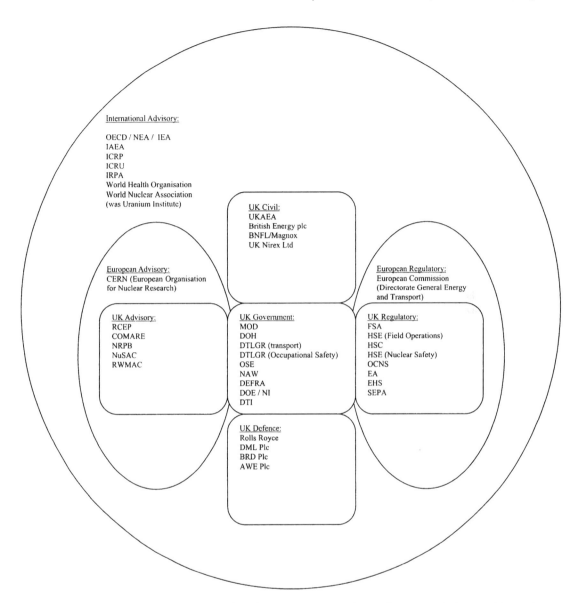

International Advisory:

OECD / NEA / IEA
IAEA
ICRP
ICRU
IRPA
World Health Organisation
World Nuclear Association
(was Uranium Institute)

UK Civil:
UKAEA
British Energy plc
BNFL/Magnox
UK Nirex Ltd

European Advisory:
CERN (European Organisation
for Nuclear Research)

European Regulatory:
European Commission
(Directorate General Energy
and Transport)

UK Advisory:
RCEP
COMARE
NRPB
NuSAC
RWMAC

UK Government:
MOD
DOH
DTLGR (transport)
DTLGR (Occupational Safety)
OSE
NAW
DEFRA
DOE / NI
DTI

UK Regulatory:
FSA
HSE (Field Operations)
HSC
HSE (Nuclear Safety)
OCNS
EA
EHS
SEPA

UK Defence:
Rolls Royce
DML Plc
BRD Plc
AWE Plc

Figure 1-5. Nuclear Decommissioning and Radioactive Waste: Policy, Advice, Regulation, and Operation in the UK, Europe, and Worldwide.

operation in the UK, Europe, and Worldwide is given in Figure 1-5.

1-6. The Kyoto Protocol and OSPAR (Oslo Paris Convention)

1-6-1. *The Kyoto Protocol*

The Kyoto Protocol, which legally binds industrialised countries to cut their emissions of greenhouse gases (mostly pollutants caused by burning coal, oil, and other hydrocarbon fuels), was signed in 1997. The protocol has been adopted by 159 countries and sets aggregate reduction targets of some 5.2% targets from 1990 levels during the years 2008–2012. The European Union heads the group that is required to make an 8% cut. The US and Japan had initially agreed to reduce emissions by 7 and 6%, respectively.

This is seen as a major part of the developed world's response to global warming mechanisms and, therefore, places emphasis on future electricity generation from nuclear and renewable energy sources, together with improvements in energy efficiency and savings. Pronuclear groups say that, without nuclear power, the EU will not meet its Kyoto commitments to reduce CO_2 levels. Antinuclear groups say that the gap can be filled with renewable energy sources (primarily wind power). However, before any resurgence in nuclear power generation is likely (in Western Europe at least), it will be necessary to demonstrate the safe and secure, environmentally acceptable, cost effective, and publicly acceptable decommissioning, waste management, and site environmental restoration of redundant nuclear facilities, including a long term sustainable solution to nuclear waste disposal. In the UK, aging Magnox reactors are coming to the end of their useful life, and clear policy decisions will have to be made to allow future stations to be built or alternative energy sources to be found.

Flexibility mechanisms have been built into the Kyoto Protocol for those countries with emission limitations or reduction commitments involving:

- *Bubbles*: to allow grouping of developed nations to pool their emission reduction targets and distribute necessary measures internally.
- *Tangible emission permits*: to allow developed nations with high compliance costs to buy permits from those countries with lower costs.
- *Joint implementation*: such that projects may be funded completely or partially by one developed country with credits for reducing emissions to be shared between participants.

In addition, a clean development mechanism has been initiated such that host countries, which do not have emission limitations or reduction targets, may generate emission reduction credits as a result of their endeavors, and these credits or permits may then be used in circulation amongst developed countries.

The European Union position is broadly that each nation should achieve the major part of its emission reduction targets by reducing emissions from its own industries. International trading of permits, creation of credits within the clean development mechanism, and the use of carbon sinks (such as forestry) should only be considered as top-up measures. The US and others have argued for a more unrestricted use of the Kyoto mechanisms and carbon sinks.

EU countries are reasonably close to their Kyoto targets — Britain thanks to its "dash for gas" as a replacement for coal fired peak electricity generation in the late 1980s and early 1990s, and Germany thanks to its closure of many inefficient Eastern area polluting industries. The rest of the EU is hardly more likely to meet the Kyoto targets than the US — in France the perverse reason being

its already high reliance on clean nuclear power generation making further reductions more difficult. The US would like developing nations such as China and India to be included.

1-6-2. *OSPAR (Oslo/Paris) Convention*

The Convention for the Protection of the Marine Environment of the North-East Atlantic ("OSPAR Convention") was opened for signature in 1992 and entered into force on 25 March 1998. Contracting parties agreed to a strategy for radioactive substances. The objective is to prevent pollution of the marine area, as defined under the Convention, from ionising radiation through progressive and substantial reductions of discharges, emissions, and losses of radioactive substances. The ultimate aim is to achieve concentrations in the environment near to background levels for naturally occurring radioactive substances and close to zero for artificial species (see Figure 1-6).

It should be noted that responsibility for the production of the latest Government Policy White Paper on Radioactive Waste Management (Cm 2919, 1995) was prepared by the Department of the Environment (now DEFRA). Further, that it is the DTI that is responsible for nuclear power policy and its Nuclear Industries Directorate (NID) for the surety and probity of expenditure by UKAEA and BNFL for their civil nuclear clean-up programs in the UK. In essence, there is a clear distinction between the regulatory roles of the Environment Agencies (EA and SEPA) and the Nuclear Installations Inspectorate (NII), as described in Chapter 18.

The International Atomic Energy Authority (IAEA) has set out a system for the establishment of a radioactive waste management system, and Government Policy essentially mirrors this framework listed here:

- identification of the parties in the different steps of radioactive waste management, including waste generators and their responsibilities;
- a rational set of safety, radiological, and environmental protection objectives from which standards and criteria may be derived within the regulatory system;
- identification of existing and anticipated radioactive wastes, including their location, radionuclide content and other physical and chemical parameters;
- control of radioactive waste generation;
- identification of available methods and facilities to process, store, and dispose of radioactive waste on an appropriate timescale;
- taking appropriately into account inter-dependencies among all steps in radioactive waste generation and management;
- appropriate research and development to support the operational and regulatory needs; and

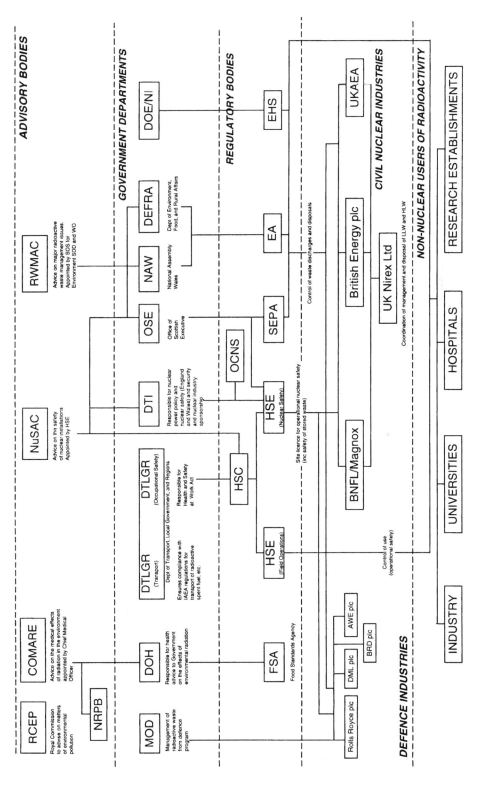

Figure 1-6. Nuclear Decommissioning and Radioactive Waste: Policy, Advice, Regulation, and Operation in the UK.

• the funding structure and the allocation of resources that are essential for the radioactive waste management, including decommissioning and, where appropriate, maintenance of repositories and post-closure surveillance.

1-7. Waste Production

First, it is necessary to put nuclear waste production into perspective. Table 1-6 shows the relative magnitude of nuclear waste produced per year compared to domestic waste volumes in the UK.

In volumetric terms, radioactive waste therefore represents only a tiny fraction of the total wastes, and less than 1% of the toxic wastes, produced each year in the UK. If an average family is considered, then they may consume some 8 MWh of electricity in the UK per annum. This, in turn, would represent a production of ~1 liter of all nuclear waste categories if all their electricity was derived from nuclear power generation. In contrast, if their power was totally supplied from a conventional coal-fired power station, then this would produce some 400 liters ash/toxic waste and some 4,000,000 liters of CO_2 greenhouse gas.

The older types of nuclear reactor were not only relatively inefficient in terms of modern nuclear power plant fuel usage, but they also produced higher quantities of nuclear wastes of a type that are difficult to deal with. Table 1-7 gives a comparison of UK reactor types by efficiency and waste production.

In compliance with IAEA recommendations, the UK Government Department of Transport, Local Government, and Regions together with UK Nirex Ltd. produces an inventory of nuclear wastes which is updated every 4 years. This produces a snapshot in time of current stocks and projects future arisings. It details the chemical and physical parameters and quantities of the wastes. The wastes are broken down into waste streams, and the inventory covers the range of nuclear wastes from LLW to HLW. Such information is an essential input for scoping

Table 1-6. Annual Radioactive Waste Production in the UK Compared With Normal Domestic Waste

Radwaste	vs	Normal Domestic Waste
50,000 m^3/year total^a	vs	40,000,000 m^3/year total of which some 3,000,000 m^3/year is poisonous solid waste and which does not necessarily decay in toxicity over time

^{a}Of which: (i) Some 90% is Low Level Waste ~45,000 m^3/year (LLW); (ii) Remainder is largely Intermediate Level Waste ~5,000 m^3/year (ILW); and (iii) only a small fraction is High Level Waste (HLW), but highly active and long-lived.

Table 1-7. Comparison of UK Reactor Types in Terms of Waste Production and Efficiency

	Magnox	AGR	PWR
Waste volume per GW year (m^3 conditioned)	1200	520	70
Power output per tonne of fuel	5,000 MW days per te	25,000 MW days per te	45,000 MW days per te

Note: te — metric tonnes.

the technical nature of any future proposed deep waste repository.

Figures 1-7a and 1-7b are histograms showing the projected UK waste volumes in unconditioned and conditioned form to 2030. Note the major advantage achieved in reduced volumes for disposal from size reduction associated with LLW arisings.

Figure 1-8 shows the projected cumulative build up of UK LLW and ILW stocks over time. A further increase in waste arisings beyond 2060 is projected to occur from reactor Stage III decommissioning wastes (primarily LLW rubble and contaminated ground).

1-8. Acronyms and Abbreviations

See also *A Dictionary of Nuclear Power and Waste Management* [12]. Some useful terms not specifically described so far in this chapter but which are covered elsewhere in this book are:

Actinide: An element following Actinium (Ac, Atomic Number 89) to Lawrencium (Lr, Atomic Number 103) in the Periodic Table. Many of the Actinides are long-lived alpha-emitters, examples are uranium and plutonium.

Activation Product: Radionuclides induced by the absorption of radiation, usually neutrons. Some significant activation products are Cobalt-60 (derived from iron-60) and tritium (from water, especially deuterium in heavy water, and lithium in concrete). Plutonium-239 is produced from Uranium-239 — refer to Annex 5.

AGR: Advanced Gas-cooled Reactor. The second generation of nuclear reactors built in the UK. Using slightly enriched uranium dioxide clad in stainless steel as fuel and operates at much higher temperatures than the earlier Magnox plants from which the design was developed.

ALARA: As Low as Reasonably Achievable. Radiological doses or risks from a source of exposure are as low as reasonably achievable when they are consistent with the relevant dose or target standard and have been reduced to a level that represents a balance between

(a)

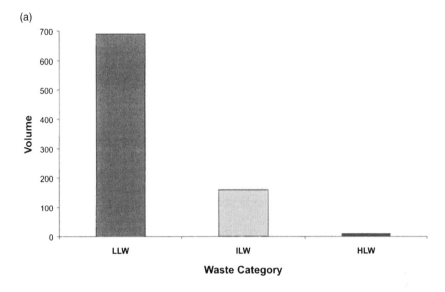

(b)

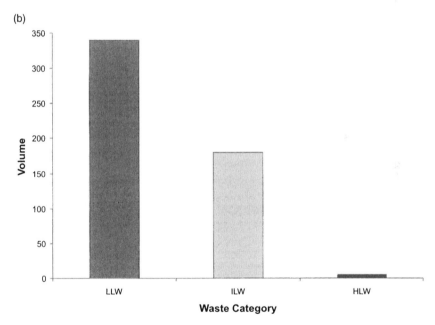

Figure 1-7. UK Waste Volumes to 2030, (a) Unconditioned: 1000 m³, (b) Conditioned: 1000 m³.

radiological and other factors, including social and economic factors. The level of protection may then be said to be optimised.

ALARP: As Low as Reasonably Practicable. To satisfy the ALARP principle, measures necessary to reduce risk are undertaken until or unless the cost of these measures, whether in money, time or trouble, is disproportionate to the reduction in risk.

BAT: Best Available Techniques.

BATNEEC: Best Available Techniques Not Entailing Excessive Cost (see Chapter 18).

BNFL: British Nuclear Fuels Ltd.

BPEO: Best Practical Environmental Option. The out-come of a systematic consultative and decision-making procedure which emphasises the protection and conservation of the environment across land, air,

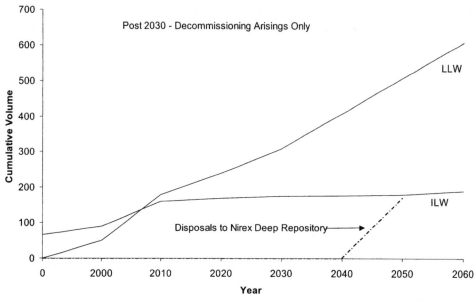

Figure 1-8. UK ILW and LLW Cumulative Arisings (Conditioned: 1000 m^3).

and water. The BPEO procedure establishes, for a given set of objectives, the option that provides the most benefits or least damage to the environment as a whole, at acceptable cost, in the long term as well as the short term.

BPM: Best Practical Means. Within a particular waste management option, the BPM is that level of management and engineering control that minimises as far as practicable, the release of radioactivity into the environment whilst taking account of a wider range of factors, including cost effectiveness, technological status, operational safety, social and environmental factors. In determining whether a particular aspect of the proposal represents BPM, the Inspectorates will not require the applicant to incur expenditure, whether in money, time or trouble, which is disproportionate to the benefits likely to be derived.

Bq: Becquerel. The standard international unit of radioactivity equal to one radioactive transformation per second (also GBq, kBq, etc.). The unit itself tells you nothing about the radiation associated with the particular transformation and, therefore, is not a direct measure of possible harm arising from the radioactivity as is the effective dose. The unit, when divided by weight (e.g., GBq/te) provides a measure of the concentration of the radioactivity.

BSS: Basic Safety Standards. A European Directive as applicable to exposure levels from nuclear wastes in terms of dose.

COMARE: Committee on Medical Aspects of Radiation in the Environment. To assess and give advice to Government on health effects of natural and man-made radiation in the environment and adequacy of data and need for further research.

Critical Group: For a given source, the critical group comprises of those members of the public whose exposure is reasonably homogeneous and is typical of people receiving the highest dose from the given source.

Criticality: The accidental (or sought after occurrence when in the core of a nuclear reactor) of a self sustaining fission chain reaction in fissile material. Hence, a "criticality incident."

Decommissioning: The process whereby a nuclear facility, at the end of its economic life, is permanently taken out of service and its site made suitable for other purposes. In the case of nuclear power stations, the IAEA defines three different stages:

- immediately after the final closure, radioactive material such as nuclear fuel and operational wastes are removed;
- the buildings surrounding the reactor shield are then dismantled; and
- finally the reactor itself is dismantled.

Delicensing: Under NIA 65, whereby the licensee has to demonstrate "no danger" from the decommissioned facility prior to removal of regulatory controls. Incorrectly driven by radioactivity levels in the UK instead of more technically correctly by dose as in the US.

Disposal: In the context of solid waste, disposal is the emplacement of waste in a disposal facility without intent to retrieve it at a later time; retrieval may be possible but, if intended, the appropriate term is storage. Disposal can also refer to the release of airborne or liquid wastes to the environment (i.e., emissions or discharges).

DoH: Department of Health (UK).

Dose: A measure of the radiation received. Various forms of dose are commonly referred to, including equivalent dose, effective dose, and the absorbed dose. Measured in Sieverts (Sv).

Dose constraint: A restriction on annual dose to an individual from a single source such that, when aggregated with doses from all sources, excluding natural background and medical procedures, the dose limit is not likely to be exceeded; the dose constraint places an upper bound on the outcome of any optimisation study and will, therefore, limit any inequity which might result from the economic and social judgments inherent in the optimisation process.

Dose limit: For the purposes of discharge authorisation, in the UK (since 1986) applied limit of 1 mSv/y to members of the public from all man-made sources of radiation (other than from medical exposure).

DTI: Department of Trade and Industry (UK).

EA: Environment Agency.

Effective dose: Effective dose relates to exposure of the body as a whole. This quantity takes account of the relative effectiveness of different types of radiation in causing tissue damage, and the relative sensitivity of different organs to increased cancer risk from radiation. Measured in Sieverts (Sv) and often quoted at the milliSievert (mSv) level.

EHS: Environment and Heritage Service.

Exemption levels: Radioactivity level below which wastes may be disposed of with ordinary household waste in land-fill sites. Typically 0.4 Bq/gm.

Fabrication: Fabrication is the process whereby plutonium and/or uranium in purified form (from reprocessing or processing) are converted either into fresh fuel for nuclear reactors or into other useful products (e.g., targets for the production of medical isotopes). This operation generates small quantities of low level waste and manufacturing scraps which are processed in order to recycle the nuclear material.

Fission Product: A radionuclide formed by the splitting of a heavy nucleus, usually into two nearly equal-by-mass fragments. Many fission products are very short-lived; after a few years strontium-90 and caesium-137 may be the most dominant fission products both with half-lives of ∼30 years. Over very long timescales, small quantities of long-lived fission products such as chlorine-36 and nickel-63 tend to dominate.

Half-life: The half-life of a radioactive nuclide is defined as the time taken for half of the number of atoms to disintegrate. Half-lives vary from less than a millionth of a second to thousands of millions of years, depending upon the stability of the nuclide involved. Refer to Annex 5.

HSC: Health and Safety Commission.

HSE: Health and Safety Executive. A distinct statutory body with day-to-day responsibility for making arrangements for the enforcement of safety legislation. The Executive is the statutory licensing authority for civil nuclear installations in the UK — a function which it delegates to senior officials within the Nuclear Installations Inspectorate (NII) which is part of HSE's Nuclear Safety Division.

ICRP: International Commission on Radiological Protection and associated with the publication of key radiological protection documentation such as ICRP 26 (1977) and ICRP 60 (1990).

ILW: Intermediate Level Waste.

IRAC: Ionising Radiation Advisory Committee. To consider all matters concerning protection against exposure to ionising radiation that are relevant to the work of the HSC.

IRR: Ionising Radiation Regulations (e.g., IRR 1985).

LLW: Low Level Waste.

LMA: Liabilities Management Authority. Yet to be constituted, but recently (2002) proposed in a Government White Paper entitled "Managing the Nuclear Legacy." The Authority will take ownership of the assets and liabilities on UK civil nuclear sites. It will then competitively tender for the decommissioning and waste management operations of the redundant facilities. The annual spend on such work in the UK is currently some £1bn per annum.

Magnox: The first generation of gas-cooled nuclear reactor, used for electricity generation at power stations constructed in the 1960s. Takes its name from the magnesium-based alloy in which the natural uranium metal fuel is contained.

MoD: Ministry of Defence.

NDA: Nuclear Decommissioning Authority. Alternative name for LMA.

NIA 65: Nuclear Installations Act 1965 — under which the HSE NII operate and ensure safety of civil nuclear facilities in the UK.

NII: Her Majesty's Nuclear Installations Inspectorate.

Nirex: UK Nirex Ltd — Nuclear Industry Radioactive Waste Executive — responsible for provision of a UK ILW (and some long-lived LLW) repository. Gives sound waste packaging advice to waste producers in lieu of formal conditions for acceptance for a possible eventual UK repository.

NRPB: National Radiological Protection Board.
To give advice, conduct research, and provide technical services in the field of protection against ionising and non-ionising radiation.

NuSAC: Nuclear Safety Advisory Committee.
To advise HSE on safety of nuclear installations.

OCNS: Office of Civil Nuclear Security.

OECD: Organisation for Economic Co-operation and Development.

Processing or Recovery: Processing is the treatment of unirradiated plutonium and/or uranium materials (e.g., unused fuel (in some cases the fuel materials may also contain thorium) or manufacturing scraps), that may or may not be contaminated with other materials, in order to recover the plutonium and/or uranium in a purified form for reuse. Although trace quantities of fission products and actinides can be present from historical irradiation, high level waste is not generated. Small quantities of low level waste are produced together with even smaller quantities of intermediate level waste, if at all.

PWR: Pressurised Water Reactor. The most recent widely utilised reactor design to be constructed in the UK and France. Derived from submarine propulsion reactor types. Uses a slightly enriched uranium dioxide clad in Zircalloy as fuel.

Radiological risk: The probability of harmful consequences of radiation in a given period of time. This term is usually used to refer to the product of the probability of a potential occurrence and the probability of developing either a cancer or hereditary effects.

Radiological safety assessment: An analysis to predict the performance of a system or subsystem, where the performance measure is radiological impact or some other global measure of impact on safety.

Radionuclide: General term for an unstable nuclide (or isotope) that emits ionising radiation (e.g., Caesium 137 ~30 year half life, Cobalt 60 ~5 year half life, Strontium 90 ~29 year half life, etc.).

RADREM: Radioactivity Research and Environmental Monitoring Committee.

RADWASS: Radioactive Waste Safety Standards.

RCF: Rock Characterisation Facility.

Reprocessing: Reprocessing is the treatment of irradiated nuclear fuel (i.e., fuel which has been used in reactor operations) to separate the plutonium and/or uranium from the high level fission product waste. The fission products have been formed by the splitting (fissioning) of the uranium or plutonium in the nuclear reactor. The products of reprocessing are plutonium and/or uranium (which are capable of being recycled as new fuel), the high level waste, which will be converted into solid glass, some intermediate level waste and low level waste.

Risk: The product of probability × consequence arising from a particular activity or scenario. For example, the radiological risk arising from a radioactive disposal facility being the probability that an individual will suffer a serious radiation induced health effect as a result of the presence of the facility. The associated risk target being a level of risk to a member of the critical group from a single disposal facility which provides a numerical standard for assessing the long-term performance of the facility.

RSA 93: Radioactive Substances Act 1993.

RWMAC: Radioactive Waste Management Advisory Committee.
Source of independent advice to Government on civil radioactive waste management.

SEPA: Scottish Environmental Protection Agency.

Sievert: The standard international unit of dose.

Source: A facility, or group of facilities, which can be optimised as an integral whole in terms of radioactive waste disposals. Also used to refer to radioactive sources (e.g., ^{60}Co).

THORP: Thermal Oxide Reprocessing Plant — located at Sellafield in Cumbria, UK. Owned and operated by BNFL.

TOR: Tolerability of Risk and associated with the ALARP principle as explained in a 1992 HSE publication covering the way in which risks from nuclear installations in the UK are regulated. 300 micro Sieverts/year being equivalent to a 10^{-5} risk as a constraint and 10–30 micro Sieverts/year being equivalent to a 10^{-6} risk and a target below which no remedial action to further lower the dose is considered essential.

Transuraniac Elements: Elements above uranium, with an atomic number greater than 92, in the Periodic Table. The 13 transuranic elements discovered to date include plutonium and americium.

Tritiated wastes: Low and intermediate level waste containing the radionuclide tritium. Tritium has a 12 year half life and is of low radioactivity, but is highly mobile and, therefore, difficult to contain.

UK: United Kingdom.

UKAEA: United Kingdom Atomic Energy Authority.

UN: United Nations.

VLLW: Very Low Level Waste, which can be safely disposed of with ordinary refuse (dust-bin disposal) and defined as < 400 kBq beta/gamma activity per $0.1 m^3$ waste or 40 kBq beta/gamma activity per single waste item. More applicable for small volumetric quantities rather than, say, bulk contaminated land.

VLRM: Very Low Radioactive Material defined as < 40 Bq/gm beta/gamma and < 1 Bq/gm alpha activity. Associated with relatively short-lived radionuclides from contaminated ground with very low levels of long lived alpha activity from traces of uranium, plutonium, and actinides.

Waste form: The physical and chemical form in which the waste will be disposed of, including any conditioning media, but excluding the container.

Waste package: The waste form and its container, as prepared for disposal.

1-9. References

1. British Nuclear Industry Forum, Whitehall House, 41 Whitehall, London SW1A 2BY.
2. International Atomic Energy Agency. *The principles of Radioactive Waste Management*, IAEA Safety Series No. 111-F, IAEA, Vienna (1995).
3. International Atomic Energy Agency. *Classification of Radioactive Waste, A Safety Guide, A Publication in the RADWASS Programme*, IAEA Safety Series No.11-G-1.1, IAEA, Vienna (IAEA Safety Series No. 111-S-1, IAEA1), Vienna (1995).
4. NEA/OECD. *Future Financial Liabilities of Nuclear Activities*, NEA/OECD, Paris (1996).
5. ww.europa.eu.int/comm./energy/en/nuclearsafety/synopses.htm#18185.
6. International Atomic Energy Agency. *TECDOC 1021*, IAEA, Vienna (1998).
7. www.iaea.org.
8. www.icrp.org.
9. www.nea.fr.
10. www.europa.eu.int.
11. Lau, F.-S. *A Dictionary of Nuclear Power and Waste Management*, Research Studies Press, ISBN 086380 051 3, 1987, Exeter, UK, John Wiley & Sons Inc.

Chapter 2
Ionising Radiation and its Control

2-1. Introduction

This chapter discusses the nature of radiation, how it can be measured, its effects on humans and the measures which are taken to protect people and the environment from its harmful effects. The capabilities and limitations of the instrumentation available for detecting and measuring radiation and radioactivity are also briefly described.

It is often said that the public fear of radiation is due to the fact that it cannot be detected by the normal senses. Whilst this is obviously true, it is in fact relatively easy to detect radiation down to very low levels with real time measuring devices. This is in contrast to some chemical hazards, such as asbestos or beryllium, which require samples to be taken and sent to a laboratory for analysis — with a time delay of some hours or even days before the results are returned.

Since there are many sources of information available on the science of radiation and its detection and measurement, it is not intended to present a detailed account on this complex topic here (see References). However, the chapter includes an overview of the International Commission on Radiological Protection (ICRP) recommendations that are relevant to radioactive waste disposal and decommissioning. It also describes the role of the UK's National Radiological Protection Board (NRPB), which is directed to advise Government on the acceptability and application of such international standards as ICRP Publication 60.

2-2. The Properties of Radiation

Shortly after the discovery of X-rays, their diagnostic potential was recognised. X-ray apparatus, for example, was used on wounded troops in Europe in the later stages of the First World War. The appearance of acute undesirable effects (such as hair loss and erythema) soon made hospital staff aware of the need to avoid overexposure. General radiation protection recommendations were proposed in the UK in the early 1920s, and the First International Congress of Radiology was held in 1925. The International Commission on Radiological Protection (ICRP) was formed in 1950, and it has published a series of recommendations since then, reflecting the increased understanding of the biological basis of radiation-induced tissue damage. These recommendations contain advice on good working practice, the quantities to be used in radiological protection, and recommend dose limits. The most recent set of recommendations, the 1990 recommendations of the ICRP, were published in 1991 [1–3]. The basic structure of the report is shown in Table 2-1.

The term radiation is used to describe a range of electromagnetic waves and particles. Radiation which causes the formation reactive ions in matter through which it passes is called "ionising radiation." The principal forms of ionising radiation which are likely to be encountered in nuclear decommissioning and radioactive waste management are:

- alpha particles, comprising the nuclei of helium atoms,

Table 2-1. Contents of ICRP 60

ICRP 60 — Section components	ICRP 60 — Corresponding contents
Introduction	History, development
Quantities	Basic and subsidiary quantities
Biological aspects	Biological effects, detriment, tissue weighting factors
Conceptual framework	Framework, system, practices, intervention
System for practices	Occupational, medical, public, potential
System for intervention	Public, remedial action, emergencies
Implementation	Responsibility, regulation, compliance, planning, exemption and exclusion
Summary	
Annexes	Quantities, biological effects, significance of radiation effects, publications

- beta particles, which are fast-moving electrons (positive or negative),
- gamma radiation, which is electromagnetic radiation similar to X-rays but more energetic, and
- neutrons, which are neutral particles emitted by atoms undergoing fission.

Radiation is emitted by the nuclei of unstable atoms which undergo decay, i.e., the spontaneous transformation into another type of atom. Several types of radiation may be emitted by the decay of a single atom. Thus, gamma radiation invariably accompanies the emission of an alpha or beta particle. The decay of an unstable nucleus through the emission of radiation may result in another unstable nucleus. There may be a series of decays, known as a *decay chain*, before a stable nucleus is reached (see Annex 5).

The ICRP organisational structure is set up with four sub-committees covering the following work areas:

Committee 1 — Radiation effects.
Committee 2 — Secondary limits.
Committee 3 — Protection in medicine.
Committee 4 — Application of the Commission's recommendations.

An unstable nucleus which emits ionising radiation is called *radioactive*. Radioactive nuclei are also known as *radioisotopes* or *radionuclei*. Each radioisotope can be characterised by a *half-life*, which is the time taken for half of the radioisotopes present to decay. Over 1000 radioisotopes are known and their half-lives vary from fractions of a second to millions of years. In two half-lives, the radioactivity is reduced to a quarter of its original level and in 10 half-lives to about one thousandth.

Radionuclides can also be characterised by the type and energy (measured in mega electron volts, or MeV) which they emit. Table 2-2 lists some of the more important radionuclides encountered in nuclear decommissioning.

Alpha and beta particles lose energy by colliding with the nuclei of any matter they pass through, causing ionisation. The heavier and more highly charged alpha particles

can be stopped by a thin sheet of paper or plastic. The lighter beta particles can be stopped by a thin sheet of metal. Gamma radiation and neutrons interact with matter to a much lesser degree and can, therefore, penetrate greater distances. The thickness of shielding required to stop gamma radiation and neutrons varies depending on the energy and intensity, but can be several metres of concrete.

It should be noted that caesium-137 does not itself emit gamma radiation. It decays to an unstable barium-137 isotope, which in turn immediately decays (half-life of 2.5 minutes) by emitting 0.66 MeV gamma radiation. The quantity of barium-137 is in a state of dynamic equilibrium with its parent and, hence, its gamma radiation can be used as a characteristic marker for caesium-137.

2-3. Basic Concepts and Units

The ICRP recommendations deal only with ionising radiation and with the protection of man. The Commission emphasises that ionising radiation needs to be treated with care rather than fear and that its risks should be kept in perspective with other risks. All those concerned with radiological protection have to make value judgments about the relative importance of different kinds of risk and about the balancing of risks and benefits. The 1990 Recommendations propose a "System of Radiological Protection," which is intended to cover all situations, that is:

- normal operations;
- situations where there is a probability of exposure (accidents and disposal of solid radioactive wastes); and
- situations where the source is not under control (e.g., radon in homes).

The principal dosimetric quantities used in radiological protection (absorbed dose, equivalent dose, effective dose, committed effective dose and collective effective dose (see Chapter 1, Section 1.9) are described in the report.

Table 2-2. Some Important Radioisotopes Encountered During Decommissioning

Isotope	Half-life (years)	Mode of decay	Principal particle energies (MeV)	Principal gamma energies (MeV)
Tritium (^{3}H)	12.3	β^-	0.019	—
Carbon-14	5,730	β^-	0.156	—
Cobalt-60	5.27	β^-	1.49, 0.67, 0.32	1.17, 1.33
Strontium-90	28.8	β^-	0.55	—
Caesium-137	30.2	β^-	1.17, 0.51	0.66
Plutonium-239	24,000	α	5.16, 5.15, 5.11	0.013, 0.03
Americium-241	432	α	5.48, 5.43	0.02, 0.06

2-4. The Measurement of Radiation

The International Commission on Radiation Measurement and Units (ICRU), set up by the First International Congress of Radiology in 1925, has developed internationally-agreed quantities and units of radiation and radioactivity. A comprehensive treatment can be found in ICRU publications. The most significant quantities and units are summarised in Table 2-3.

In Europe, SI units have been adopted as standard. However, the original units are still found in the literature and tend to be common currency in the US.

The unit of radiation *exposure*, the *roentgen*, was the earliest unit for measuring radiation, and was originally defined as the quantity of X-radiation which produced one electrostatic unit (esu) of charge (0.3E-9 coulomb) in a cubic centimeter of air at standard temperature and pressure. This was later changed to refer to coulombs per unit mass of air (C/kg). It is strictly only applicable to X-rays and low energy gamma rays. For other radiations, a more complex quantity, known as *kerma,* has been introduced to describe the processes which occur when ionising radiation imparts energy to matter. However, discussion of kerma is beyond the scope of this book.

The unit of *radioactivity* was originally defined as the *curie* (Ci), defined as the disintegration rate of the quantity of radon gas in equilibrium with one gram of radium. This was later set precisely at:

$$1\,\text{Ci} = 3.7 \times 10^{10} \text{ disintegrations per second.}$$

The modern SI unit for radioactivity, the *bequerel* (Bq), is simply the amount of a substance which decays at a rate of 1 disintegration per second. Whilst it is conceptually simple, the bequerel is an inconveniently small unit.

The relationship between radioactivity and exposure depends on the interaction between radiation and air. For the purposes of determining the effect of radiation on other materials, it is necessary to define a unit of *absorbed dose*, which is a measure of the energy deposited in joules per kilogram (J/kg) by the radiation in the material which absorbs it. The original unit is the *rad*. The SI unit, the *gray* (Gy), is 1 J/kg, equivalent to 100 rad.

For X-rays, gamma rays, and electrons, the damage caused to biological tissue is approximately proportional to the energy deposited, i.e. absorbed dose. However, this proportionality does not hold for more heavily ionising radiations such as alpha particles. The correction for this effect depends on the ionisation energy per unit length of radiation path, which will vary for different points along the path of an individual particle. However, as an approximation, a weighting factor has been introduced to modify the absorbed dose to define the *dose equivalent*. This dimensionless factor was originally called the *quality factor*, Q, which is related to the *linear energy transfer* (LET) for the radiation concerned. The *dose equivalent* was measured in *rem*:

$$1\,\text{rem} = 1\,\text{rad} \times Q$$

where $Q = 1$ for electrons and all electromagnetic radiation, $Q = 10$ for fission neutrons and protons, and $Q = 20$ for alpha particles and other heavy particles.

ICRP now recommends radiation weighting factors (w_R) based on the type and the nature of the radiation (whether an external field or radiation from an internally deposited radionuclide). Hence, the *equivalent dose* is defined as:

$$1\,\text{sievert} = 1\,\text{gray} \times w_R.$$

In practical terms, there is no difference between w_R and Q for alpha, beta, and gamma radiations; for neutrons, w_R varies in the range 5–20, depending on the energies of the neutrons involved.

In many practical situations, only part of the body may be exposed to radiation, or the exposure may vary between different tissues. To deal with this problem, ICRP recommends *tissue weighting factors*, w_T, by which the equivalent dose to individual organs should be multiplied to give the *effective dose*:

$$\text{effective dose} = \Sigma(w_T \times \text{equivalent dose to organ } T)$$

where Σ denoted summation over all the organs concerned. Table 2-4 gives the weighting factors recommended by ICRP for the most significant organs.

2-5. The Biological Effects of Radiation

Ionising radiation causes two basic types of harmful effects, called "deterministic" effects and "stochastic" effects.

Table 2-3. Summary of Radiation Units

Quantity	Name	Unit	Definition
Radiation exposure	roentgen	R	2.58E-4 C/kg air
Radioactivity	curie	Ci	3.7E10
	bequerel[a]	Bq	disintegrations/sec
			1 disintegration/sec
Absorbed dose	rad	rad	0.01 J/kg
	gray[a]	Gy	1 J/kg (= 100 rad)
Equivalent dose	rem	rem	rad × w_R[b]
	sievert[a]	Sv	Gy × w_R (= 100 rem)

[a] SI units, i.e. International System of Quantities and Units.
[b] w_R = radiation weighting factor (explained in text).

Table 2-4. ICRP Tissue Weighting Factors

Tissue	w_T
Gonads	0.20
Red bone marrow	0.12
Colon	0.12
Lung	0.12
Stomach	0.12
Bladder	0.05
Breast	0.05
Liver	0.05
Oesophagus	0.05
Thyroid	0.05
Skin	0.01
Bone surfaces	0.01
Remainder	0.05

Table 2-5. ICRP Risk Factors for Stochastic Effects

Detriment	Adult workers	Whole population
Fatal cancer	4.0×10^{-2}	5.0×10^{-2}
Nonfatal cancer	0.8×10^{-2}	1.0×10^{-2}
Severe hereditary effects	0.8×10^{-2}	1.3×10^{-2}
Total	5.6×10^{-2}	7.3×10^{-2}

Deterministic effects have a threshold dose and, above that threshold, the frequency and severity of the effect increases with increasing dose. Examples are erythema and hair loss. Stochastic effects have a simple proportional relationship between dose and probability of occurrence (which implies that these types of effect can never be eliminated, only the occurrence can be minimised). Examples are fatal cancer and severe hereditary diseases in offspring.

One of the most difficult tasks for the ICRP and others has been to quantify these harmful effects and to create a measure of the overall risk from ionising radiation, which they called the health detriment. Since much of the data on the effects of ionising radiation on man is based upon studies of Japanese survivors of atomic bombs, it is appropriate to high dose rates. Therefore, this data had to be extrapolated to low doses and low dose rates. In addition, the lifetime cancer risk for the Japanese survivors had to be estimated, since not enough time had yet elapsed for all the cancers to have been expressed. Following an extensive review of the Japanese and other data, the ICRP produced a new set of risk factors for irradiation of a number of organs and tissues. These were used to derive the rounded tissue weighting factors for effective dose. They then calculated the health detriment from a combination of the incidence of fatal cancers, nonfatal cancers and severe hereditary effects, each weighted for severity, and when they occur in the irradiated person. For normal operations and optimisation, effective dose is seen as an adequate surrogate for health detriment. However, in practice we are generally concerned with much smaller doses of radiation where the acute *deterministic* effects are negligible. At low doses, we are concerned with *stochastic effects*, i.e., enhanced risks associated with the induction of cancers and leukemia (*somatic effects*), and damage to genes and chromosomes transmitted to subsequent generations (*hereditary effects*).

In summary, the overall health detriment following exposure to low doses of radiation amounts to $\sim 7.3 \times 10^{-2}$ Sv^{-1} (see Table 2-5). The risk factor for an exposed working population, aged 18–64 years, is slightly less, at 5.6×10^{-2} Sv^{-1}. A recent review of the risk of radiation induced cancer at low doses and dose rates has confirmed that the linear, no threshold (LNT) dose response model is the most appropriate and supported the ICRP dose rate reduction factor of 2.

The acute effects of radiation on humans for a single whole body dose are established to be:

- at ~ 1 Gy, symptoms of radiation sickness will be apparent, but the patient will almost certainly recover (but with an enhanced risk of later, stochastic effects);
- at ~ 4 Gy, there is a 50% chance of death;
- at ~ 8 Gy, death will occur within 2 months, due to bone marrow failure;
- at ~ 15 Gy, death will occur within 2 weeks, due to gastrointestinal tract failure; and
- at ~ 40 Gy, death will occur within 2 days, due to central nervous system failure.

Higher total doses can be tolerated if they are delivered in fractions over a period of time which allows the body repair mechanisms to function between each fraction. These estimates depend on the data projection method used. Other sources give different estimates, but higher estimates are inconsistent with the absence of any detectable effect due to variations in the natural background of radiation. It should be noted that no hereditary effects of radiation have been observed in human populations at any dose level, even among the children of survivors of Hiroshima and Nagasaki. Estimates of hereditary effects on humans, therefore, depend on data from animal experiments, studies of cell cultures, and theoretical models.

There is continuing debate as to how the data for stochastic effects at relatively high doses should be extrapolated to the much lower dose levels associated with the operation of nuclear licence sites and the regulated uses of radioactive sources. The mostly widely used assumption, recommended by ICRP, is that there is no dose threshold for the onset of stochastic effects (somatic or hereditary), and that the chance of these effects occurring is linearly dependent on radiation dose at low levels. In the

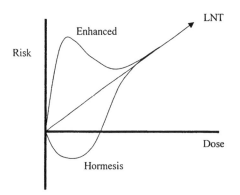

Figure 2-1. Alternative Risk–Dose Curves for Low Level Radiation.

absence of unequivocal proof to the contrary, this *linear, no threshold* (LNT) hypothesis is a safe, conservative assumption.

There are, however, dissenting voices [1–3]. On one side of the argument, there are those who claim that low levels of radiation can have a beneficial effect — known as *radiation hormesis*. On the other side, there are claims that low levels of radiation have an enhanced risk which can be perhaps as high as 100 times that predicted by the LNT extrapolation. These competing theories are shown graphically in Figure 2-1.

When radiation is absorbed by biological tissues, the ionisation it causes results in changes at a molecular level, which leads to damage at the cellular level and ultimately to the organ and whole body. In irradiated cells, damage can be detected to chromosomes, the packages in which DNA is contained within the cell. This damage has been attributed to breaks in the DNA chain, followed by rejoining of the broken fragments in a different way. Complex repair processes respond to this damage so that the vast majority of damaged cells do not lead to lasting effects. In principle, however, it is conceivable that a single broken DNA chain caused by a single photon or particle of radiation has a finite probability of leading to a stochastic effect. This is the basis of the argument that there is no threshold. On the other hand, organisms have sophisticated immune systems which respond positively to a challenge, thus becoming more efficient. This is the basis of viral immunisation; a similar mechanism could explain radiation hormesis. Proponents of the opposite theory, that low levels of radiation cause enhanced risk, point to mechanisms which only operate at low levels and rapidly become saturated as the radiation dose increases.

Recent laboratory studies have observed two effects which illustrate the complexity of the response of biological systems to radiation damage:

• *Genomic instability*: colonies of stem cells grown from a single surviving irradiated cell show aberrations

which vary from cell-to-cell and become expressed only several generations after the irradiated cell.
• *The bystander effect*: cells which have been in the environment of an irradiated cell but not themselves irradiated exhibit chromosome damage and genomic instability. Some form of signaling between the irradiated cells and the unirradiated ones seems to be the only explanation.

The linear no-threshold hypothesis remains the accepted and most credible basis for estimating radiation effects at low doses.

2-6. Radiological Protection Principles

2-6-1. *Introduction*

There is a conceptual framework for radiological protection which has been set out by the ICRP and forms the basis of legislation in most countries. The overall objective is to prevent the occurrence of deterministic effects by keeping doses below the relevant thresholds, and to ensure that all reasonable steps are taken to reduce the induction of stochastic effects. In the past, ICRP has quantified this through an optimisation process based on classical cost-benefit analysis. This meant calculating collective dose arising from a particular practice involving the use of radiation and equating the cost of mitigating that dose with the benefit in terms of lives saved. This approach, however, does not provide sufficient protection for each individual exposed to the dose. The 1990 Recommendation added a restriction by the introduction of the concept of a *constraint*. The constraint is a criterion that can be applied to a single source in order to ensure that the most exposed individual is not exposed to excessive risk.

To achieve this objective, ICRP has developed a framework based on three elements:

• justification,
• dose limits for protective action, and
• optimisation of protection.

2-6-2. *Justification*

The first principle is that a practice involving exposure of people to radiation should do more harm than good. This procedure implies a quantified balance of costs and benefits, but in practice decisions are made in a qualitative way. For example, domestic smoke detectors contain around 40 kBq of Americium-241. However, the very small risk involved in exposure to this radioactivity is far outweighed by the benefit in terms of avoiding loss of life through early detection of fires. On the other hand, the risks arising from the use of radioactive luminised watch dials and signs is not considered to be outweighed by the benefits, since there are other means of illuminating dials. Consequently, this use of radioactivity is no longer allowed in most countries.

In the case of exposure for medical purposes, where the radiation doses can be substantial, the practice has to be justified primarily in terms of the exposure conferring more good than harm to the patient. However, the detriment to radiological staff and members of the public must also be considered. The relative significance of such doses can be placed in context by noting that many nuclear facilities have experienced incidents where radiation alarms were triggered by staff returning to work following medical radiation therapy.

2-6-3. *Dose Limits for Protective Action*

The exposure of individual(s) from all practices involving exposure to radiation should be subject to dose limits. Dose limits are aimed at ensuring that no individual is exposed to radiation risks that are judged to be unacceptable in normal circumstances. In general:

- Occupational dose limits, and intervention levels for the public either in emergencies or for radon in homes, are set at about 10 times the natural background, i.e., ~20 mSv per year.
- Added doses arising from discharges of radioactivity to the environment are kept to about one tenth of background, i.e., <300 μSv per year.
- In many countries, exemption from regulatory control is allowed if doses are below about one hundredth of background, i.e., typically <10–20 μSv per year.

These dose limits can be considered as establishing a minimum level of health protection (see Table 2-6). However, it should be noted that in many countries there is public pressure to reduce dose limits to much lower levels. For example, both the Irish and Norwegian Governments have protested to the UK Government about discharges to the Irish Sea from the Sellafield reprocessing plant, even though the doses to the most exposed individuals arising from artificial radioactivity in the Irish Sea (as determined by the Radiological Protection Institute of Ireland) amounts to only a few μSv per year, which would normally be considered as harmless.

2-6-4. *Practices and Intervention*

The ICRP considers that the most effective way to control exposure is at the source. Two types of human activity are considered by the ICRP. Those activities that increase the overall exposure to radiation are called "practices,"

Table 2-6. ICRP Dose Limits

Application	Dose limit	
	Occupational	Public
Effective dose	20 mSv per year, averaged over 5 years	1 mSv in a year
Equivalent dose to lens of eye	150 mSv	15 mSv
Equivalent dose to skin	500 mSv	50 mSv
Equivalent dose to hands and feet	500 m mSv	50 mSv

ICRP 60 "practices" (i) Occupational — constraints and limits.
(ii) Medical — constraints.
(iii) Public — critical groups, constraints and limits.

whereas those activities that decrease the overall exposure are called "intervention." Examples of a practice are routine discharges, the use of contaminated marine sediment for landfill and the change of use of contaminated land. Intervention will usually apply to public exposures from natural sources of radiation and from environmental contamination following an accident. The system of protection for intervention is based on the general principles that any intervention must do more good than harm and the scale of the intervention should be such that the net benefit, less the cost, is as large as reasonably achievable. The dose limits for practices do not apply to intervention.

Before a program of intervention is initiated, it should be justified and optimised, i.e., there should be a net benefit (including allowance for anxiety) from the adopted action, and the benefit should be maximised by settling the details of that action. The two main examples of intervention are, first, the need to reduce high levels of radon gas in homes and, secondly, for a potential or actual accidental release of radioactive materials to the environment. In neither case does the ICRP make new recommendations about numerical levels, but task groups have been set up to provide guidance in due course. In the meantime, the recommendations in ICRP Publication 39 [4] and ICRP Publication 40 [5] remain valid. ICRP Publication 63 [6] updates and extends Publication 40 and includes quantitative guidance on intervention levels. Advice on radon at home and at work is given in Publication 65 [7].

ICRP has divided exposures in practices into three categories: occupational, medical, and public. Occupational exposures are those incurred at work and include exposures to natural sources at work. Medical exposures are those incurred by patients as part of their diagnosis or treatment, those incurred willingly by individuals helping patients, and those incurred by volunteers in a program of biomedical research. Public exposure encompasses all other exposures.

2-6-5. *Optimisation of Protection*

After the application of Protective Dose Limits, there is an additional requirement that residual doses should be kept "as low as reasonably practicable" (ALARP). Essentially, this means that if there is scope for reducing doses further, even if they are already compliant with dose limits, at a reasonable cost, then this should be done. Once again, this can formally involve the use of cost-benefit analysis, but in practice it is more likely to be done qualitatively through the application of "common sense."

The most important part is the requirement to optimise the protection requirements and the ICRP have introduced the concept of a dose constraint for use in the optimisation procedure. A "constraint" is an upper bound on the individual dose from a single source. This differs from a dose "limit" which relates to the total dose an individual receives from all relevant sources. ICRP has specified dose limits for workers and for the public (see Table 2-6). Risk limits and dose constraints have not been set.

ICRP is currently discussing changes to the system which will shift the emphasis further from collective to individual dose. It is widely believed that the use of collective dose in situations where extremely small doses to a very large number of people leads to a distortion of the process. An extreme example is the doses across Europe arising from the Chernobyl accident. It is likely, therefore, that ICRP will in future focus on keeping individual doses both below a defined action level and as low as reasonably achievable. The ALARA requirement would not be linked to collective dose. If the risk of harm to the most exposed individual is trivial, then the total risk is trivial — irrespective of how many people are exposed.

2-6-6. *The Control of Occupational Exposure*

The control of occupational exposure is achieved by the use of dose constraints and dose limits, as given in Table 2-6. For women there are two alternatives: if the woman is not pregnant then the basis for control of occupational exposure is the same as that for men; if the woman is or may be pregnant, then extra controls are needed to protect the unborn child. ICRP have published dose coefficients for intakes of workers in Publication 68 [8], and these are available on CD ROM.

2-6-7. *The Control of Medical Exposure*

The control of medical exposure is achieved by the use of dose constraints and optimisation. ICRP recommend that dose limits should not be applied to medical exposures and that medical exposures should not be included when considering compliance with the dose limits applied to occupational or public exposures. ICRP 73, published in 1996 [9], clarifies how the system recommended in Publication 60 should be applied in medicine.

2-6-8. *The Control of Public Exposure*

The control of public exposure is achieved by the use of dose constraints and dose limits. The dose limits are given in Table 2-6. It is often convenient to class together individuals who form an homogeneous group with respect to their exposures to a single source. If such a group is typical of those most highly exposed by the source, it is known as a "critical group." In optimisation, the dose constraint should be applied to the mean dose in the critical group from the source. ICRP have published dose coefficients for intakes by members of the public, including a number of different age groups, in Publication 72 [10] (also available on CD ROM).

2-6-9. *Potential Exposures*

Potential exposures should be considered as part of the system of protection applied to practices. However, the exposures, if they occur, may lead to intervention. Therefore, there are two objectives: prevention (reduction of probability of occurrence) and mitigation (limitation and reduction of exposures). In theory, potential exposures could be controlled by the use of risk constraints and risk limits, by analogy with the use of dose constraints and dose limits for actual exposures. However, the techniques for assessing risk are still being developed, and ICRP give no figure for a risk limit at this time. In general, ICRP recommend that dose and risk constraints should be treated separately. However, if the doses, should they occur, are below dose limits, ICRP consider that it is adequate to use the product of the expected dose and its probability of occurrence as if this were a dose that was certain to occur. ICRP Publication 64 [11] shows how the fundamental safety principles can be applied to all potential exposure situations.

2-7. Practical Advice on Radiation Protection Implementation

The ICRP gives advice on the regulation of practices, regulation in the context of potential exposures, and stresses the need for a safety based attitude in everyone. ICRP consider that it is helpful to use a set of reference levels or values of measured quantities above which some specified action should be taken. They include:

• recording levels, above which a result should be recorded, lower values being ignored;

Table 2-7. Approximate Timescales over which Prevention of Harmful Releases of Radionuclides into the Environment from a Nuclear Waste Disposal Facility must be Considered

Years	Past historical events over such timescales	Possible future events over such timescales	$t_{1/2}$ [a]
10^2	Discovery of radioactivity	"Greenhouse" effects	
10^3	Norman conquest	Large ecological changes, e.g., lakes fill with weeds	
	Egyptian pyramids	Mineral and energy resources exhausted?	^{14}C
10^4	Discovery of agriculture		
	Last glaciation of Northern Europe	Next glaciation	^{239}Pu
	Use of fire and tools by humans		
10^5	Emergence of Neanderthal man	Time between major glaciations	^{99}Tc
10^6	Emergence of Homo sapiens	Stable geological formations remain relatively unchanged	^{237}Np
10^7	Evolutionary branching between humans and apes	Appearance of new families of species?	^{129}I
10^8	Dinosaurs populated the earth	Large-scale movements of continents (thousands of kilometers)	
10^9	Appearance of multi-cellular organisms	Significant probability of "nearby" supernova, or meteorite impacts	
		Increase in solar intensity sufficient to erase life on earth	^{238}U
	Age of the earth to date	Sun becomes red giant	

[a] $t_{1/2}$ represents the approximate half-lives of some significant radionuclides in solid waste.

- investigation levels, above which the cause or the implications of the result should be examined; and
- intervention levels, above which some remedial action should be considered. The two practical application topics that are of wide interest are the classification of workplaces and exemption levels.

2-8. The Role of NRPB

The NRPB was set up in the UK in 1970 by an Act of Parliament with the following functions: to provide advice, to conduct research, and to undertake technical services on the protection of mankind from radiation hazards. One major responsibility is to advise on radiation protection standards. This advice can be to Government, industry, or the public. Government frequently, but not always, incorporates NRPB advice in subsequent legislation. NRPB is also very involved with international standard setting organisations, such as the ICRP, EC, and IAEA.

2-9. Practical Advice on Principles for Solid Radioactive Waste Disposal

Following the 1990 ICRP Recommendations, the NRPB recognised that there was need for advice on the application of radiological protection principles to the disposal of radioactive waste on land, and they published advice and guidance in 1992 [12]. The three basic principles recommended by NRPB for the protection of the public

following the disposal of solid radioactive wastes are as follows:

(i) Individuals and populations who might be alive at any time in the future should be accorded a level of protection at least equivalent to that which is accorded to individuals and populations alive now.

(ii) In order to ensure that individual members of the public are not exposed to unacceptable risks, the radiological risk to an average typical member of the critical group, attributable to a single waste disposal facility, shall not exceed the risk constraint of 10^{-5} y^{-1}.

(iii) The radiological risks to members of the public should be as low as reasonably achievable; economic and social factors being taken into account (ALARA).

For the purpose of these objectives, risk is defined as the overall probability that a serious deleterious health effect will occur as a result of exposure to ionising radiation. NRPB recommended that calculations to predict radiological risks in the future should take due account of the uncertainties inherent in such predictions. The level of detail in the calculations should reflect the reliability of the information available, and should, therefore, change according to the length of time into the future being considered. Table 2-7 gives, for perspective, a chronological list of a number of historical and (predicted) future events. For times up to about 100 years after the closure of the site, it may be assumed that some form of institutional control over the site will remain. During this period, the system of dose limitation should be applied. For times greater than

100 years or so, but less than about 10,000 years into the future, the NRPB considered that the risk to members of the critical group should be estimated for comparison with the risk constraint. Assumptions about the human environment and human behavior more than 10,000 years or so into the future will necessarily become increasingly arbitrary and, therefore, should be replaced by more general ones. For simplicity, the NRPB recommended that general assumptions should be applied after about 10^4 years. The NRPB considered that individuals who might be alive beyond 10^4 years will be adequately protected if calculations indicate that suitably chosen, hypothetical reference communities would not be exposed to unacceptable risks.

Furthermore, any predictions about the natural environment more than 10^6 years or so into the future are highly speculative, and, therefore, NRPB considered that risk calculations should not be continued beyond this time. Qualitative arguments should be used, however, to show that the likelihood of any sudden, significant increases in risks after this time is low. The specified risk constraint should, therefore, apply from the time institutional control of the site is assumed to be lost (100 years or so after closure) until such time as risk calculations cease to be valid, taken to be 10^6 years or so from the present day. Low probability events which, should they occur, could lead to the exposure of individuals to doses or dose rates high enough to cause serious deterministic health effects should be treated separately. Steps should be taken in the selection and design of a disposal facility to ensure that the probability of such events occurring is ALARA. The total probability of such events occurring as a result of natural events and processes should be below a specified constraint of 10^{-6} y^{-1}.

Calculations to predict radiological risks should include estimates of the uncertainty in these predictions due to incomplete or inadequate knowledge of the system being modeled and the environmental behavior of radionuclides. The stages that could be included are as follows:

- Sensitivity analyses, field studies, and natural analogs to address conceptual and modeling uncertainty.
- Uncertainty as to the future evolution of the site by means of a series of distinct scenarios, representing qualitatively different possibilities. Central value risk calculations may be performed for each scenario.
- Uncertainty analysis to address parameter uncertainty, giving a probability distribution of possible outcomes (i.e., risks).

Whilst recommending that all risks should be kept ALARA, the NRPB recognised the difficulties involved in carrying out detailed optimisation studies for solid waste disposal facilities; in particular, the difficulty in obtaining reliable estimates of the total risk over long timescales,

and the extensive resources often required to carry out such studies (especially when this involves study of a number of possible disposal sites). The NRPB, therefore, recommends that, if the risk to an average member of the critical group, attributable to a single waste disposal facility, does not exceed the specified design target of 1 in 10^6 per year, then the optimisation requirement should be relaxed for that site. The design target represents a level of individual risk which is widely regarded as acceptable, and which is rarely taken into account by individuals in making decisions as to their actions.

The critical group concept needs to be modified for use in the context of solid waste disposal. Hypothetical critical groups should be assumed to exist at the time and place where environmental concentrations of radionuclides are predicted to be highest. The habits of these groups should broadly represent the habits of observed present-day critical groups, but should not be based on the most extreme examples. The critical group for times beyond 10^4 years should, in general, be a reference subsistence community with habits broadly typical of those of subsistence communities in the present day. The habits of the community should be consistent with their status, and extreme habits should not be used.

All risks, to individuals and populations, should be kept as low as reasonably achievable, economic and social factors being taken into account. Predictive calculations of collective dose (or societal risk) for input to optimisation studies, particularly those extending far into the future, are unlikely to be reliable, and, therefore, such calculations are not, in general, recommended. When individual risk is used as an input for optimisation studies, separate consideration should be given to the probability and dose elements of risk.

ICRP Publications 77 [13] and 81 [14] recognised the problems of estimating collective dose over long periods of time in the future and of assessing the risk from future human intrusion into a repository. There was a growing consensus that it is not possible to assign a meaningful probability to such events, as there is no scientific basis for predicting the nature or probability of future human actions. Other issues included what assumptions should be made about future biosphere conditions and about the habits of future critical groups.

There is also the important question of optimisation of protection in the context of a solid waste disposal system. Conventionally, collective dose has been an input into optimisation procedures, but estimates of collective doses to future populations from disposal of long-lived wastes are surrounded by considerable uncertainty. This may make any estimate of collective dose essentially unusable. Furthermore, the current judgments about the relationship between dose and detriment may not be valid for future populations. The dose or risk constraints should

Table 2-8. Analytical Approaches Recommended under ICRP/NRPB for Assessment of Nuclear Waste Repository Performance over the Long Timeframes Involved

Timeframe under consideration	Analytical approach
Up to 100 years	Site under controls and monitored, discharges assessed upon dose limitation
10^2–10^4 years	Assessment to consider risks to members of the critical group(s)
10^4–10^6 years	Consideration of risks to hypothetical reference community
Beyond 10^6 years	Qualitative reviews over a range of scenarios

Constraints for single site: 10^{-5} y^{-1}; Design target for site: 10^{-6} y^{-1} (compliance with possible relaxation upon optimisation); Probability constraints for deterministic effects: 10^{-6} y^{-1} (as applicable to natural events only with ALARA considerations for other eventualities).

increasingly be considered as reference values for the time periods farther into the future, and additional arguments should be duly recognised when judging compliance. Two broad categories of exposure situations should be considered: natural processes and inadvertent human intrusion. Doses or risks arising from natural processes should be compared with a constraint of 0.3 mSv per year or its risk equivalent of about 1×10^{-5} per year. With regard to inadvertent human intrusion, the consequences from one or more plausible stylised scenarios should be considered in order to evaluate how robust the repository design is to such events.

Examples of deep waste disposal systems being planned or in use throughout the World are given in Annex 1. Table 2-8 indicates the analytical concepts to be considered when appraising repository performance in respect of the possible detrimental return of radionuclides to the environment over the considerable timeframes involved.

2-10. Exemption of Sources from Regulatory Controls

The ICRP consider that there are two grounds for exempting a source from regulatory control. One is that the source gives rise to small individual and collective doses in both normal and accident conditions. The other is that no reasonable control procedures can achieve significant reductions in individual and collective dose. Exemption is, therefore, the limit of what is considered to warrant supervision on the part of the competent authority. The radiological basis for exemption from regulatory control has been reviewed by IAEA [15], who concluded that an annual individual dose of "a few tens of microSieverts" or

less provided a basis for exemption. Furthermore, to take into account exposures from more than one exempt practice, it was recommended the critical group exposure from one such practice should be of the order of 10 microSieverts per year. The IAEA also require the collective dose to be ALARA and suggest that this may be assumed if it is below 1 manSv per year of practice. Annex 1 of the European BSS contains levels for exemption from the reporting requirement. These levels are intended for small quantities of radioactive material that do not need to enter the regulatory system. They are not applicable to material leaving a licensed site. The relevant quantity in that case is the clearance level, and EC has produced guidance on clearance levels for metals, building rubble, and general clearance levels [16–18]. This is summarised in Figure 2-2.

2-11. Chronic Exposures

These have been defined as those exposures that persist in time. Specific interest is in exposure of the public from land that has been contaminated by past practices or previous events. For example, early luminising operations with Ra-226, testing of nuclear weapons, and long term contamination following an accident. These situations do not always fit readily into the categories of "practice" and "intervention." NRPB has published some advice on this topic [19], and the UK government has commissioned some research on the criteria for designation of contaminated land already occupied: the intervention situation. Although the criteria for redevelopment of land for new use have not yet been addressed by government, the nuclear industry has initiated the Safegrounds project (see Chapter 23), which aims to produce guidance on good practice for the cleanup of contaminated land.

2-12. Methods of Radiation Detection

There is a wide range of methods available for detecting and measuring radiation, which can be sensitive over an extremely wide range of intensity, from single particle events to the flash produced by a nuclear weapon. Table 2-9 summarises the principal methods in common use.

The first discovery of penetrating radiations from natural radioactive ores by Bequerel stemmed from the observation that the radiation darkened some photographic plates stored nearby. The darkening of photographic emulsion is still used in some radiation dosimeters. The extent of darkening is proportional to the degree of exposure and can be measured by the extent to which the exposed film attenuates a beam of light (a technique known as densitometry).

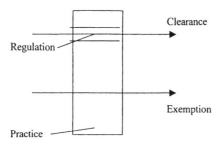

Notes: (i)   Exemption levels described in terms of Bq and Bq/gm in
Annex 1 of BSS.

     (ii)   Clearance levels described in terms of Bq/gm, metals to RP89, buildings and
rubble to RP113, and general waste to RP122.

     (iii)   See Chapter 23 concerning contaminated ground and the "safegrounds"
project.

Figure 2-2. Exemption and Clearance Levels Associated with Radioactive Wastes.

Table 2-9. Nuclear Radiation Detectors

Name	Primary interaction	Medium	Mode of use
Ionisation chamber	Ionisation	Gas	Primary ionisation measured as current pulse or mean current
Proportional Counter Geiger counter	Ionisation	Gas	Primary ionisation increased by gas multiplication and current pulses registered electronically
Scintillation counter	Excitation of electronic levels and emission of photons on return to ground state	Gas, liquid or solid	Light pulses measured with photomultiplier
Photoluminescence, thermoluminescence	Excitation of electronic levels, subsequently released by UV light or heat to emit light	Solid	Photo emission measured as total integrated light output using photomultiplier
Semi-conductor counter	Production of electron-hole pairs	Solid	Current pulses amplified electronically
Cerenkov counter	Production of photons by Cerenkov effect	Gas, liquid or solid	Light pulse measured with photomultiplier
Cloud chamber	Ionisation	Gas	Tracks photographed
Bubble chamber	Ionisation	Liquid	Tracks photographed
Dielectric detector	Ionisation	Solid	Tracks developed by etching

However, most techniques for detecting and measuring radiation uses the ability of the radiation to cause ionisation in materials which absorb it. This is illustrated by Figure 2-3. A beam of ionising particles enters a gas chamber with parallel plates. A potential difference, V, applied across the plates gives rise to a uniform electric field. As the particles slow down in the chamber, they ionise gas atoms by ejecting electrons and leaving positive ions behind. If the electric field is weak, the electrons and ions will recombine, but a few will drift apart, and a small current will flow in the circuit. If the potential difference is increased, the increasing field strength will cause more ion pairs to separate, until a point is reached where all the ions are collected on the plates and the current reaches a plateau where the saturation value remains at I_0 when $V > V_0$.

Early experimenters such as Marie Curie used simple electroscopes and electrometers. The scope of gas ionisation detectors was increased by Rutherford and Geiger by applying high enough electric fields to accelerate the primary ions produced to an energy where they caused further, secondary ionisation. This gas multiplication technique can generate sufficiently large ion currents to detect a single ionising particle and the technique has been developed into the modern *Geiger-Mueller* counter.

Ionisation in gases can also be used to observe visually the tracks of ionising radiation in the *Wilson Cloud Chamber*. This device uses air which is saturated with

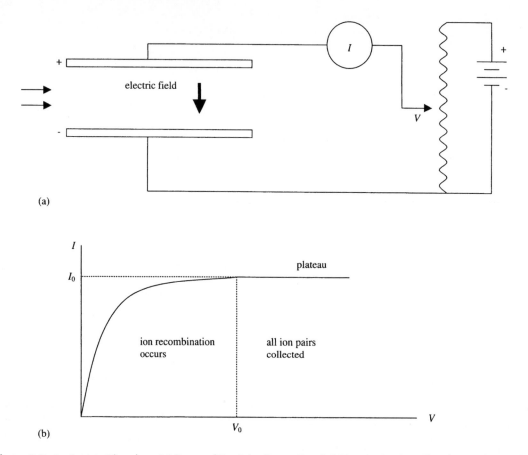

Figure 2-3. Ionisation Chamber. (a) Beam of Particles Enters Parallel Plate Ionisation Chamber with Variable Potential Difference V Applied Across Plates. (b) Plot of Current I vs. Potential V.

water vapor. The passage of a radiation particle causes condensation of water along the track of the particle. This track can be illuminated and photographed. A more recent variation of this concept is the *Bubble Chamber*, which uses liquid gases which produce a line of bubbles along the track of an ionising particle.

The *scintillation counter* uses the property of certain materials to emit a flash of light when energised by the passage of a radiation particle. A light-sensitive photomultiplier tube can detect very tiny flashes of light from a scintillation screen and register it electronically. The earliest application used zinc sulfide screens for counting alpha particles. Large sodium iodide crystals are used for high-efficiency counting of high energy gamma rays. Liquid and plastic scintillators are used for many types of radiation and give a fast response.

Other luminescence phenomena used for detecting and measuring radiation are photoluminescence and thermoluminescence. These methods are particularly suited for integrating exposure doses over long periods of time. The amount of exposure is measured by exposing the material to ultraviolet light (photoluminescence) or heating it (thermoluminescence) and measuring the amount of visible light emitted by the material. The technique is often used as an alternative to the older method of photographic film in personal dosimeters worn by radiation workers.

The passage of radiation through very pure single crystals of semiconductor material creates electrons and complementary electron holes, which are mobile and can be collected under the influence of an applied electric field, similarly to gas ionisation detectors. The two most commonly used materials for these solid state ionisation detectors are silicon and germanium. Silicon detectors can be used at ambient temperatures, but germanium requires to be cooled to liquid nitrogen temperatures.

2-13. Choosing Detection Equipment

There are many types of radiation detector available commercially, ranging from small hand-held units to large

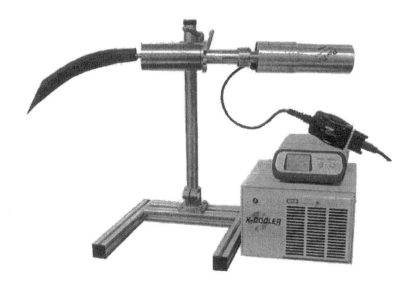

Figure 2-4. Gamma Ray Measuring System (Photograph courtesy of ORTEC^{TM}).

fixed devices for personnel monitoring. Hand-held monitors usually provide a reading of total alpha or gamma radiation (separate monitors are required for each type). A hand-held monitor gives a real-time reading, i.e., there is no delay involved in sending a sample to a laboratory for analysis. Alpha monitoring is difficult, because alpha particles are readily shielded by a thin layer of paint or wet soil, and only surface activity is detectable.

In order to identify specific radionuclides, it is generally necessary to use gamma spectrometry. The measurements tend to be slow and expensive. Whilst portable equipment is available, it is more usual to send samples to a laboratory. In order to make more effective use of the readings from a portable monitor, it is often possible to identify a "fingerprint" for the radioactivity being measured. The use of a fingerprint assumes that, for a given set of circumstances, the composition of the radioactivity does not vary and the full inventory can be inferred from a single measurement of, say, total gamma radiation. Thus, for example, if the radioactivity is known to come from irradiated fuel elements, the measurement of gamma radiation primarily from caesium-137 can be related to the quantity of strontium-90 and plutonium, which are more difficult to measure directly. The fingerprint is established by taking samples and carrying out complete radionuclide assays under laboratory conditions. This assay can then be used to calibrate the portable monitor used for real-time field surveys. However, care must be taken using the fingerprint method, because differential rates of migration of different radionuclides can cause the fingerprint to vary.

Figure 2-4 shows a typical gamma-ray measurement system, supplied by ORTEC^{TM}. This system consists of a germanium detector, liquid nitrogen or mechanical cooling system, preamplifier, detector bias supply, linear amplifier, analog-to-digital converter (multichannel storage of the spectrum), and data readout device.

The detector is housed in shielding to reduce the background from sources other than the sample. The sample is placed in the shielding at some distance from the detector. Gamma photons emitted by the sample interact with the Ge crystal to produce as pulse. The amplitude of the pulse is proportional to the energy of the photon absorbed by the crystal. Each pulse is registered according to its pulse height to produce a spectrum in the form of an histogram (counts per unit energy) of the incident photons. The system can be calibrated so that the nuclides giving rise to the peaks in the spectrum can be identified.

2-14. Practical Aspects of Radiation Protection

2-14-1. Introduction

The control of radiation is a major factor in the design of nuclear plant and its operation, including decommissioning and waste management. There are four principal elements in this control:

- *Shielding*. This absorbs radiation. The material used and its thickness depends on the nature and intensity of the radiation. The most common materials are concrete, lead, and steel. For a reactor or high level radioactive waste, several meters of reinforced concrete may be used, whilst laboratory experiments may require only a few centimeters of lead. Water is very commonly used, particularly for the storage of irradiated fuel elements.

• *Containment.* This prevents the spread of radioactive material, particularly liquids, gases, or dusts. The containment volume can vary from relatively small gloveboxes to the complete envelope of a large building. Containment is most commonly achieved by use of a ventilation plant to maintain a reduced air pressure within the contained volume. The extracted air is filtered through high efficiency particulate filters. Where a process involves radioactive gases or vapors, it may also be necessary to use some form of physical or chemical process (e.g., scrubbing or absorption on molecular sieve material) to remove the activity. Radioactive isotope sources are frequently hermetically sealed to a very high standard in metallic canisters to prevent any airborne hazard.

• *Distance.* In many practical situations, it is sufficient to take advantage of the inverse square law for the attenuation of radiation with distance (for a point source; for an extended source, attenuation follows an inverse linear relationship). Radioactive sources may often be handled safely with long tongs, and the boundary fence of a facility may be set at a sufficient stance to reduce the maximum levels of radiation at the boundary to an acceptable level for exposure to the public.

• *Time limitation.* Unnecessary accumulation of radiation dose can frequently be avoided by minimising the amount of time that workers spend in an area of elevated background radiation. For example, an area where waste drums are being temporarily stored before collection might display a notice advising staff not to loiter in this area.

2-14-2. *The Designation of Controlled and Supervised Areas*

An important aspect of managing doses to workers and the general public is a system for designating areas where there is a risk of exposure. A risk assessment should be carried out to establish the nature of the risk and to identify the measures necessary to restrict the exposure.

Under UK statutory regulations (IRR 99) [20,21] an employer must designate as a *controlled area* any area where:

• it is necessary for any person who enters or works in the area to follow special procedures to restrict significant exposure to ionising radiation in that area;

• prevent or limit the probability and magnitude of radiation accidents or their effects; or

• any person working in the area is likely to receive an effective dose greater than 6 millisieverts a year or an equivalent dose greater than three-tenths of any relevant

limit specified in IRR 99 for employees aged 18 years or above.

The procedures for working in a controlled area will vary and be determined by local rules applicable to the area. Areas designated as controlled areas must be shown by warning notices, their boundaries suitably demarcated, and a description of them included in local rules. In general, a controlled area will involve controlled access via a physical barrier. Personnel crossing the barrier may be required to don personal protective clothing. Depending on the circumstances, this may simply involve gloves, overshoes, and a lab coat or overalls. If a full change of clothes is required, then special change rooms will have to be provided. Personal dosimeters must be worn by staff entering the area. Staff leaving the area will be required to monitor hands, feet, and clothing. Hand washing and shower facilities will normally be required.

An employer must also designate as a *supervised area* any area under his control, not being a controlled area:

• where it is necessary to keep the conditions under review to determine whether the area should be designated as a controlled area; or

• in which any person is likely to receive an effective dose greater than 1 millisievert per year or an equivalent dose greater than one-tenth of any relevant limit specified in IRR 99 [20,21] for employees aged 18 years or above.

Supervised areas must be signified by warning notices indicating the nature of the radiation sources and the risks arising from them. A supervised area will be routinely surveyed for radiation and contamination, and access will be controlled to ensure that only authorised personnel enter the supervised area. However, the requirement for protective clothing and radiation monitoring of personnel leaving the area is likely to be significantly less than for a controlled area.

The following paragraphs describe the system used by UKAEA for designating and managing controlled and supervised areas. An area is designated as a controlled area if:

• the external dose rate in the area under normal planned operations exceeds, or is likely to exceed, 3 microsieverts per hour when averaged over the working day;

• the work with ionising radiation is such that there is a significant risk of spreading contamination outside the working area;

• the hands of an employee can enter an area in which the time average dose rate exceeds, or is likely to exceed, 75 microsieverts per hour;

• it is necessary to prevent, or closely supervise, access to the area by employees who are unconnected with the work with ionising radiation while that work is under way;

Table 2-10. Equivalent Dose Limits for Employees aged 18 years and Above

Organ	Limit
Lens of the eye	150 millisieverts in a calendar year
Skin	500 millisieverts in a calendar year
Hands, forearms, feet and ankles	500 millisieverts in a calendar year
Abdomen of woman of reproductive capacity	13 millisieverts in any consecutive 3 month period

- employees are liable to work in the area for a period sufficient to receive an effective dose in excess of 6 millisieverts per year or an equivalent dose greater than three-tenths of any relevant limit specified in Table 2-10; or
- the area is the subject of a local contingency plan designed to restrict exposure following a radiation accident in that area.

An area is designated as a supervised area if:

- it is necessary to keep the conditions in the area under review to determine whether the area should be designated as a controlled area; or
- it is an area in which any person is likely to receive an effective dose in excess of 1 millisievert per year or an equivalent dose greater than one-tenth of any relevant limit specified in Table 2-10.

2-14-3. *The Categorisation of Controlled Areas*

Areas designated as controlled areas on the basis of the criteria specified above are categorised according to the level of the radiological hazard associated with them and whether it arises from internal or external exposure. This categorisation is designed to facilitate operational control, having regard to the range of radiological conditions associated with plant and operations under UKAEA management.

Hazard rating

The radiological hazard rating system to be applied within UKAEA is as shown in Table 2-11.

Each area is designated as a controlled area in accordance with the rating system above. In determining risk category, all relevant factors are considered, including the nature of the work with ionising radiation to be carried out, the risk assessments relating to such work, and the range of radiological conditions liable to arise in the course of planned operations. The radiological risk is expressed in terms of gamma radiation dose rates and the levels

Table 2-11. UKAEA's Hazard Rating System for Designated Areas

Degree of hazard	Risk category	Nature of area
High	Cat H	Exclusion
Moderate	Cat M	Restricted
Low	Cat L	Operational

Table 2-12. Criteria for Categorising Designated Areas

Airborne activity (% DAC over 8 hours)	Surface contamination (becquerels per cm^2)		Category
	Alpha^a	Beta^b	
>50	>2	>20	H
<50	<2	<20	M
>10	>0.4	>4	
<10	<0.4	<4	L

^{a}The alpha criteria apply provided that the risk assessments and radiological surveys confirm that alpha emitting radionuclides of high toxicity form an insignificant proportion of the total alpha contamination present, or likely to arise, in the area. If this is not the case, the categorisation of the area will require special consideration in consultation with the Radiological Protection Adviser (RPA).

^{b}The beta criteria refer to contamination by an unspecified mixture of beta emitting radionuclides. If the risk assessments and radiological surveys confirm that the beta contamination present, or likely to arise, in an area is predominantly due to beta emitting radionuclides of low toxicity it may be appropriate, with the agreement of the relevant RPAs, to apply less restrictive criteria.

of airborne and removable surface contamination anticipated in the course of planned operations. Each area is then categorised in accordance with the criteria set out below.

Categorisation on internal radiation hazard

Areas designated as controlled areas because of an internal radiation hazard are categorised in accordance with the following criteria, which refer to the conditions anticipated under normal planned operations, as seen in Table 2-12.

Categorisation on external radiation hazard

Areas designated as controlled areas because of an external radiation hazard are categorised according to the following criteria, which refer to the conditions anticipated under normal planned operations, as shown in Table 2-13.

The following additional points need to be considered:

- The criteria for designation relating to concentrations of removable surface contamination refer to the concentrations of such contamination averaged over

Table 2-13. Doserate Criteria

Dose rate (microsieverts per hour)^a	Category
>25	H
<25 but >10	M
<10	L

^{a} Averaged over a working day.

areas not exceeding $1000\,cm^2$ for walls, floors, and ceilings, and not exceeding $300\,cm^2$ for other cases.

- In determining the extent of any controlled or supervised area, it is permissible to take account of physical boundaries such as walls and fixed partitions around the area to be designated.
- If it is considered more convenient to delineate the designated area in terms of such boundaries, this may be done provided that they are not too remote from the working area to enable proper control to be exercised.
- Any area included within a designated area for reasons of convenience is subject to all the requirements applying to the designated area including those relating to local rules, restrictions on access, control of access, control of contamination, and monitoring.
- When determining whether or not an exposure is likely to be significant in the context of this standard, it should be noted that HSE advice on the interpretation of IRR 99 indicates that "significant dose" is taken to mean "a dose of the order of 1 millisievert."

2-14-4. *Personal Protective Equipment*

There is a range of equipment available for the protection of personnel working in a radiological environment:

- *Protective clothing.* This can range from simple overshoes, gloves, and overalls to complete pressurised suits. It also covers standard equipment for protection from industrial hazards, such as steel toe-capped shoes, hard hats, and heavy-duty gloves which can protect from corrosive chemicals. A list of British Standard Specifications for PPE is shown in Table 2-14.
- *Respiratory protection.* For work in areas where there is a risk of minor air-borne contamination, respirators should be worn. These can be either passive respirators (gas masks) where the wearer breathes through a canister of suitable filter material, or positive pressure respirators, which provide a supply of filtered air from a battery operated pump to a face mask. Simple gauze face masks (surgical masks) offer little protection and are not recommended. Pressurised suits carry their own air supply, which can either be via an airline or in the form of pressurised air bottles in a back-pack.

- *Personal electronic dosimeter.* In addition to the standard photographic film badge or TLD worn by radiation workers (which are typically read on a weekly or monthly basis), it is often advisable for workers in an area of enhanced radiation to wear a personal electronic dosimeter. This advice gives an instant read-out of radiation dose, and can also be set to give an audible warning when a threshold level is reached. These devices allow doses to be measured on a task-by-task basis, and are often used in conjunction with a system of dose budgeting. This involves making an estimate of the dose to be received in advance of carrying out a task, and ensures that ALARP assessments of the work methods.
- *Alpha-in-air monitors.* These small portable devices use a battery operated pump to pass a stream of air over a filter material which absorbs airborne contamination. The filters are monitored in a laboratory to measure any alpha contamination. These devices do not give a real-time alarm of airborne contamination, for which separate building alpha-in-air monitors are required.

2-15. Summary

- The term *Ionisation Radiation* encompasses various particles and electromagnetic waves which cause ionisation in substances which absorb them. The principal types of radiation relevant to nuclear decommissioning are alpha and beta particles, gamma rays, and neutrons.
- The measurement of radiation is a complex topic. The most commonly used units are *becquerels*, which measure radioactivity, and *sieverts*, which measure radiation dose.
- The effects of radiation on people can be understood in terms of *acute effects*, i.e., illness which results from exposure to high doses, and *stochastic effects*, which relate to the enhanced risk of cancer or hereditary effects due to low doses.
- Radiation protection principles assume that there is no threshold for stochastic effects. Any exposure of people to sources radiation must be: (i) justified by some benefit, (ii) subject to protective action limits, and (iii) notwithstanding protective action limits, as low as reasonably achievable (ALARA).
- There is a wide variety of instruments available for measuring radiation, including hand-held monitors which give real-time measurements.
- Personnel and the general public are protected by many methods, including shielding, containment, designation of controlled and supervised areas and the use of personal protective equipment (PPE).

Table 2-14. British Specifications for Protective Personal Equipment

BS 697	Specification for rubber gloves for electrical purposes
BS 1397	Specification for industrial safety belts, harnesses, and safety lanyards (current for a transitional period)
BS 1542	Specification for equipment for eye, face, and neck protection against non-ionising radiation arising during welding and similar operations
BS 1651	Specification for industrial gloves
BS 1870	Safety footwear
BS 2653	Specification for protective clothing for welders
BS 4275	Recommendations for the selection, use, and maintenance of respiratory protective equipment
BS 5462	Specification for lined rubber boots with protective midsoles
BS 5845	Specification for permanent anchors for industrial safety belts and harnesses
BS 6159	Polyvinyl chloride boots
BS 6408	Specification for clothing made from coated fabrics for protection against wet weather
BS 6858	Specification for manually operated positioning devices and associated anchorage lines for use with industrial belts and harnesses
BS 7028	Selection and maintenance of eye protection for industrial and other uses
BS 7184	Recommendations for the selection, use, and maintenance of chemical protective clothing
BS EN 132	Respiratory protective devices — definitions
BS EN 133	Respiratory protective devices — classification
BS EN 134	Respiratory protective devices — nomenclature of components
BS EN 136	Full face masks
BS EN 137	Self-contained open circuit compressed air breathing apparatus
BS EN 138	Fresh air hose breathing apparatus with full face mask, half mask, or mouthpiece assembly
BS EN 139	Compressed air line breathing apparatus with full face mask, half mask, or mouthpiece assembly
BS EN 140	Half masks and quarter masks
BS EN 141	Gas filters and combined filters
BS EN 143	Particle filters
BS EN 145	Self-contained closed circuit breathing apparatus
BS EN 146	Power assisted filtering devices incorporating helmets or hoods
BS EN 147	Power assisted filtering devices incorporating full face masks, half masks, or quarter masks
BS EN 149	Filtering half masks against particles
BS EN 166	Specification for eye protectors for industrial and nonindustrial purposes
BS EN 169	Filters for welding and related techniques
BS EN 170	Ultra-violet filters used in personal eye protection
BS EN 171	Infra-red filters used in personal eye protection
BS EN 269	Power assisted fresh air hose breathing apparatus incorporating a hood
BS EN 270	Compressed air line breathing apparatus incorporating a hood
BS EN 271	Compressed air line or powered fresh air hose breathing apparatus incorporating a hood for use in abrasive blasting operations
BS EN 340	General requirements for protective clothing
BS EN 341	Descender devices
BS EN 344	Requirements for safety, protective, and occupational footwear
BS EN 345	Safety footwear
BS EN 346	Protective footwear
BS EN 347	Occupational footwear
BS EN 352	Hearing protectors
BS EN 353	Guided type fall arresters
BS EN 354	Lanyards
BS EN 355	Energy absorbers
BS EN 358	Work positioning systems
BS EN 360	Retractable type fall arresters
BS EN 361	Full body harness
BS EN 362	Connectors
BS EN 363	Fall arrest systems
BS EN 364	PPE against falls from a height — Test methods
BS EN 365	PPE against falls from a height — Instructions for use and for marking
BS EN 374	Protective gloves against chemicals and micro-organisms

Continued

Table 2-14. Continued

BS EN 379	Filters with switchable or dual luminous transmittance for personal eye protectors used in welding or similar operations
BS EN 397	Industrial hard hats — heavy duty
BS EN 421	Protective gloves against ionising radiation
BS EN 464	Protective clothing for use against liquid and gaseous chemicals
BS EN 471	Specification for high visibility warning clothing
BS EN 812	Industrial hard hats — light duty
BS EN 60903	Specification for gloves and mitts of insulating material for live working

2-16. References

1. International Commission on Radiological Protection. "1990 Recommendations of the Commission. ICRP Publication 60," *Annals of the ICRP*, 21(1–3) (1991).

2. International Commission on Radiological Protection. *Recommendations of the International Commission on Radiological Protection, ICRP 60*, Pergamon (1991).

3. Higson, D. "Resolving the Controversy of Risks from Low Levels of Radiation," *The Nuclear Engineer*, 43(5):132–137 (September–October 2002).

4. International Commission on Radiological Protection. "Principles for limiting exposure of the public to natural sources of radiation. ICRP Publication 39," *Annals of the ICRP*, 14(1) (1984).

5. International Commission on Radiological Protection. "Protection of the public in the event of major radiation accidents: principles for planning. ICRP Publication 40," *Annals of the ICRP*, 14(2) (1984).

6. International Commission on Radiological Protection. "Principles for intervention for protection of the public in a radiological emergency. ICRP Publication 63," *Annals of the ICRP*, 22(4) (1991).

7. International Commission on Radiological Protection. "Protection against Radon-222 at home and at work. ICRP Publication 65," *Annals of the ICRP*, 23(2) (1994).

8. International Commission on Radiological Protection. "Dose coefficients for intakes of radionuclides by workers based on the 1990 Basic recommendations and on the 1994 respiratory tract model. ICRP Publication 68," *Annals of the ICRP*, 24(4) (1994).

9. International Commission on Radiological Protection. "Radiological protection and safety in Medicine. ICRP Publication 73," *Annals of the ICRP*, 26(2) (1996).

10. International Commission on Radiological Protection. "Age dependent doses to members of the public from intakes of radionuclides; Part 5, Compilation of ingestion and inhalation dose coefficients. ICRP Publication 72," *Annals of the ICRP* (1996).

11. International Commission on Radiological Protection. "Protection from potential exposure: a conceptual framework. ICRP Publication 64," *Annals of the ICRP*, 23(1) (1993).

12. NRPB. "Board statement on radiological protection objectives for the land-based disposal of solid radioactive wastes," *Documents of the NRPB*, 3(3) (1992).

13. International Commission on Radiological Protection. "Radiation protection policy for the disposal of radioactive waste. ICRP Publication 77," *Annals of the ICRP*, 27 (Supplement) (1997).

14. International Commission on Radiological Protection. "Radiation Protection Recommendations as Applied to the Disposal of Long-lived Solid Radioactive Waste. ICRP Publication 81," *Annals of the ICRP*, 28(4) (2000).

15. International Atomic Energy Agency. *Principles for exemption of radiation sources and practices from regulatory control*, IAEA Safety Series No 89, IAEA, Vienna (1988).

16. CEC. *Recommended radiological protection criteria for the recycling of metals from the dismantling of nuclear installations*, CEC RP 89(1988).

17. CEC. *Recommended radiological protection criteria for the clearance of buildings and building rubble from the dismantling of nuclear installations*, CEC RP 113 (1999).

18. CEC. *Practical use of the concepts of clearance and exemption — Part 1: Guidance on general clearance levels for practices*, CEC RP 122 (2001).

19. NRPB. "Board Statement on radiological protection objectives for land contaminated with radionuclides," *Documents of the NRPB*, 9(2) (1998).

20. Stationery Office. *Work with Ionising Radiation, Approved Code of Practice and Guidance to the Ionising Radiation Regulations 1999, (IRR 99)*, Stationery Office 1999) London, UK.

21. Stationery Office. *The Ionising Radiations Regulations 1999*, SI 1999/3232, Stationery Office (1999), London, UK.

Chapter 3

Decommissioning — Introduction and Overview

3-1. Definition and Scope

The term "decommissioning," as used within the nuclear industry, means the actions taken when a facility has reached the end of its useful life, in order to ensure that it is managed safely in a manner that protects workers, the general public, and the environment. These actions can range from simply closing the facility (with minimum works to remove radioactive material coupled with continuing maintenance and surveillance), to the complete dismantling of the facility and restoration of the site for unrestricted use. In the case of a UK nuclear licensed site, achieving an end-point of unrestricted use involves a process of delicensing, which is addressed in Chapter 23.

3-2. Stages of Decommissioning

It is internationally accepted that there are three distinct stages of nuclear decommissioning, originally defined by the IAEA [1]. These stages may be separated by extended periods of Care and Maintenance (C & M) with the appropriate security and radiological surveillance, or they may follow directly one after the other in a continuous, systematic, and progressive manner. The state of a facility at the end of each of the three stages is described below and illustrated in Figure 3-1.

Stage 1. Reactors are completely defueled and the fuel usually shipped away from the reactor. All heat transport fluids and readily removable contaminated materials are removed. For nonreactor facilities, all radioactive sources and readily removable equipment are removed. The containment is maintained intact, and the atmosphere inside the containment building and enclosures are controlled. The ventilation system may be operated as required. Access to the inside of the containment building is controlled by physical barriers and administrative procedures. Periodic measurements and visual checks are

carried out to ensure that contamination control systems continue to function properly.

Stage 2. Contaminated areas are decontaminated to the extent appropriate. Remaining areas with unacceptable residual radioactivity levels (e.g., reactor core structures) are sealed to prevent unauthorised access. Contaminated parts that are easily dismantled are removed and transferred off-site or into plant areas that are to be sealed. Ventilation plant and other active safety systems are no longer needed. Some monitoring equipment will remain operational, depending on specific circumstances. Some parts of the plant or site could be converted to new uses or released with certain constraints for uses not involving other radioactive sources. Surveillance around the restricted area is required, but is less extensive than in Stage 1.

Stage 3. All materials, equipment, and structures in which radioactivity levels exist above prescribed limits are removed to an approved storage or disposal site. The site and any remaining equipment and materials may be released for other purposes without any radiological protection restrictions. No further surveillance, inspection, or tests are necessary.

It should be noted that, although these stages of decommissioning were originally defined by the IAEA and are widely used internationally, they are no longer recommended by the IAEA. The Agency now defines "phases" in a facility's life as follows:

- Operational phase.
- Shut-down transition phase (defuelling & postoperational clear-out known as POCO).
- Preparation for safe enclosure.
- Safe enclosure period.
- Final dismantling.

Of course, if decommissioning proceeds directly from the shut-down phase to final dismantling, the safe enclosure (sometimes known as "safestore") phase is not relevant.

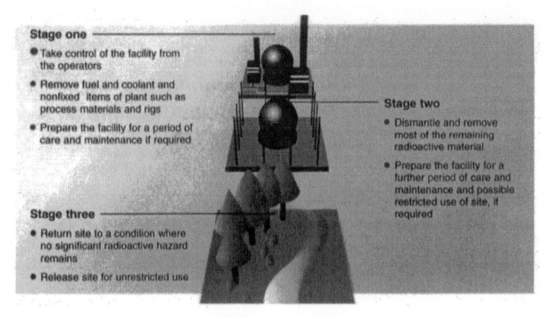

Stage one
● Take control of the facility from the operators
● Remove fuel and coolant and nonfixed items of plant such as process materials and rigs
● Prepare the facility for a period of care and maintenance if required

Stage two
● Dismantle and remove most of the remaining radioactive material
● Prepare the facility for a further period of care and maintenance and possible restricted use of site, if required

Stage three
● Return site to a condition where no significant radioactive hazard remains
● Release site for unrestricted use

Figure 3-1. The Three Stages of Reactor Decommissioning.

Different countries approach the detail of this reactor decommissioning framework in different ways. In the US, detailed regulatory guides and rules which define decommissioning have been issued by the US Nuclear Regulatory Commission (NRC) and describe decommissioning as "safely shut down the Nuclear Power Plant (NPP) and reduction of the radioactivity inventory to a level which allows the normally not restricted use of the plant or remaining parts of the plant." Decommissioning strategies include DECON (Decontamination), SAFSTOR (Safe Storage), and entomb (Entombment). Periods of Care and Maintenance (C&M) between decommissioning stages are perhaps up to 60 years.

In the UK, regulatory regimes are less prescriptive and the Regulator requires to be satisfied of the adequacy of the decommissioning strategy within the framework of Government policy and guidance that they issue to their inspectors. In general, the "process of reactor decommissioning should be undertaken as soon as it is reasonably practical to do so, taking account of all relevant factors" (including the type of facility, the nature of its radioactive inventory, cost and overall financial, economic and resource issues). Such wording is inevitably open to interpretation. Regulators in the UK press for the "systematic and progressive reduction of hazards" on the earliest possible timescales [2]. Conversely, operators may wish to give more consideration to relevant economic factors and balance the real advantages in the safety and simplicity arguments of decommissioning after allowing for a period of radioactive decay (less

dose uptake to workers and lower cost should remote radioactive waste handling not be required). An attempt is made to minimise total discounted costs, including infrastructure costs, with Stage 2/Stage 3 decommissioning C & M periods therefore extending to perhaps 100 years.

In Sweden and Germany, for example, more immediate total decommissioning is the favored strategy. However, BWRs have relatively less difficult waste treatment issues than those associated with the older generation UK Magnox type reactors.

3-3. Drivers Determining Decommissioning Plans and Programs

Later chapters deal in more detail with Government policy and the regulatory framework. At this point, it is enough to note that, in general, Stage 1 decommissioning or Post Operational Clear Out (POCO) should be undertaken as soon as possible after closure. As indicated above, subsequent stages of decommissioning may follow on directly or be deferred for a period, depending on circumstances. The timing for the decommissioning of different facilities must be considered on its merits, so as to achieve the most appropriate safe and secure, environmentally, and publicly acceptable approach offering value for money. Some relevant factors to be taken into consideration are listed in Table 3-1.

Table 3-1. Some Relevant Factors to be Taken into Consideration when Considering Decommissioning Timescales

- The potential hazards to public, workers, and the environment.
- The availability of waste routes.
- Corporate memory retention and availability of suitably qualified and experienced personnel (SQEP).
- The time required to plan the work and develop decommissioning techniques and equipment.
- Radioactive decay:
  - the benefit from decay of cobalt 60 in reactor steels, and
  - in-growth of americium 241 in Plutonium Contaminated Material (PCM).
- Retention of the structural integrity of the facility.
- Maintenance of the organisation.
- Changes in regulatory requirements.
- Changes in the real value over time of costs and benefits.
- The time value of money (discounting effects — see Chapter 13).
- The impact on support and infrastructure costs.

3-4. Risk Versus Hazard

It is a basic principle of radiological protection that any practice involving the exposure of people to radiation should be justified in terms of a positive net benefit. In the case of nuclear power generation, the benefit derived from the electricity produced goes some way to justifying the risk involved in producing it. However, once the reactor has been shut-down, there is no longer any benefit being produced, and as long as the reactor can be retained in a safe and secure condition with minimum risk to the public then there would be far less reason to get on with the decommissioning; especially since this activity in itself may give a small dose uptake to workers. For this reason, UK decommissioning policy is framed in terms of "systematic and progressive reduction of hazard."

The distinction between hazard and risk is difficult to grasp at first sight, but is nevertheless fundamental. Risk involves the probability of some event happening multiplied by the consequences of that event. Hazard is an intrinsic property of an object, whether it is a can of petrol or a can of nuclear fuel. In the case of the can of petrol, the size of the hazard depends on the amount of petrol and its volatility. A 200 liter drum is more hazardous than a 1 liter bottle. The risk associated with the petrol is determined by the potential for it to be dispersed and subsequently ignited, ingested, or even just absorbed on the skin. This can be determined by establishing a set of fault sequences that involve identifying *pathways* and *receptors* that might be exposed, all of which can be quantified in probabilistic terms. The risk associated with a hazard can be reduced by engineered safeguards — for example by minimising the amount that is stored (inventory), by ensuring that it is stored in a robust container, by avoiding potential sources of ignition should a release occur and by installing fire detection and suppression systems. All of these measures reduce the risk. Only the first of them (minimising inventory) reduces the hazard. In the jargon of modeling risk and consequences, the hazard is described as the *source term*.

When considering the risk and hazard associated with radioactivity, measures that effectively reduce these can be considered (see Figure 3-2). For example, vitrified fission product clearly has intrinsically less risk than the same amount of radioactivity in the form of highly active liquor. The probability of a dangerous accident leading to a release of radioactivity to the environment is less (all other things being equal) with the waste in vitrified form. However, the vitrified waste could still present a hazard, as it could be vaporised (for example, in an intense fire), and it, therefore, retains its intrinsic property to do harm. It is legitimate to refer to reduced hazard when it involves measures that are robust to any conceivable event that could occur to the hazard, without outside intervention.

Later chapters will discuss decommissioning in terms of hazard reduction. This includes measures to condition radioactive waste in *passively safe forms*, i.e., where the radioactivity is packaged or immobilised in a form that is physically and chemically stable and which minimises the need for control and safety systems, maintenance, and monitoring.

3-5. Contrasting Reactor Decommissioning With Other Facilities

Different facilities present different hazards (see Figure 3-2). Safety, environmental, and economic considerations normally require at least some work to be undertaken immediately after closure (i.e., Stage 1 decommissioning). Beyond this, the exact scope of work undertaken at each stage, and the length of time between stages, is determined on a case-by-case basis. Normal practice is summarised below.

Reactors

Reactors are normally defuelled immediately after closure and the coolant removed. This removes typically 99% of the radioactivity, and substantially reduces the hazard presented by the facility. The majority of the remaining radioactivity is normally embedded in the structure in the form of activation products. Delaying the later stages of decommissioning allows the radiation levels and the quantities of radioactive waste to fall as a result of radioactive decay.

Reactor structures are normally robust and can be maintained with a high level of safety over an

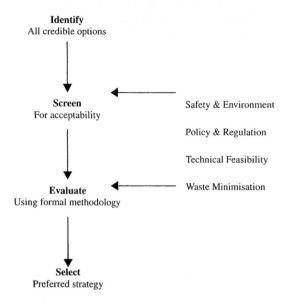

Figure 3-2. Selecting a Decommissioning Strategy.

extended period. Under such circumstances, it will be as satisfactory to retain decommissioning wastes *in situ*, pending a disposal route, as it will be to decommission the reactor and place the wastes in interim storage.

The timing of Stage 3 decommissioning of reactors depends on a range of factors, a number of which support some deferral. Amongst the most important of these are:

* The reduction in waste volume and operator dose resulting from radioactive decay.
* The availability of waste management and disposal routes.
* Economic factors, including the alignment of final decommissioning with the closure of the nuclear licensed site.

The benefits of allowing radioactive decay to reduce waste volumes and the potential doses to decommissioning operators are well recognised. In the UK, if a national repository for ILW becomes available, it will be possible to dispose of stored wastes, and thus close the site; at this point, there are strong economic drivers to complete decommissioning and site restoration to save infrastructure costs. If a national repository is not available before the closure of the current UK Drigg Low Level Waste Facility, it may nevertheless be advantageous to ensure that Stage 3 decommissioning of reactors is complete. For example, if a reactor will have largely decayed to LLW by around 2040 and will require LLW disposal, then it is best to decommission prior to the planned closure of such an existing disposal facility rather than rely on only the possibility of an alternative waste route being

available. Hence, the latest stage for commencing Stage 3 decommissioning is around 2035–2045 (depending on the expected duration).

The "DIDO" Materials Test Reactor is at UKAEA Harwell, Oxfordshire. It has undergone Stages 1 and 2 decommissioning, and is currently in a state of Care and Maintenance (C&M) awaiting the benefits derived from the natural radioactive decay of reactor materials before final Stage 3 decommissioning (see Figure 3-3).

Plutonium Facilities

Facilities which have been used for handling plutonium do not normally benefit from radioactive decay between Stages 1–3, because of the long half lives involved. The optimum strategy for such a facility requires prompt decommissioning, which minimises the dose-rate to workers arising from the in-growth of the gamma emitter americium 241 and the deterioration of the plant due to damage by alpha radiation (see Annex 4, Section A.4.6). Normal practice is to remove process equipment, dismantle glove-boxes, and wash out vessels during Stage 1 decommissioning. The timing of subsequent stages of decommissioning takes account of the integrity of the building structure and plant (see Figure 3-4).

Caves and Cells

Caves and cells, e.g., postirradiation examination (PIE) facilities, are generally robust concrete structures which provide a high degree of containment. Mobile activity is normally removed from caves and cells immediately after closure, in order to reduce the hazard presented by the facility. The hazard presented by these structures after the removal of mobile activity is normally low, so later stages of decommissioning can safely be deferred if appropriate.

Other Facilities

Other facilities, including waste treatment plants, waste stores, and laboratories, will undergo Stage 1 decommissioning immediately after closure. The timescales for the later stages of decommissioning will be determined on a case-by-case basis, taking account of the overall site strategy and other related factors (see Figure 3-5).

3-6. Availability of Guidance and Reference Information

There are numerous sources of information on decommissioning. Official guidance documents are available

From Reactor Operation   to   Care and Maintenance

Figure 3-3. From Reactor Operation to Care and Maintenance.

Before   After

Figure 3-4. Alpha Laboratories (Plutonium Handling Facilities, UKAEA, Harwell) from Laboratory Use to Complete Decontamination/Decommissioning for Alternative Use. (Currently in Unrestricted Use as Offices.)

- Shielded cells for remote handling, e.g., post-
  irradiation examination (PIE) of fuel
- Waste treatment plants and stores
- Active laboratories

Figure 3-5. Examples of "Other Facilities": Shielded Cells and Active Laboratories.

in the UK from HSE / NII, EA and SEPA, IAEA, and the Commission of the EU. Companies such as BNFL and UKAEA have their own internal management systems which set out policy and guidance [3-7].

3-7. References

1. International Atomic Energy Agency. *Decommissioning of Nuclear Power Plants and Research Reactors*, IAEA Safety Guide No. WS-G-2.1, Vienna (1999).

2. Stationery Office. *UK Government White Paper Managing the Nuclear Legacy – A strategy for action*, Cm 5552, The Stationery Office, London (July 2002).

3. Stationery Office. *Review of Radioactive Waste Management Policy*, Cm 2919, The Stationery Office, London (1995).

4. HSE. *Decommissioning on Nuclear Licensed Sites. Guidance for Inspectors*, HSE, London (2001).

5. UKAEA. "Radioactive Decontamination," *Atomic Energy Code of Practice AECP1057*, Issue 4, UKAEA, Harwell, Oxfordshire (1992).

6. UKAEA. "Care and Maintenance Guidance Document," *Atomic Energy Code of Practice AECP1085*, Issue 1, UKAEA, Harwell, Oxfordshire (1998).

7. Cumo, M., I. Tripputi, and U. Spezia. *Nuclear Plants Decommissioning*, University of Rome, Italy (2002).

Chapter 4
Typical Government Policy on Decommissioning

4-1. Introduction

This chapter briefly describes how and why Government is involved in the decommissioning of nuclear facilities in the UK. It goes on to describe some of the main issues from a Government perspective, current issues, and the challenges ahead. The chapter also describes decommissioning within the current European Commission research framework programs.

4-2. How and Why is Government Involved?

4-2-1. *Historical*

In the 1940s, 1950s, and 1960s, the UK Government and the nuclear industry focused on the development and application of nuclear technology for civil and weapons purposes. Harwell, in Oxfordshire, was the first UK nuclear site, but Sellafield, Dounreay, Aldermaston, and most other nuclear sites in the UK date from that period. Much was achieved and the nuclear industry now makes a significant contribution to the British economy. However, in the early years the industry created substantial liabilities in the form of wastes that needed to be treated and plants that needed to be decommissioned [1]. These liabilities are often referred to as the "legacy." The early emphasis was on developing the new technology for military and civil purposes rather than on consideration of the most appropriate means of decommissioning redundant facilities in the future. The plethora of bodies involved in the administration of nuclear affairs in the UK is illustrated in Figure 1-8 in Chapter 1.

Government has interests in nuclear safety, security, decommissioning and waste management policy, assistance to Eastern Europe and the former Soviet Union, and maintaining a relationship with the industry (e.g., through the British Nuclear Industry Forum (BNIF)).

4-2-2. *Safety*

The Secretary for Trade and Industry is accountable to parliament for safety at nuclear power stations and other licensed civil nuclear sites in the UK. The Secretary of State is advised on nuclear safety issues by the independent Health and Safety Commission (HSC), which has a statutory responsibility for ensuring that there is an adequate framework for the regulation of health and safety across most industry sectors, including the UK nuclear industry.

4-2-3. *Regulatory Policy*

Government is responsible for nuclear regulatory policy and the execution of that policy. It is committed to the principles of sustainable development. In connection with nuclear site remediation, this involves taking responsibility for the forward clean-up program now (since it is the current generation that has gained from the technology) rather than leaving this to future generations (see UK Government White Paper Cm 4345 [2]).

4-2-4. *Security*

The Department of Trade and Industry (DTI) participates in the international debate about standards of physical protection, and ensures, through national regulations and guidance on security measures, that the appropriate measures are taken in relation to the likely threat against the facilities. Since October 2000, the DTI has been the security regulator for the UK's civil nuclear industry. It is responsible for setting the standards and enforcing compliance.

4-2-5. *Decommissioning and Waste Management*

The Government's priority is to ensure that the legacy is managed safely, securely, and cost effectively in a way that ensures protection of the environment [1].

Radioactive Waste Management Policy

Overall policy for radioactive waste management is the responsibility of the Department of Environment, Food, and Rural Affairs (DEFRA) and the devolved administrations. The DTI is primarily concerned with ensuring that the views of the nuclear industry and certain other industries are represented in the decision-making process determining radioactive waste management policy.

Coordination and Best Practice

DTI supports a Liabilities Management Group (LMG), which draws together those public sector organisations in the UK with nuclear liabilities. It is made up from industry members (BNFL, including BNFL Magnox Generation, UKAEA, MoD, and DTI). It has task forces looking into best practice in the areas of safety, research and development, procurement and performance. LMG guidance documentation covering these areas is available from the member organisations [3].

4-2-6. *National Economic Benefits*

Government is inevitably involved in nuclear power, because it continues to own a substantial fraction of the industry that grew up in the public sector. It is financially responsible for the majority of the legacy, and it recognises that the scale of spend is very significant. Indeed, it was realised in 2001 that the costs of decommissioning the BNFL liabilities were so large that they could not be covered by BNFL's reserves and anticipated income streams from BNFL's commercial work and historic customers alone. Such a large expenditure, therefore, contributes significantly to the economy. In addition, Government wishes to ensure that the nation benefits from the skills held in the nuclear area.

4-2-7. *The Consequences of Failure*

Government business is political in nature, and Ministers are acutely aware that dealing with an industry with potentially high hazards leads to public concern. Nuclear issues generate considerable public and parliamentary interest. Programs of work involving public money have to be justified against other calls on public expenditure in a meticulously open and transparent way, such that it can be demonstrated that the reduction of nuclear legacy hazards gives a wider benefit to society. The consequences of failure are, therefore, too great to be left entirely up to market forces.

4-3. Some of the Key Drivers for Government

4-3-1. *The Costs Involved*

There are approximately £50bn (undiscounted) of public sector civil nuclear liabilities in the UK. The BNFL Sellafield site alone accounts for some 65% of the total. The annual UK expenditure on nuclear clean up is currently some £1bn per annum. Government has underwritten the costs of all UKAEA liabilities (some £8bn) and funds UKAEA at about £270m per annum to maintain safety and security and to achieve progress on its environmental site restoration program. The costs of this work are therefore of key consideration to Government.

Estimated reactor decommissioning costs in other countries are given in Table 4-1. Such estimates need to be carefully understood before being used for comparative purposes, taking into account some of the typical factors below [4]:

- Their scope (Are the costs to the end of Stage 3 or just for defuelling? Do the costs include return of the site for unrestricted use back to the environment, etc.?);
- The timing of the decommissioning (Is the level of radioactivity at the time of decommissioning such that robotic handling is required?);
- Technical factors (Are the costs associated with one of a series of reactors or is this an early one-off specialist reactor with potential difficulties such as arising from the core fire in Pile 1 at Windscale in the UK?);
- Waste management issues (Are waste routes available? Are the treatments and conditioning requirements for the decommissioning wastes understood? Do the costs include fuel and decommissioning waste conditioning to recognised end points and associated disposal costs?);
- The decommissioning program (Is the program continuous or does it involve periods of Care and Maintenance?);
- Administrative factors (Are the requirements stemming from Government policy and the Regulatory system prescriptive and well understood or open to debate and uncertainty?); and
- Financial and economic factors (Are the figures all quoted in the same money of the year values? Are they discounted or undiscounted estimates? What discount rate has been used, etc.?).

Table 4-1. Estimated Reactor Decommissioning Costs

Type of reactor	Estimated decommissioning costs	Notes
Prototype Fast Breeder, PFR, Dounreay, UK, 270 MW	~£500m (2002 money values, undiscounted)	Not designed with decommissioning in mind. Difficult liquid sodium coolant.
Generic PWR, USA	US$ 290m (1999, Lower threshold value)	NRC minimum value estimates to licence release (not to "green field" site).
Generic BWR, USA	US$ 370m (1999, Lower threshold value)	NRC minimum value estimates to licence release (not to "green field" site).
Average BWR, USA	US$ 420m (1998 money values to licence termination)	NEI (Nuclear Energy Institute) study based on 30 BWRs from 540 to 1140 MW with and without full disposal and site remediation costs.
Average PWR, USA	US$ 368m (1998 money values to licence termination)	NEI (Nuclear Energy Institute) study based on 60 PWRs from 500 to 1095 MW with and without full disposal and site remediation costs.
Caorso BWR, Italy, ~850 MW	~US$ 500m (2000 money values)	Société Gestione Impianti Nucleari (SOGIN).
Trino PWR, Italy, ~250 MW	~US$ 280m (2000 money values)	
For specific reactor types, costs per reactor:	Million Euro	1997–2000 UNIPEDE study covering 12 countries (10 European plus South Africa and Canada). Note spreading between lowest and highest estimated costs for:
Belgium	548	• Overall — factor of 6
Canada	403	• Project Management, Planning, and Licencing — factor of 3
France	498	• Waste Management — factor of 10.
Germany	601	• Safestore strategy in UK negates direct comparison.
Hungary	459	
Italy	466	
The Netherlands	562	
South Africa	340	
Spain	323	
Sweden	273	
Switzerland	458	
UK	293	

There is an increasing cost trend over time for reactor decommissioning associated with increasing waste disposal costs.
Decommissioning to final site restoration has been estimated to add some US$19.3m to costs (2000 money values) in the US.

4-3-2. National and International Responsibilities

To the public, the nuclear industry appears to pose a unique hazard and is seen by some as involving an unwanted risk in their lives. The nuclear industry is controversial and the Government has to balance rational assessment of the risks against the costs. Government, then, has national and international responsibilities for ensuring that a framework is in place that addresses society's concerns about safety, security, and protection of the environment. Since the present generation has benefited from nuclear power and technology over the last 50 years, the principle of sustainable development suggests that this generation should address the problem of decommissioning the legacy rather than pass it on to future generations. However, the timing of decommissioning

has to be taken into account for the factors described in Chapter 3, including arrangements for the disposal of waste. Government, therefore, seeks to achieve this reasonable balance between the necessary and sufficient *precautionary* measures associated with the risks involved and the *proportionality* of the costs and impact on society of reducing these risks to a tolerably acceptable level.

4-3-3. Business Potential

Dealing with the nuclear legacy contributes to the UK's (and that of other countries) standing and export business potential in the world. Potentially, world-wide, there are over 400 power reactors, each of which could cost up to £500m to fully decommission, including site remediation to unrestricted future use. In addition, there are almost 700 research reactors, over half of which are more than 20 years old. Nuclear decommissioning is therefore big business. National firms have the technology and ambition to be successful in winning a significant share of this world-wide opportunity if based upon a good home market record of achievement. The DTI has an "International Nuclear Safety Programme" (NSP) currently standing at some £84m over 3 years to improve safety in Eastern Europe and the Former Soviet Union (FSU).

4-4. Current Developments

4-4-1. *Structural Issues*

Government wishes to ensure that the decommissioning is undertaken effectively and that public money is, therefore, being well spent. In the UK, Government uses a Quinquennial Review (QQR) process to assess public services. Such reviews first look at whether such a public sector service is indeed required, and then look at how its remit, ways of working, and overall efficiency may be improved.

A recent 2000–2002 QQR of the UK civil nuclear industry has concluded that a new body is required to oversee the management of all UK publicly funded civil nuclear liabilities (i.e., for managing decommissioning, waste management, and site environmental remediation). It has concluded that the public sector cannot discharge these liabilities alone. It requires the help of the private sector and wishes to establish a competitive market for site restoration. Government is clear that it is only by managing the liabilities as a whole that the necessary focus, strategic control, and direction can be achieved. Government will, therefore, set up a Liabilities Management Authority (LMA) to let contracts on a competitive basis to, and work in partnership with, site licensees for the discharge of decommissioning programs. The LMA will seek to develop a strong supply chain and a skills base capable of sustaining the clean-up program over the long time scales involved [1]. The proposals are described in Government White Paper, Managing the Nuclear Legacy, Cm 5552, July 2002 and as reviewed by the House of Commons Trade and Industry Committee [5].

4-4-2. *Skills Issues*

The successful discharge of the forward nuclear decommissioning program requires the continued need for nuclear "know-how." There is a recognition of the "graying" of staff (staff age profiles show a large proportion of staff aged 45 and over and a lack of younger persons being attracted to the industry). At the university degree level,

there are less and less students taking technical courses in the UK and across Europe. Decommissioning will, however, require large numbers of staff with engineering and technical skills as well as associated generic skills such as project management, planning, safety assessment, and risk management. Following a national forum in 2001, the DTI has set up a Nuclear Skills Group. This will coordinate human resource planning by:

- identification of the skills gaps through a skills audit,
- development of solutions in conjunction with stakeholders, and
- stimulate initiatives to encourage workers to the industry.

This has already spawned a nuclear Decommissioning, Waste Management and Site Environmental Remediation Post-graduate course at the University of Birmingham in the UK. The essential need for staff to be able to demonstrate that they are suitably qualified and experienced persons (SQEPs) under UK site licence conditions, as described in Chapter 1, is a driver for such initiatives at all technical levels throughout the workforce.

4-4-3. *Regulatory Issues*

The public requires that the regulatory environment will be effective and that it will not allow an unacceptable level of risk. The Regulatory authorities are, therefore, rightly demanding, and Government is committed to further improvements in Regulation. This requires of Government transparency, accountability, consistency, and targeting. The Health and Safety Executive and the Environment Agencies continuously review their processes in the light of these improvement themes. In addition, the Government encourages a vigorous industry/regulator dialogue.

First, Government is ensuring adequately coordinated interaction between the Safety and Environmental regulators [6]. Government is also looking at regulatory areas where policies, such as "delicensing," need to be reviewed or updated. Of critical importance to nuclear environment regulation over the next few years will be the Department of the Environment, Food, and Rural Affairs (DEFRA) policy documents on "Statutory Guidance" and "UK Discharge Strategy" (related to UK aspects of the 1998 OSPAR agreement). Overall, the task is one of ensuring "joined up Government."

4-4-4. *Waste Issues*

Government and the devolved administrations are currently looking into the most appropriate ways of managing solid radioactive wastes; especially to defined "end points" including disposal. For reasons of transparency and engagement of the public, this is being handled through a public consultation exercise entitled "Managing Radioactive Waste Safely." This is a slow business so as to avoid past experiences leading to lack of public confidence. The first stage of engagement with the public is, therefore, to set the scene and not to try and resolve issues. Views are sought on specific questions such as the options for the disposal of radioactive wastes, factors to be considered when planning for decommissioning, and the status of recovered plutonium and uranium. An independent body is then proposed to advise Government on the initial conclusions. The whole consultation process may take up to 7 years.

4-5. Decommissioning Research Framework Programs of the European Community

Since 1979, European Government has spent more than 60 million Euros on:

*   the development of decontamination and dismantling techniques for different types of nuclear installations;
*   technologies for waste minimisation, such as melting of steel components;
*   the development of decommissioning strategies and management tools;
*   the development of remote handling systems for high activated components (the TELEMAN program); and
*   development of planning and management tools for decommissioning projects.

As a result of such Research Framework Programs, the EC considers that most of the dismantling techniques and technologies involved in the decommissioning process have now reached the industrial stage, and a large number of reports are available [7].

The emphasis in European funded Research and Development (R&D) is, therefore, changing in the 5th Framework Research Programme (FP-5, 1998–2002) from technology research to:

*   development of management tools;
*   coordination of member countries' requirements;
*   collection of the practical results from member countries' decommissioning programs;
*   dissemination of experience and training requirements (including training programs); and
*   dissemination of the results of former research programs.

The EC has proposed a European Networks of Excellence around specialist areas such as nuclear decommissioning

and, in the 6th Framework Research Programme (FP-6, 2002–2006), the EC will support a Thematic Network on Decommissioning [8]. (The EC DB Tool is a database used to collect technical performance data and the EC DB COST is a database used for collecting waste arisings, dose uptake, etc.)

Of particular importance are the results from five pilot decommissioning projects which have been sponsored by the EC since the early 1990s. They have been chosen to cover different aspects of decommissioning or types of plant and are described below:

*   *Fuel Reprocessing Plant* (AT1 at La Hague, France). Successfully completed and the plant being cleaned up for alternative use.
*   *Windscale Advanced Gas Reactor* (WAGR at Windscale, UK). A textbook case for the future dismantling of graphite cored reactors. Innovative dismantling techniques included: (i) use of gamma cameras to detect and sort radioactive hot spots, (ii) decontamination by use of lasers, (iii) ultrasonic cleaning of filters and surfaces, and (iv) stereoscopic cameras for the control of remote handling machines.
*   A *pressurised water reactor* (PWR - BR3, Belgium). Development of dry and underwater cutting techniques on highly active reactor internals.
*   A *boiling water reactor* (BWR - KRB-A, Gundremmingen, Germany). Dismantling of the heat exchangers, the core internals, concrete bioshield, and the reactor pressure vessel successfully completed.
*   A *VVER type reactor* (Russian design, Greifswald, Germany). One of the largest reactor decommissioning projects in the World, commencing with Stage 3 dismantling of five VVER-440 reactors and one VVER-70 reactor. Remote controlled dismantling of the first reactor pressure vessel commenced in 2001.

FP-5 work also includes:

*   Standardised Decommissioning Cost Estimation as a benchmark exercise from the VVER reactor work;
*   Production of a compendium on the state-of-the-art in decommissioning; and
*   Documentation on innovative remote dismantling techniques (IRDIT).

In conclusion, it can be seen that the above initiatives when coupled with the considerable EC work on radiation safety during decommissioning (Directive on Basic Safety Standards) [9], the environmental impact of decommissioning [10], waste treatment and the unconditional release of dismantling waste [11], together with work on public perceptions [12] all greatly assist the decommissioning challenges ahead.

4-6. The Challenges Ahead

The challenges for Government may be summarised as:

- To maintain the importance of nuclear decommissioning within Government budget allocations on the basis that it has strong sustainability and environmental restoration credentials;
- To explain that much has been achieved, but that much more has still to be done;
- To demonstrate that the goal of efficient, effective, and successful decommissioning is achievable and can be delivered; and
- To support National, European, and International initiatives which assist the effectiveness of decommissioning programs.

However, it must do this within a framework that protects public interests and engages stakeholders in an open and transparent fashion rather than using a "decide and defend" approach. In this way, Governments may then exploit their national capabilities in the very substantial home and overseas markets. Essentially, "World-class engineering skills are essential . . . , from the dismantling of closed nuclear facilities, to the construction of new plant for the processing of nuclear materials and waste, and from the decontamination of land to the development of safety and environmental protection systems" [13].

4-7. References

1. Extract from a Statement by Patricia Hewitt, Secretary of State for Trade and Industry, on Civil Nuclear Liabilities (28 November 2001), prior to the publication of Government White Paper, Cm 5552 (July 2002).

2. UK Government White Paper, Cm 4345, Sustainable Development, Stationery Office, London, UK.

3. Procurement practices in liabilities management; Procurement performance measures in liabilities management; An assessment of the records and documents required for efficient liabilities management; The "safegrounds" project, dealing with land management guidance for nuclear licensed and defence sites, www.safegrounds.com.

4. UNIPEDE. *OECD/NEA Cost Estimates for Decommissioning of Nuclear Reactors. Why do they differ so much?* UNIPEDE 1998-211-0002 (1998).

5. House of Commons, Trade and Industry Committee, Managing the Nuclear Legacy, Comments on the Government White Paper, Fifth Report of Session 2001–02, HC 1074-1, July 2002, London Stationery Office (Government White Paper "Managing the Nuclear Legacy — A Strategy for Action," Cm 5552 (July 2002)).

6. HSE and EA. "The working relationship between HSE and EA on nuclear safety and environmental regulatory issues," *HSE & EA Statement of Intent* (Aug 2001).

7. EC Nuclear Energy Work Programme, 1998–2002, *Official Journal L26*, 1999.

8. www.ec.tnd.net. The EC DB TOOL described in this website is a database used to collect technical nuclear facility decommissioning performance data and the EC DB COST is a database used for collecting waste arisings, dose uptake, etc. and associated cost data.

9. Directive 96/29/EURATOM.

10. *Environmental Impact Assessment for Decommissioning of Nuclear Installations, EIA* (Scotland) Regulations, 1999.

11. EC Radiation Protection 89, *Recommended radiation protection criteria for the recycling of metals from the dismantling of nuclear reactors*, EC Radiation Protection.

12. www.ec-decom.be.

13. DTI, Energy Minister, Brian Wilson, *New Civil Engineer*, special report on Nuclear Decommissioning, Thomas Telford Press, UK (2001).

Chapter 5

The Transition from Operations to Decommissioning

5-1. Introduction

The transition phase between operations and decommissioning begins when the plant has been declared or forecast to become redundant. It continues until the decommissioning plan is firmly in place and being implemented. Successful decommissioning depends on careful planning, before the plant shuts down, to ensure a smooth transition from the end of operations to the start of decommissioning. Ideally, adequate notice should be given of the intention to shut a plant — up to 2 years is required to carry out the required planning work. In reality, the decision to shut a plant is often precipitated by adverse commercial circumstances, which may leave less time to plan the transition.

The physical activities carried out on the plant during the transition phase are typically:

- removal of spent fuel and other hazardous materials and wastes
- post-operational clear-out (POCO) to reduce the hold-up of hazardous material,
- changes in the configuration and status of systems reviewed against a safety assessment, and
- installation of barriers to prevent the spread of contamination, where necessary.

Shut down of a plant also involves a major organisational change, often with a major reduction in staff numbers. This will be accompanied by significant cultural change as the nature of the work changes, with a much greater focus on project management approach. A smooth transition process therefore needs to:

- consider measures to identify and preserve key skills and knowledge, and
- mitigate the impact of the changes on staff morale.

Perhaps most importantly, during the transition phase there is a need to ensure that the funds required for decommissioning are allocated in a timely manner.

A smooth transition will not be achieved if there is:

- insufficient time to plan for it, due to a sudden decision to close the facility,
- indecision about the decommissioning strategy,
- lack of clarity about regulator requirements,
- lack of suitable infrastructure, e.g., waste management facilities,
- loss or demotivation of personnel, and
- insufficient funds.

With regard to the last point, bear in mind that all organisations have a budgeting process that requires funds to be allocated sometime in advance of commitment — this could be several years.

5-2. Preparing for the Transition

A project team should be established well in advance of the planned shut-down to prepare for the transition phase. This team will:

- prepare a decommissioning strategy and plan (or update any that already exist),
- identify options for spent fuel and/or management of other radioactive materials and wastes,
- identify routine care and maintenance requirements through the transition phase and into decommissioning,
- identify manpower requirements,
- estimate costs and secure sources of funding,
- evaluate project risks, and
- prepare safety documentation.

A typical decommissioning project team during the final operational phase of a plant might involve the functions shown in Figure 5-1.

The management structure at the start of the transitional phase will be that which ended the operational phase. The structure will then evolve as the transition progresses. The transition plan should address the changes and additions in roles that will be required.

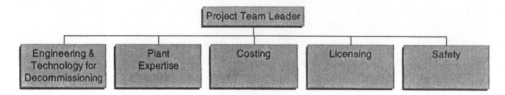

Figure 5-1. A Typical Small Decommissioning Project Team Structure During Plant Operation.

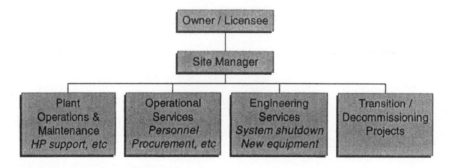

Figure 5-2. A Notional Functional Organisational Structure Suitable for the Transition Phase.

Two different types of activity can be identified during the transition phase:

- conversion of the facility from an operational configuration to a safe shut-down state and subsequent preparation for either long-term care and maintenance (Safestore) or immediate dismantling, and
- preparation of a detailed decommissioning plan requiring current information on the condition of systems, structures, components, and materials.

A typical organisational structure during the transition is shown in Figure 5-2.

5-3. Human Resource Issues

It is necessary to prepare staff for the technical and organisational changes that accompany the shut-down of a facility. The most obvious change is the reduction in staff numbers that usually occurs. This affects staff morale and commitment. To mitigate this effect, it is necessary to consult with staff and other stakeholders. Staff should be informed in a timely way of who will be retained and who will be displaced. During the last months of operation of a nuclear power plant or waste management facility it is necessary to put measures in place to enhance the co-operation of the operational personnel with those planning the future decommissioning works.

The transition planning team will identify a staff reduction profile. This will take account of the need to use experienced operating staff to carry out at least the initial decommissioning tasks, such as fuel removal, POCO, and

system reconfiguration. A policy is required to identify what work should be contracted out.

It is also important to identify those staff who have key skills and knowledge that must be retained. Often these will prove to be the individuals best placed to find alternative employment and they may need special incentives to ensure that they stay.

The basic skills required to support the transition are the same as those needed for operation of the facility. However, there is a key difference. Decommissioning work places a strong emphasis on project management principles. This involves:

- a specification for the scope and the end-point for the works required,
- safety assessment and safety case preparation,
- cost estimating and budgeting, and
- flexible working teams.

Staff will need to be given training on the culture of working within a project management environment.

5-4. Information Requirements

Planning for decommissioning requires good information on the radiological status of the facility and on the hazardous materials which may be present. A reliable database is needed, specifying the quantity, type, distribution, and physical and chemical form of hazardous materials which are to be expected. This information will be collected from existing records and data, *in situ* measurements, and/or sampling and analyses.

Prior to shut-down, it is important to identify those records which need to be kept to support decommissioning and ensure that they are stored in an accessible form. Criteria for selecting records for retention are typically:

- technical and safety information required to assist dismantling and/or periods of long-term care and maintenance (Safestore),
- compliance with statutory and regulatory requirements,
- records of historical or cultural interest, and
- records which may be needed in support of legal defense against possible litigation.

5-5. Implementation Issues

The activities to be carried out during the transition phase are principally:

- removal of spent fuel and/or radioactive materials used during operations,
- system clear-out operations (POCO),
- treatment, conditioning storage, and/or disposal of operational wastes,
- decontamination, or fixation of contamination, and
- reconfiguration of systems.

Removal of Spent Fuel, etc.

Removal of the spent fuel from a reactor or, in the case of other facilities, the inventory of radioactive materials, typically reduces the radioactive hazard by around 99%. This not only has safety implications, but involves additional costs in managing the facility, for example through continuous shift manning. Any delay in removal of the radioactive inventory leads to potential problems through loss of suitably qualified and experienced personnel (SQEP) and degradation of the infrastructure for handling fuel and other materials. There is also a risk that transport of radioactive materials will become increasingly controversial and expensive.

POCO

All fluids (such as coolant, heavy water moderator, hydraulic fluids, solvents, etc.) should be removed from retired systems and disposed of while experienced personnel are available. Small items of contaminated equipment and hazardous material such as sodium or chemicals should also be identified and removed where possible. After removal of the contents, systems should be flushed or decontaminated to meet specified end-points.

Treatment of Operational Wastes

The wastes generated during the transition phase are similar to those during plant operation and maintenance. The hazard within the plant should be reduced as far as possible by processing or disposing of accumulated hazardous wastes. If a disposal route is available, use it. If not, the waste should be conditioned to a passively safe form (e.g., immobilise sludges).

Decontamination

Consider reducing operational exposure during subsequent decommissioning or care and maintenance phases by decontaminating circuits, tanks, and containers to remove loose activity from inner surfaces. The method and extent of decontamination requires assessment using multiattribute or cost-benefit analysis. The ALARP principle applies. If decontamination is too difficult, consider fixation to reduce airborne contamination. (Chapter 10 deals with decontamination techniques.)

System Reconfiguration

Once the facility is no longer operational, some of the systems will no longer be needed and can be shut-down, thereby saving on operating and maintenance costs. There is a need to decide which systems need to be kept operational and for how long. For example, fuel handling equipment is required only for as long as there is fuel remaining to be handled. Clearly, there is a need to identify those systems that must be kept to ensure that safety requirements are met and to enable care and maintenance operations to be carried out. The assessment of requirements need to consider:

- costs of fuel, power, and C&M requirements,
- the scope for replacing complex or worn out systems with simpler new ones, and
- retention of equipment for possible use during future decommissioning operations.

5-6. Costs of Transition Activities

In preparing a budget for transition activities, it is necessary to identify those systems that are no longer required after the shut-down of the operations. By reconfiguring or retiring these systems, financial savings can be made as described above. Costs can also be reduced by a review of purchasing and spare parts policy. The same standards may no longer be applicable to consumables and services on a shut-down facility. It is possible that spare parts holdings may be reduced. In general, consideration needs to be given to the following:

- continuing operations and maintenance of systems,
- characterisation of radioactive/hazardous inventory,
- removal of spent fuel and/or other radioactive inventory,
- system reconfiguration (including design and installation of new systems),
- waste management and treatment,
- decontamination and immobilisation of residual contamination, and
- project management.

In conclusion, early planning is the key to a smooth cost effective transition from operational activities to the decommissioning phase. Planning requires the timely allocation of resources to a dedicated decommissioning team whose activities will include hazard identification, cost reduction initiatives, the production of simplified waste management plans, as well as human resource initiatives to maintain skilled staff and motivation amongst the work force. As a matter of policy, relevant data and records should be collected while plant operators are still on hand to assist in the retention of the "Corporate memory" associated with the facility. A radioactive and hazardous materials inventory should be compiled together with the production of clear decommissioning objectives, costed options, planning, and safety documentation keeping stakeholders fully informed.

Chapter 6

Reactor Decommissioning — The Safestore Concept

6-1. Introduction

There are real technical and safety benefits arising from the deferral of the later stages of conventional (AGR, BWR, PWR, etc.) reactor decommissioning. This chapter describes the studies which indicate this. At the end of generation and Stage 1 decommissioning (see Chapter 4 — Stage 1 decommissioning involves fuel and coolant removal, together with nonfixed items of plant, thereby removing some 99% of the radioactivity from the reactor, together with preparation of the facility as a safestore for a period of care and maintenance) by deferring the dismantling of the safestore for at least 85 years for an AGR and some 50 years for a PWR, both the potential radiation dose to decommissioning workers and the volumes of radioactive wastes are significantly reduced.

This chapter shows that there are also real and tangible economic arguments in favor of inserting periods of Care and Maintenance (C&M) into the decommissioning process which arise from a reduced radioactive waste inventory, waste minimisation, and the subsequently reduced waste disposal costs. Some consider that there are also the more theoretical economic benefits arising from a view on the time value of money and the subsequent lessening of discounted decommissioning costs if the works are deferred. Chapter 13 covers such purely financial and economic appraisal considerations.

There are also arguments against deferral. These are associated with social responsibility issues for the current generation and the interpretation of the principle of sustainable development. In addition, there may also be high infrastructure costs associated with the care and maintenance of a facility which negate the financial drivers over the long time scales (perhaps 100 years) involved while waiting for the benefits of radioactive decay to accrue.

This chapter describes such issues in the context of the Safestore strategy applied to British AGR decommissioning.

6-2. Decommissioning and Radioactivity

6-2-1. *Decommissioning Strategy and Option Selection*

Technically, the timing of decommissioning is driven by radioactive decay considerations in the context of:

- dose uptake to workers, and
- the quantities of radioactive waste generated.

However, equally important to the electricity generating company and owner of the redundant facility are the cost and financing arguments that, in part, stem from these technical factors. Table 6-1 illustrates the decommissioning sequence involved with the safestore concept.

6-2-2. *Activation Inventory*

It is normal practice to produce an activation inventory of specific and total activity covering some 24 nuclides of concern over a total of some 145 component types or elements within or surrounding the reactor core. This information allows an assessment of the reduction in radioactivity from natural decay. Figure 6-1 illustrates point dose rate results for various points within a partially dismantled reactor.

6-2-3. *Worker Dose Modeling*

The various dose rates at the different locations within the reactor can then be assessed in conjunction with decommissioning workforce modeling to derive a view on the likely dose uptake to workers under different decommissioning scenarios and methodologies. Figures 6-2 and 6-3 illustrate examples from a typical analysis from such work in the UK for the Advanced Gas-cooled Reactor (AGR). Based upon restricting worker dose to a design safety guideline limit of 10 mSv per annum, the technical

Table 6-1. The Safestore Strategy

Defueling	Plant dismantling	Safestore construction	Care and maintenance	Reactor dismantling	Site clearance and release
This is the first major activity following the end of generation. Fuel is removed from the reactor in much the same way as routine defueling and, after a period of cooling, is sent to reprocessing or storage.	The non-radioactive plant and buildings, such as the turbine hall, circulating water system, and ancillary buildings will be dismantled. Scrap materials will be recycled and the visual impact of the site will be considerably reduced.	To allow the radioactivity in the reactor buildings to reduce naturally, they will be modified to form a durable robust structure that will provide protection to the radioactive plant and structures for many decades. This is called the "Safestore Structure" and will be designed to be passively safe, secure and intruder resistant.	The safestore is designed to provide passively safe stable storage with minimal maintenance.	During this phase, the reactors and all remaining plant and systems will be dismantled. The radioactive waste materials will be transferred to a waste management facility where they will be treated and packaged into a form suitable for disposal or further interim storage.	The end point of decommissioning a power station will be the eventual and complete clearance and delicensing of the site.
Defueling removes over 99% of the radioactivity from the power station.	Potentially mobile radioactive wastes, such as sludges and resins, will be treated and packaged into a stable form suitable for interim storage or disposal.		This does not mean that it will simply be left. A comprehensive inspection and surveillance program will be put in place to ensure the safestore continues to do the job it was set up to do — providing safe storage of the remaining radioactive facilities.		Following that dismantling, an environmental monitoring program will be undertaken to check for the presence of any residual radioactivity on the site. Any contaminated land issues will be dealt with and the site will then be confirmed clean before being delicensed and made available for reuse.

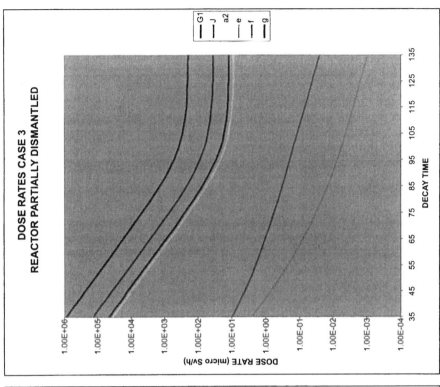

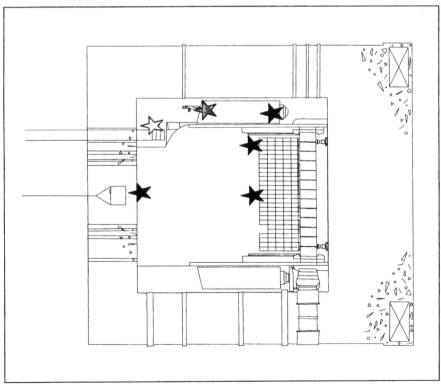

Figure 6-1. Partially Dismantled Reactor Core Point Dose Rate Locations and Results.

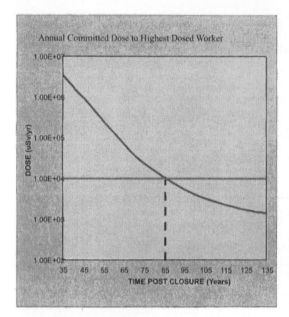

Figure 6-2. Typical Worker Dose Modeling Results (Annual Committed Dose to the Highest Dosed Worker).

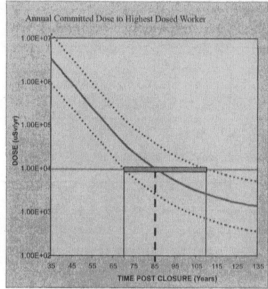

Figure 6-3. Typical Worker Dose Modeling Results (Annual Committed Dose to the Highest Dosed Worker).

conclusions are that reactor decommissioning should be deferred for approximately 85 years, so as to significantly reduce worker dose uptake and approximately 100 years to achieve the 5 mSv annual dose target. An uncertainty band around such calculations indicates that reactor decommissioning could be safely undertaken with a minimum of remote handling following a 70–110 year period after the end of generation.

6-2-4. Radioactive Waste Minimisation Modeling

The costs of radioactive waste disposal are significant, and any reduction in the inventory will have a direct impact on decommissioning costs. This is not only a volumetric consideration. Free release material is, of course, cheaper to dispose of in a conventional landfill site than VLLW or LLW. LLW is, in turn, considerably cheaper to dispose of than ILW. The ratio of such disposal costs in the UK is typically:

- 1–1.5 (Free release) to;
- 2.5–7.5 (Controlled burial, for VLLW) to;
- ~125 (LLW disposal) to;
- ~25,000 (ILW disposal),

respectively, per unit volume including an average allowance for conditioning, packaging, and transport to the waste disposal site. In the case of ILW, no disposal facility currently exists in the UK. Therefore, an allowance has to be included in any financial analysis for interim ILW storage as a planning assumption until such time that such a facility might become available.

The activation inventory is used to establish the quantities of packaged wastes arising from the reactor dismantling, taking into account shielding requirements. This is then used, in turn, to identify the optimum time for reactor dismantling on the basis of waste minimisation. Figure 6-4 illustrates the projected reduction in packaged ILW volumes arising from reactor decommissioning with different safestore periods. Applying this analysis to the UK AGRs indicates that the emplaced volumes of ILW in a future possible national repository or surface store would reduce by more than half from ~56,000 cum to ~21,000 cum.

6-2-5. Arguments Against Deferral

Any particular decommissioning option has to be justified. There comes a time when the cost of such justification may exceed the savings envisaged from following a proposed

Numbers of ILW Boxes

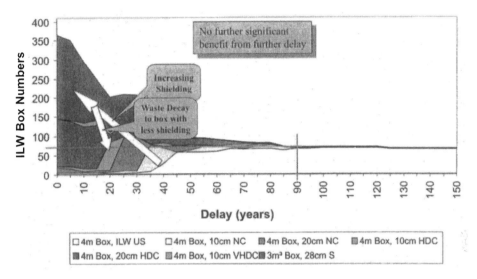

Figure 6-4. Projected Packaged ILW Volumes Plotted Against Decommissioning Deferral Periods.

decommissioning option. The justification may require a huge investment in desktop studies, so as to satisfy the Company management, Regulators, and other stakeholders. Such studies do not, of course, actually add to the end goal that all parties are looking for, namely decommissioning progress. Typical issues that have to be addressed in order to confirm the appropriateness of the safestore concept include:

- confirmation of the retention of the structural integrity of the reactor containment over the timescales involved;
- confirmation that the structures will not degrade over the deferral period and require large maintenance or refurbishment costs;
- that an appropriate risk management scheme is in place;
- that Corporate knowledge is not lost (including adequate record retention);
- the sensitivity of the economic arguments are soundly based (escalation of cost of capital);
- retention of a competent workforce knowledge;
- achievement of passive safety and the adequate cost modeling of intervention during the deferral period;
- increasing Regulatory requirements over the Care and Maintenance (C&M) period, making final decommissioning more expensive than originally envisaged; and
- the likelihood of continued availability of waste routes and sensitivity of increasing costs of waste disposal over the timescales involved.

6-3. Decommissioning Activities

Tables 6-1 and 6-2 detail the activities involved in reactor decommissioning. Figures 6-5 to 6-13 illustrate Stage 1 to

Table 6-2. Decommissioning Activities

	Decommissioning activities	Stage of decommissioning
1.0	Pre-closure planning	Stage 1, Figures 6-5–6-7
1.1	Defueling	Stage 1
1.2	Decommissioning engineering preparation works	Stage 1
1.3	Management of potentially mobile operational wastes	Stage 1
2.0	Plant decommissioning[a]	Stage 2, Figures 6-8–6-10
2.1	Safestore construction[b]	Stage 2
2.2	Site surveillance[c] and Care and Maintenance (C&M)[d]	Stage 2
3.0	Preparation for reactor dismantling	Stage 3, Figures 6-11–6-13
3.1	Vault waste management	Stage 3
3.2	Reactor dismantling	Stage 3
3.3	Site clearance and release	Stage 3

[a] Plant decommissioning includes:

· Installation of new services where necessary,
· Dismantling of nonradioactive plant and systems,
· Dismantling of nonradioactive buildings, and
· Strip out and dismantle radioactive ancillary buildings.

Table 6-2. (continued)

[b]Safestore development:

· Modification of existing buildings to safestore structures,
· Safestores to remain safe, contain hazard, be weatherproof, and secure,
· Provision of stable storage conditions for radioactive plant and materials, and
· Achievement of passive safety.

[c]Safestore Surveillance and Care and Maintenance objectives include:

· Maintenance of safe and stable passive storage,
· Minimal site maintenance,
· Infrequent access inside safestore structures,
· Environmental monitoring program,
· No permanent site presence,

· Remote security surveillance, and
· Waste management to Regulatory requirements.

[d]Monitoring and Surveillance Equipment includes:

· Monitoring of sump levels with alarms,
· Installation of smoke detection systems in critical areas,
· Temperature and humidity measurements in the safestore and reactor,
· Corrosion measurements in safestore and reactor,
· Ground contamination measurements in boreholes around the site,
· Intruder detection systems, and
· Solid, liquid, and aerial environmental discharge monitoring.

With all of the above measurements telemetered to a central off-site control room during the C & M period which includes a planned regime of inspection and maintenance.

Stage 3 decommissioning activities, including a safestore period of Care & Maintenance.

6-4. Paying for Decommissioning

The financial effect of these technical advantages in both cash and discounted cost terms is illustrated in Figure 6-5.

British Energy use a 3% discount rate in their analysis to derive a Net Present Value (NPV) reactor decommissioning cost figure. Obviously, the higher the discount rate, the more seemingly advantageous (in discounted cost terms) deferral appears to be. However, it is essential that a rigorous approach to Care and Maintenance is analysed before jumping to such a conclusion [1,2].

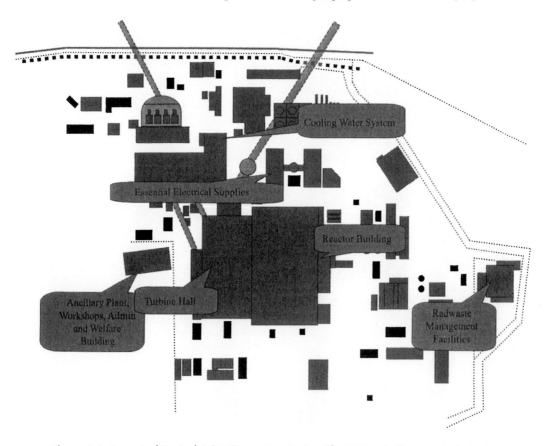

Figure 6-5. Layout of Typical AGR Generating Station Plant (Prior to Decommissioning).

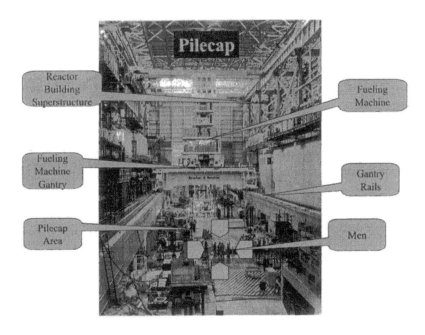

Figure 6-6. Internal View of Typical AGR Generating Station (Prior to Decommissioning).

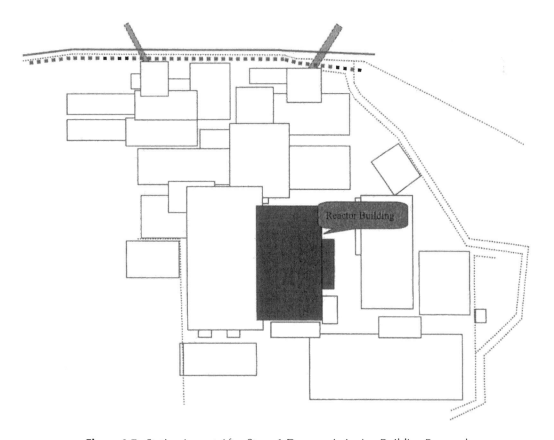

Figure 6-7. Station Layout After Stage 1 Decommissioning Building Removal.

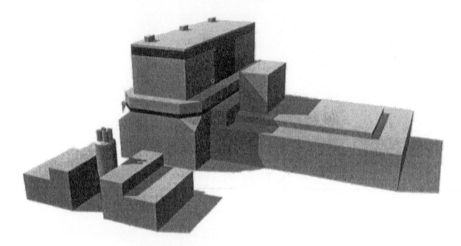

Figure 6-8. Stage 2 Decommissioning Hinkley Point B.

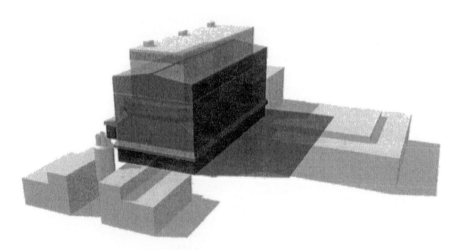

Figure 6-9. Stage 2 Decommissioning Hinkley Point B Safestore Construction.

Figure 6-10. Stage 2 Decommissioning Hinkley Point B Conceptual Safestore.

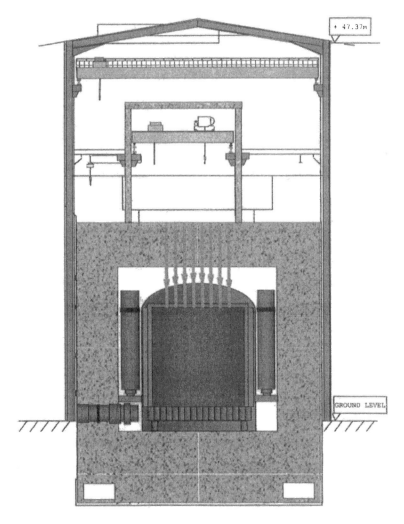

Figure 6-11. Stage 3 Decommissioning Preparation.

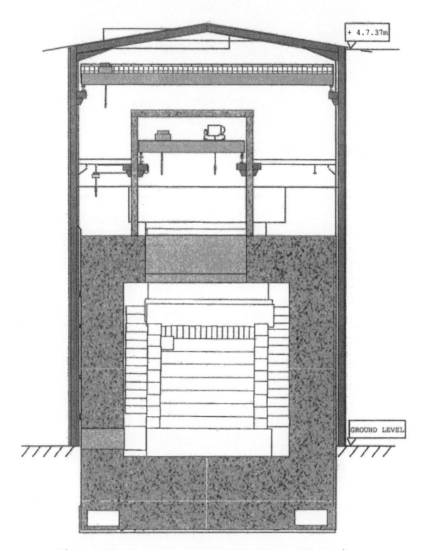

Figure 6-12. Stage 3 Decommissioning Reactor Dismantling.

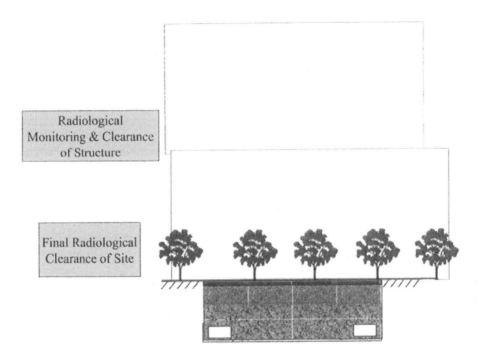

Figure 6-13. Stage 3 Decommissioning: Site Clearance and Release.

In the UK, British Energy (a private nuclear power generating Company) funds the initial Stage 1 activities (defueling and removal of potentially mobile operational wastes) from provisions in the Company's balance sheet [3] (see Chapter 12). The subsequent decommissioning work is funded through a separate (segregated) decommissioning fund held and administered by the independent Nuclear Trust. This was set up at the time of privatisation of British Energy (BE) in 1996. The fund performance is periodically reviewed to ensure that the fund, including annual contributions from BE, will be sufficient to meet the decommissioning liabilities.

Based upon current estimates of station operating lives and lifetime output predictions, Table 6.3 shows, in 1999/2000 year money values, the likely undiscounted and discounted (3% per annum discount rate) liabilities costs for the decommissioning of the eight British AGRs together with the accrued value of the segregated decommissioning fund to March 2000. The difference between the discounted and undiscounted costs reflects the fact that the decommissioning costs concerned will not fall due for payment until some considerable time into the

Table 6-3. AGR Decommissioning Liabilities

Undiscounted AGR liabilities (8 reactors)	£4.6bn
Discounted AGR liabilities (3%)	£0.9bn
Segregated fund accrual to end FY 1999/2000	£0.9bn

future and primarily after the end of the safestore period (see Figures 6-6–6-13).

6-5. References

1. UKAEA. *Atomic Energy Code of Practice AECP 1085,* Issue 1, UKAEA Harwell, Oxfordshire, UK.
2. Bayliss, C. "Practical Applications of UKAEA's Decommissioning & Liabilities Management Toolbox," *Ibc 7th International Conference & Exhibition on Decommissioning Nuclear Facilities,* London (30/31 October 2000).
3. Bayliss, C. "Decommissioning — Choosing and Prioritising the Right Options," *Ibc 6th International Conference & Exhibition on Decommissioning Nuclear Facilities,* London (15/16 June 1999).

Chapter 7
Decommissioning PIE and Other Facilities

7-1. Introduction

This chapter describes the decommissioning activities and techniques involved in the remediation of a variety of facilities other than reactors. These include:

- Postirradiation Examination (PIE);
- Fuel fabrication;
- Fuel reprocessing;
- Waste processing; and
- Research and Development (R&D) laboratories.

All of these have a diversity of equipment, structures, and inventories. The actual approach adopted needs to be adapted to meet the individual circumstances, but forward planning is the key to success. Experience indicates that the results from practical inspection and analysis may not exactly match the anticipated radioactive materials inventories, records, and supposedly "as built" facility drawing data. Therefore, a degree of flexibility has to be built into the decommissioning programs to cater for such things as a higher degree of contamination than expected and the associated increased decommissioning program durations.

7-2. Key Issues to be Considered

In a similar manner to reactor decommissioning, and in addition to the required Regulatory related paperwork, the following issues need to be addressed and included for in the decommissioning program:

- Production of a radiological inventory;
- An investigation into the operational history of the facility which might shed light on the location and extent of the likely contamination;
- The gathering of structural information;
- Production of an integrated decommissioning program that includes for the provision of available waste routes to a recognised end point such as interim storage in a surface facility or a waste disposal facility;

- A consideration of the most suitable decontamination methods to be utilised (see Chapter 10); and
- The dismantling methods to be adopted (see Chapter 11).

Experience indicates that when older facilities have been left unused without an adequate Care and Maintenance (C&M) regime, the services — general small power and lighting, fire alarms, ventilation systems, in-cell equipment, etc. — all deteriorate, and it is an expensive and time consuming program to upgrade these services which may be required before the actual facility may be decommissioned and knocked down. In a similar way, old wastes may have been left to accumulate without adequate records. Laboratories may have toxic chemical hazards as well as radioactive inventories to consider. The spread of contamination into the foundations of the building and then possibly into the groundwater under the facility will add greatly to the remediation costs and lengthen the program. In particular, hot (gamma) cells involve shielding walls of a massive structural nature which are not easily dismantled. The lower worker productivity brought about when using protective gear (respirators, air hoods, and full pressurised suits) needs to be taken into account in the program. Figure 7-1 shows personal protective equipment being worn during decommissioning of an alpha laboratory.

7-3. Alpha and Gamma Radiation Working

When carrying out assay work, it must be recognised that alpha (and soft beta) radiation is more difficult to detect when screened from view below surfaces. Radionuclides that have penetrated concrete surfaces may, therefore, be screened to some extent by the depth of cover through which they have migrated. In such cases, the actual costs and time taken to remove the contamination will be greater than anticipated unless some allowance is made or a more extensive, and probably intrusive, survey is carried out before decommissioning work commences.

Ingestion of small quantities of uranium or plutonium by breathing, or from entry into the blood stream

• Respirators

• Air hoods

• Full pressurised suits

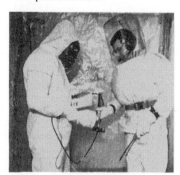

Figure 7-1. Full Pressurised Suit Working During Decommissioning of an Alpha Laboratory.

Nuclear Engineered Advanced TeleRobot

Figure 7-2. The NEATER Robotic Arm.

through cuts or abrasions in the skin, can cause eventual death. Therefore, alpha facilities need a high degree of containment from the outside environment, which is usually achieved using HEPA filtered ventilation systems. Worker decommissioning operations may, therefore, necessitate the use of pressurised suits so as to further negate the possibility of particulate entry beyond the physical barrier of normal protective clothing.

For radionuclides which emit very penetrating gamma radiation, shielding is required. Such radiation is relatively easy to detect, but the need for shielding makes worker movement difficult if not impossible and may negate workers entering such environments. Hence, depending on the worker dose uptake, remote operations using manual or fully automatic robotic arms are needed.

Figure 7-2 shows a typical Nuclear Engineered Advanced TeleRobot (NEATER) and Figure 7-3 the use of robots in the decommissioning of an irradiated fuel high active handling facility.

Robotics reduce the need for high operator dose uptake and exposure to risk. Once set up, they can achieve greater productivity than the equivalent hands-on manual pressurised suit working and, thereby, reduce the decommissioning program man-hours and costs. They also have the advantage, if used correctly, of reducing the waste arisings (particularly secondary wastes). Figure 7-4 illustrates an alpha facility stage 1 Post Operational Clear Out (POCO) using a robot in a location where manual entry would only be allowable using pressurised suit working practices. Work involves mounting the glovebox on a

Figure 7-3. High Active Handling Gamma Cell Decommissioning (showing before and after decontamination and cell removal together with the use of a robotic arm).

remotely rotatable work bench, dismantling remotely by robotic manipulation and cutting, transfer of the wastes to a posting cell, monitoring, and then assaying the waste as it is placed into a standard drum container.

Manual glovebox dismantling follows a similar pattern for stage 1 POCO. The area is disconnected from normal services, modular containment is installed, and the area put under a pressurised suit working regime. The box is cut-up manually and placed in 200 liter contact handable Intermediate Level Waste (c-ILW) drums, monitored and assayed for storage prior to eventual disposal. This sequence is illustrated in Figure 7-5.

7-4. Decommissioning Examples

This section illustrates some of the practical features involved in decommissioning PIE and other facilities.

The High Active Handling Cell illustrated in Figure 7-3 was used to break down irradiated fuel elements and experimental rigs. It contained simple concrete-shielded (hot) cells with zinc bromide windows and manipulators. Waste from the cell had to be cut up (size reduced) and dispatched, together with large items such as a steel work benches. The high background dose rates, primarily associated with Cobalt 60 gamma radiation, led to use of the NEATOR TeleRobot (Figure 7-2) for dismantling purposes. An issue of interest is to ensure that

hydraulic piping and electrical cabling on such equipment does not degrade in such an environment.

Figure 7-6 illustrates a seven storey Chemical Engineering Laboratory that was built in the 1950s and decontaminated and fully demolished in 1998. It contained large contaminated radioactive ventilation systems, including duct systems and fans together with many experimental rigs and fume cupboards. In addition to the alpha, beta, and gamma radiation hazard, the building also had a large chemical inventory including sodium from fast reactor research and mercury. A stepwise approach using the processes described in this chapter was adopted.

The decommissioning of the Fuel Handling Facility was the first full alpha facility decommissioning from an operational phase through to a green field site at Winfrith in Dorset, UK. The plant produced plutonium-based fuels for experimental purposes for 35 years and contained many interconnected gloveboxes and heavy industrial scale equipment including ball mills, mixers, grinders, and furnaces. Decommissioning as illustrated in Figure 7-7, was undertaken in 2000–2001 in three stages:

- *Stage a*: size reduction and removal of gloveboxes and their contents;
- *Stage b*: removal of building equipment, ventilation plant, drains, etc., and decontamination of the structure; and

Box on tilt table

Waste posting

Cutting glovebox

Figure 7-4. Alpha Glovebox Remote Dismantling.

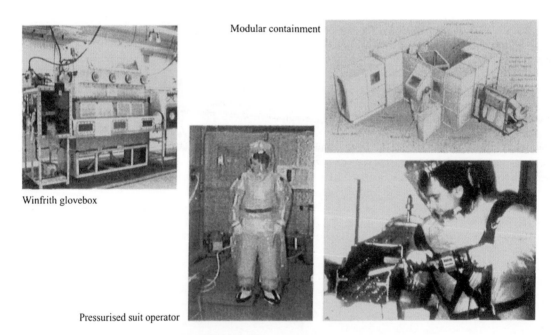

Modular containment

Winfrith glovebox

Pressurised suit operator

Figure 7-5. Manual Glovebox Dismantling.

- *Stage c*: dismantling and removal of the building structure, basement, and ground remediation.

Manual dismantling of the gloveboxes within a pressurised suit working area was adopted. The steps necessary for the ground remediation and site delicensing are described in Chapter 23.

Harwell Glovebox dismantling is illustrated in Figure 7-8. This decommissioning project involved the dismantling of some 170 gloveboxes (50 dismantled by manual methods and 120 using the NEATER robot), some of which were very large floor-to-ceiling affairs. Prompt decommissioning following the operational phase

 Early 1950s

 Fingal Cell

Internal View

 Final Stages of
Demolition

Figure 7-6. Chemical Engineering Building Strip-Out, Decontamination, and Demolition.

• Produced Pu-based fuel for 35 years
• Inter-connecting gloveboxes
• Equipment included ball mills, mixers, grinders, and furnaces

Manual dismantling of glovebox within pressurised
suit area

Remediated site following demolition and restoration

Figure 7-7. Winfrith Plutonium Fuel Manufacturing Facility Decommissioning.

- Originally 240 gloveboxes

- 50 dismantled by manual methods

- 120 dismantled by NEATER robot

- 70 still operational (AEAT)

- 10 large fixed gloveboxes ("shop windows") decommissioned

Figure 7-8. Harwell Glovebox Dismantling.

Figure 7-9. Decommissioning Liquid Effluent System Delay Tanks and Pipes to a Fully Restored Site.

Figure 7-10. Surveying and Sampling of Effluent Drainage Systems.

was adopted to avoid the build up of gamma radiation (and associated increased decommissioning costs) arising from ingrowth of americium 241 during the natural plutonium radioactive decay process. Such high alpha, low beta/gamma work requires containment, but only very limited shielding. A combination of pressurised suit working (which is expensive, very demanding on the workers, and always includes a degree of risk associated with damage to suits when used in conjunction with heavy equipment) and robotics was adopted.

Post Irradiation Examination (PIE) facilities are used to dismantle irradiated fuel elements and other highly active components for detailed examination of their physical properties. Such facilities consist of shielded cells with viewing windows, manipulators, and in-cell hoists. Heavy duty cutting and welding equipment may have to be used for decommissioning in conjunction with posting facilities and maintenance areas. The history of usage of such facilities is important when drawing up the likely radioactive inventory to be encountered during the decommissioning phase. They may well have been used for inadequately documented general purposes involving a variety of radiation sources. If records are scarce, then the decommissioning team should be encouraged to carry out interviews with ex-workers who may remember the types of work carried out and use this information gathering in conjunction with remote and intrusive surveys.

Liquid effluent systems from laboratories have the potential for requiring extensive contaminated ground remediation arising from fractured pipework systems. The pipes themselves may be decommissioned by first slitting to open up (or by sending a remote crawler with a camera and detection device along the pipe length) for monitoring purposes. They may then be cleaned using high pressure water jetting followed by trench excavation work and pipe removal. Concrete tanks may be decontaminated by scabbling and metal tanks by cutting up for volume reduction purposes. The identification and subsequent removal of contaminated soils may be assisted by the use of GPS-linked survey systems, as described in Chapters 23 and 25. Illustrations of decommissioning of effluent delay tanks and pipe work are shown in Figures 7-9 and 7-10.

Chapter 8

Preparation of Documentation for Decommissioning

8-1. Introduction

This chapter describes how to set up the necessary documentation, especially safety documentation, required to carry out a decommissioning project. The descriptions are based upon UK Regulatory requirements, but these form a systematic suite of documentation that may also apply in the International context.

The UK Site Licence Condition 35 (Decommissioning) sets out a requirement for adequate arrangements for the decommissioning of any plant or process which may affect safety. It requires the production and implementation of decommissioning programs for each plant which may be open to scrutiny by the Regulator (HSE NII). It also allows for the possibility of a staged approach to decommissioning. The typical set of documents required to fulfill this condition is:

- a Decommissioning Plan and Program — setting out what is to be done, how and when;
- a Decommissioning Safety Case — demonstrating a justification for why the Plan and Program are safe and meet regulatory requirements; and
- a Postdecommissioning Report — stating what has been done, describing the final end-point, and the lessons learnt which may be applicable to future work.

In addition, normal conventional site construction works Health and Safety Regulations (CDM Regulations) apply to demolition and decommissioning works in the UK.

8-2. Decommissioning Plan and Program

The Decommissioning Plan provides a strategic overview of the decommissioning project through to its final end-point. The end-point may be unrestricted use of the site or some other agreed condition, if complete clearance for unrestricted use is not appropriate.

The Program should provide a statement of the decommissioning tasks to be carried out, together with timescales. If detailed proposals are not available for later stages, these should be developed as decommissioning progresses and embodied during periodic reviews of the Program.

The Decommissioning Program may be described in terms of a Work Breakdown Structure (WBS). A WBS organises the project into manageable activities at different levels (overall objectives, sub-sections, individual tasks, sub-tasks, etc.). The WBS can be used to generate a project schedule, which links the various tasks together in a logical flow, taking account of interdependencies, key dates, and milestones. The project schedule can then be displayed as a simple bar chart, or a more complex diagram showing dependence criteria, such as a Gantt or Program Evaluation and Review Technique (PERT) chart.

8-3. Decommissioning Safety Case

A Decommissioning Safety Case (DSC) is used to justify the safety of the proposed methods of decommissioning. The safety case should demonstrate a logical connection between the plant condition at shut-down (identifying the hazards), the proposed decommissioning tasks, the associated risks in performing these tasks, and the safety management arrangements that will minimise the risks. In addition, the safety case should ensure that the facility is in a safe condition at the end of the decommissioning work.

Where the decommissioning processes have not been fully developed at the time of shut-down, some initial decommissioning tasks (e.g., POCO) may be carried out under the existing Operational Safety Case (OSC) or can be treated as a "modification" to the OSC. However, any significant dismantling operations are not acceptable as modifications to the OSC, and a decommissioning safety case should be put in place to cover them.

The main elements addressed in a Decommissioning Safety Case (DSC) are as follows:

- a clear definition of the scope of the decommissioning operations to be carried out,
- a clear definition of the end point for decommissioning,
- the demonstration that the facility meets the identified decommissioning safety principles,
- the suitability of safety-related items of plant or equipment,
- a detailed assessment of the hazards associated with decommissioning,
- a justified detailed safety argument, and
- the safety management arrangements associated with decommissioning.

The methods proposed for decommissioning should consider the potential hazards to workers, the general population and the environment, and should be in accordance with the ALARP principle (As Low As Reasonably Practicable — see Chapter 18). Specific dose and risk targets will be set and agreed. Formal risk assessment methods (e.g., HAZOPS) should be applied. This assessment should answer the following questions:

- What could go wrong?
- What would be the consequences?
- How can we prevent or minimise the chances of these consequences occurring?
- How would we deal with these consequences if they did happen?

The Decommissioning Safety Case (DSC) needs to consider exposure routes such as:

- release of active gases (such as tritium),
- generation of contaminated aerosols (e.g., from liquid decontaminants),
- contaminated dust,
- export of contaminated items of equipment, and
- direct radiation (from fuel, radiation sources, contaminated, and activated items).

Engineered measures (such as shielding, containment, and/or remote handling methods) should be considered to minimise doses. Where such engineered features are not practicable or are excessively costly, dose minimisation methods relying on management procedures and work instructions may be used. It may be sensible to discuss the available options informally with the regulator at an early stage before deciding on a preferred option. The Decommissioning Safety Case (DSC) needs to be supported by:

- a radioactive inventory for the plant,
- inventories of non-radioactive hazardous materials,
- engineering information (records, drawings, as-built condition), and

- operational history (including data on incidents which may have caused the spread of contamination).

Most licence holders operate a safety management system that requires tasks to be categorised in terms of their hazard potential. The categorisation will then determine the extent of the independent scrutiny and peer review applied to the Decommissioning Plan and Safety Case. Typically, the tasks are categorised as follows:

1. Potential for off-site hazard in terms of exposure or dose to the public.
2. Off-site hazard not significant but on-site hazard is possible external to the building.
3. Maximum potential hazard is limited with the building.
4. No significant nuclear safety hazard.

The hazard category applied to decommissioning a facility is generally determined by the hazard at the start of the decommissioning phase. As the decommissioning progresses and the hazard is reduced, it is possible to invoke a procedure for reducing the category of later stages of decommissioning. For example, a facility may be Category 1 at the beginning of Stage 1 decommissioning, but then be placed in long-term care and maintenance as a Category 2 facility when Stage 1 is complete.

For large decommissioning projects, it is recommended that the DSC should present an overview of the project and be followed by a series of more detailed safety reports covering individual tasks or phases of the program. Each task can be considered as a modification to the DSC. It should be stressed that the term modification when used in this context does not apply to the physical operation, but to a modification of the current safety case, which is the DSC.

The modern approach to presenting safety cases requires the preparation of a Safety Report supported by a Safety Report Support File, the intention being that the safety arguments for decommissioning are presented in a succinct and clear manner. Table 8-1 contains the proposed content of the Safety Report modified to present the suggested content of a DSC presented in the Safety Report format.

8-4. Conventional Safety Documentation Requirements

The full requirements of the CDM Regulations apply to all projects where the primary activity is construction/decommissioning. The Regulations require notification to HSE of the project and define the roles and responsibilities of key players as follows:

- the Client,
- the Planning Supervisor, appointed by the client,

Table 8-1. Content of a Typical Decommissioning Safety Case (Safety Report Style).

Introduction	Description	Hazard assessment	Safety management
Safety Report			
Scope of decommissioning; Nature of clearance sought; Summary of Arguments; Main hazards; Main safeguards; ALARP; Action Plan items	Summaries of: facility (plant and services); Decommissioning operations; KSRE/SRE; Waste management; Decommissioning issues relating to end points — plant state, inventory, etc.	Fault schedule; Normal operations; Deterministic arguments; Probabilistic arguments; Identification of safety controls; Non-radiological hazards; Environmental effects	Facility specific issues; Safety related posts; Interactions; Arrangements for controlling operations
Safety Report Support File			
Detailed description (plant, operations and services); Operational history and experience — details of either previous operations or decommissioning; Decommissioning Program; Detailed waste management discussion	Engineering substantiation; Other safety inspections; Full set of DSPs and substantiation; Design basis analysis; Identification of safety controls	HAZOP Records; Full Fault Schedule; Detailed HAZANs and methodologies; Data sources; Screening records; Generation of KSRE etc.; Post accident Recovery; Detailed environmental assessment	Generic issues; Links to EIMT Schedule; Any detail of plant specific items; Terms of reference for staff; Building Manuals; Management Systems Manual; CDM Material

ALARP: As Low As Reasonably Practicable; KSRE/SRE: Key Safety Related Equipment/Safety Related Equipment; HAZOPS: HAZardous OPerationS; HAZANs: HAZard ANalyseS; DSPs: Design Safety Principles; EMIT: Examination, Maintenance, Inspection, and Testing; and CDM: Construction and Design Management (UK, Health & Safety at Work material).

- the Principal Contractor,
- the Designer, and
- other Contractors or Subcontractors.

The documentation required by CDM Regulations includes:

- a Health and Safety Plan, and
- a Health and Safety File.

The H&S Plan is generally prepared in two stages — Pretender and Construction. Where work is carried out in-house, only the detailed "Construction phase" H&S Plan will be required.

Pretender H&S Plan

The purpose of the Pretender H&S Plan under the CDM Regulations is to convey information to contract tenderers, on the health and safety risks of the construction/decommissioning works which the Principal Contractor has to manage. The Plan provides information on the significant risks, the standards to be applied to control them, and any other requirements laid down by the client. It provides sufficient information about specific problems to enable the competent contractor to make adequate provision for health and safety resources on submitting his tender response and may refer to the DSC. Guidance on

the contents of the Plan is provided in the HSC document "A guide to managing health and safety in construction."

The Construction Phase H&S Plan

The Construction Phase H&S Plan should be prepared before decommissioning starts. The management requirements are all aspects of the safety management systems that will be discussed in the DSC. A detailed Fault Schedule (containing a description of the accident conditions associated with construction and commissioning, consequences, and engineered and administrative safeguards) will be required for the DSC. This schedule, coupled with the description of the project-specific safety management, should effectively cover the requirements to demonstrate adequate arrangements to ensure the health and safety of all workers.

8-5. Management Procedures and Quality Assurance

Management procedures describe the stages in a process, the responsibility for completing the stages, and the records of successful completion to be produced. They are usually working documents which outline processes, though are sometimes detailed documents covering how each stage is successfully completed.

Management procedures are listed in Quality or Management System manuals, QA programs, or in contents lists of procedures manuals.

QA programs are used to supplement standard quality documentation (quality manuals, procedures, working instructions) for particular plants or projects which need further amplification of the overall controlling organisation and quality management systems. This may be because of the size of the activity/project, its complexity, special management interface requirements, or important safety considerations.

A Quality plan identifies all the key steps in the decommissioning process that need approval of one or more parties. Included are schedules, checklists, data capture sheets, flow charts, networks, or any other documents which describe or identify specific QA practices, resources, and activity sequences relevant to a particular project or task.

Work Instruction is a generic term for all other forms of written instruction necessary to implement the local management system. They define work sequences, methods of the equipment used, and the controls and verifications applied.

Site Regulations are mandatory rules applicable to the licence holder's staff, contractors, visitors, and tenants. They convey mandatory instructions from site management concerning safety or security on the particular site.

Codes of Practice provide general recommendations on acceptable standards for particular topics. They typically contain operational principles, methods, design information, and data which can be used as a basis for assessing whether the licence holder meets modern standards for the topic. Codes of Practice are by definition advisory and need to be interpreted in the light of the particular planned application and changes in technology/custom and practice since publication. Codes of Practice may be published by UKAEA; regulatory bodies, e.g., HSE, IAEA; professional bodies or trade associations; or British/European/International Standards Bodies. Approved Codes of Practice published by the HSE are a special case where they are almost mandatory: the user may depart from the guidance therein, but has to be able to justify that the practice operated is at least as good as that defined in the code.

Guidance Notes are used to provide information on how procedural requirements may be discharged. They typically include factors which should be considered in making decisions on particular topics. Guidance Notes are advisory, and staff may deviate from the advice provided that they have considered the issues, have decided on an alternative approach, and accepted responsibility for the approach. Guidance Notes are subject to regular review as other procedural documentation. They may be produced in-house or come from Governments,

Regulators, or other outside bodies such as Industry associations or joint working parties.

Supporting documents are used to supplement or enhance QA procedures, programs, plans, and work instructions. Typically, they include specifications, standards, standard forms, log books, records, and reports.

A Post Decommissioning Report (PDR) is required at the completion of the decommissioning tasks for the Stage covered by a Decommissioning Program. Its purpose is to provide a report on the tasks carried out demonstrating that the tasks have been carried out and highlighting any lessons learnt from the tasks. The report should include:

- a description of the facility,
- the decommissioning objectives and radiological criteria set for the end-point,
- references to the safety case and supporting documentation prepared during decommissioning,
- a description of the work done and of any remaining buildings or equipment not decommissioned,
- a final radiological survey report,
- an inventory of radioactive materials, including amounts and types of waste generated during decommissioning and their location for storage and/or disposal,
- an inventory of materials, etc., released from radiological control,
- a summary of any unusual events and incidents that occurred during decommissioning,
- a summary of occupational and public doses received during the decommissioning,
- a summary of the costs incurred, and
- lessons learnt.

Where Care and Maintenance forms a separate sub-stage, separate reports will be required prior to and at the end of the period of deferral.

In addition, certain records are needed to be kept for typically 50 years to comply with statutory requirements. These include the decommissioning plan, safety case, licensing documents, plant drawings, health physics records, plant maintenance schedules, safety incident reports, emergency plans, training records, and authorisations.

8-6. Examples of Typical Safety Documentation

8-6-1. Materials Test Reactors to Stage 2 Decommissioning

DIDO and PLUTO are materials testing reactors at Harwell in the UK. They ceased operation in 1990 and were decommissioned to Stage 2 over the following 5 years. Fuel was removed in the first 3 months following

Figure 8-1. Aerial View of the DIDO and PLUTO Materials Test Reactors at Harwell, Oxfordshire, UK.

Table 8-2. DIDO and PLUTO: Key Decommissioning Documentation

Decommissioning safety case	Supporting documents
• Decommissioning Plan (including radiological inventory and description of all major tasks)	• Operational safety document
	• Procedures
	• Working instructions
• Decommissioning Program (schedule)	• Operational safety reports
	• Progress reports
• Quality assurance program	• Postdecommissioning report
• Safety principles and radiological standards	
• Safety justification	

shut-down, using standard equipment and procedures under the original operational safety case. An aerial view of the reactor site is shown in Figure 8-1, and the associated set of documentation generated to support this decommissioning program is shown in Table 8-2.

8-6-2. *Jason (Royal Naval College) Reactor to Stage 3 Decommissioning*

Jason was an Argonaut class reactor used for training purposes at the Royal Naval College, Greenwich, London, UK. It was decommissioned to Stage 3 between 1997–2000. One particular aspect of the decommissioning was that special equipment was needed to be designed and manufactured in order to remove the fuel from the reactor. Figure 8-2 illustrates reactor fuel removal in the heart of London and Table 8-3 shows the set of documentation used for decommissioning Jason.

8-6-3. *Site Environmental Remediation to Unrestricted Use*

The Harwell Science park is the location of the first experimental nuclear reactor in the UK. The Graphite Low Energy Experimental Pile (GLEEP) went critical in 1947. Like many such sites in the UK, it was located at an old disused aerodrome. The reactor itself was built in an aircraft hangar. The Southern Storage Area (SSA) is outside, but adjacent to, the Harwell nuclear licensed site. It was used for airfield shelters and ammunition storage, and subsequently as an interim waste (conventional, nonnuclear, and toxic) transit area. In order to remediate the site for completely unrestricted use, it has been fully remediated. Figure 8-3 is an example of an internal "intranet" based system, which allows the user to search for the environmental remediation documentation required for this decommissioning and remediation work.

In conclusion, it is essential in such a highly regulated environment to have a comprehensive set of principles

Figure 8-2. Defueling the Jason Training Reactor at Greenwich, London, UK.

Table 8-3. Jason Reactor Key Decommissioning Documentation (to Stage 3)

Principal documents	Supporting documents
• Decommissioning Plan	• Operational safety
• Health and safety plan	document
(pretender and detailed)	• Procedures
• Preliminary design and	• Working instructions
Preliminary safety report	• Operational safety reports
• Detail design	• Progress reports
• Predecommissioning safety	• Postdecommissioning
reports	report
• Preoperational safety reports	
• Postdecommissioning report	

```
⅃00 Safety Argument            ⅃10 Traffic Arrangements
⅃01 Safety Case Validation and Compliance  ⅃11 Site Security Details
⅃02 Risk Assessment            ⅃12 Emergency Instructions and Plans
⅃03 Safety Management          ⅃13 Training Plan
⅃04 Health and Safety Procedures  ⅃14 Quality assurance
⅃05 Detailed Project Design
⅃06 Monitoring Program
⅃07 Waste Handling Program
⅃08 Sample analysis techniques
⅃09 Record Keeping
```

Figure 8-3. Conventional Site Environmental Remediation Key Documentation: Computer Log Index Page.

and objectives for the production of decommissioning documentation. The "decommissioning plan" for a facility should include such a suite of documentation. The timely production of such documentation is crucial to the achievement of the overall decommissioning objective. Experience shows that failure to sufficiently allow for this at the outset will inevitably lead to program delays.

Chapter 9
Radiological Characterisation

9-1. Introduction

Decommissioning of a contaminated facility should not commence without the prior collection of as much data as possible about the radioactive inventory, i.e., the range of contaminating radionuclides and the quantities present within the facility. It is also necessary to know the physical and chemical state of contaminants and their distribution by area (floors, walls, ceilings, etc.). Knowledge of the radionuclide composition will assist in determining:

- the appropriate activity assessment methodologies,
- the methods to be used in decontamination and dismantling,
- the optimum phasing of decommissioning operations,
- the volume of radioactive waste arisings and form of packaging of wastes required to be compatible with long-term storage and eventual disposal, and
- the estimated dose to workers and identifying worker safety requirements (ALARP, see Chapter 18).

The radioactivity in a facility may originate from one of a number of processes:

- irradiated fuels (reactors and PIE facilities),
- neutron irradiation of structural material, reactor components, or shielding,
- isotopes generated for a specific purpose (e.g., medical or industrial sources) within a reactor,
- separated actinides and fission products arising from reprocessing of irradiated fuel, and
- contamination arising from loss, leakage, or spills during processing.

For example, the preparation of samples for PIE measurements will generally have involved cutting and polishing of irradiated fuel and, therefore, have given rise to highly radioactive dust particles. Work involving the dissolution of fuel or other radioactive material (e.g., for chemical analysis or separation of isotopes) will have generated secondary liquid wastes that may have spilled or otherwise spread contamination. One of the first tasks is to characterise the nature of potential radioactive contamination within the facility. This radiological characterisation may involve obtaining data from several sources:

- reviewing existing information, such as historical facility usage records and radiological survey data,
- calculations using codes for activation, nuclear fuel burn-up, and radioactive decay,
- *in situ* measurements,
- sampling and analysis, and
- documentation.

Some facilities may incorporate additional nonradiological hazards which must also be addressed. Hazards commonly associated with radioactive facilities include asbestos, beryllium, lead (both as shielding material and incorporated into paints), and other heavy metals requiring assessment and control under the UK COSHH (Control of Substances Hazardous to Health) Regulations.

This chapter deals with characterisation of redundant nuclear facilities at the beginning of a decommissioning project and at intermediate stages up to the point where a building is being prepared for demolition. The characterisation of contaminated land is covered in Chapter 24.

9-2. General Approach

Characterisation [1] is an essential step at the beginning of the decommissioning process and may need to be repeated at different stages during the decommissioning. The results will be used to plan the methods used to dismantle the facility and manage the radioactive waste. It is also needed to determine the hazards to which workers and the general public will be exposed. It is important to have access to historical records.

The characterisation needs to be carefully defined and executed, particularly with regard to choice of methods, instruments, sampling procedures, etc. The methods used and results obtained must be well documented.

The first step in carrying out a radiological characterisation is to review the existing historical information.

This will involve a search of health physics records and other records which will give an indication of the facility's operational history, such as building maintenance records. "As-built" drawings and information on the structural condition of the facility are valuable. It is particularly important to find any references to incidents during the life of the facility that may have led to the spread of contamination. The most recent occupants of the facility prior to shut-down may be unaware of things that happened 20–30 years ago. It is a good idea to identify and debrief some of the personnel (often retired) who were associated with the early history of the facility.

These initial investigations will be useful in planning the more detailed characterisation work. It would be helpful to know, for example, if the building were ever used for work involving alpha-materials. If it can be shown convincingly that no alpha materials were ever used, then the task of characterisation is greatly simplified.

9-3. Characterisation Plan

Once a review of readily available information has been carried out, a Characterisation Plan should be developed to obtain additional information to fill in gaps in the data. This will involve direct monitoring or sampling/analysis of all materials and areas which are potentially contaminated. The Plan should identify:

- the types, numbers, sizes, and locations of samples required,
- the methods and equipment to be used in collecting samples,
- the type and methods of analysis (specifying the lower limit of detection),
- the instrumentation required,
- data validation and reporting,
- the methods to be used for disposal of samples,
- radiation protection and other hazard controls during sampling and characterisation, and
- quality assurance requirements.

There are three kinds of data which may be used to estimate the radioactive inventory:

- calculated data for the radioactive content of structural materials and fuels (using computer codes such as FISPIN or ORIGEN which take into account the radioactive decay over time and resulting fission and/or activation products),
- in situ measurement of dose-rates and/or contamination levels (by manual or remote means) using real-time instruments, and
- sampling and analysis under laboratory conditions.

Computer codes for prediction of induced activities in materials can provide good estimates of the inventory to be found in the residual plant. However, to achieve this they require a detailed knowledge of the irradiated material composition, the geometry during irradiation, and the irradiation conditions. This information may not be available at the decommissioning phase.

In situ measurements of dose-rates and contamination levels within plant are an essential first step in providing useful information on the distribution and probable scale of radionuclide inventories within a facility. Portable hand-held or mobile instrumentation can be used to provide rapid mapping of activity levels over an entire site, enabling variations to be obviously detected. Such surveys also serve to detect "hot particles" or residual sources remaining within the facility. They are also required to enable worker dose restraint objectives to be set, and will determine the dismantling methods that can be used. The instruments used measure dose-rate in terms of Sieverts per hour.

Laboratory analysis can vary from simple measurements of total activity using a proportional counter (a crude but rapid measurement), to high resolution spectrometry for determining specific isotopes (lengthy and expensive but precise). The resulting activity measurements are usually expressed in terms of Becquerels per gram of material.

In many cases, the extent of low-level radioactive activation or contamination in the structure of a facility can only be determined after the removal of the bulk of the radioactivity which may be present. This means that radioactive surveys and characterisation may need to be carried out several times at different stages of a decommissioning project, before the next stage can be planned in detail.

9-4. *In Situ* Measurements

Dose-rate measurements are made using a proportional counter or similar instrument held at a fixed, convenient distance from the contaminated surface. This method will give gross radiation readings which will allow the relative activity distribution across the plant to be determined. Such surveys will generally not identify the nature and quantity of the isotopes present. However, in many circumstances it is possible to undertake limited sampling and analysis to determine a radionuclide "fingerprint" that allows the total activity to be inferred from *in situ* measurements of gamma activity. This is based on the assumption that, for a given facility or piece of equipment (e.g., ventilation ducting), the mixture of isotopes present will be approximately constant. The fingerprint is established by taking samples and carrying out detailed

30 cm^2 grid surveys

Figure 9-1. Initial and Final Laboratory Building Survey.

measurements of all the radionuclides present using laboratory counting techniques (spectrometry). The fingerprint is then used to infer the total radionuclide inventory in Bq/gm from real-time radiation measurements (Sv/hr) using a simple Geiger counter. This method can be particularly helpful in measuring alpha-activity which cannot easily be measured in real-time, by linking it to a more easily detected gamma activity. Care needs to be taken, however, to avoid errors due to variations in the fingerprint. For example, preferential plating-out of an isotope along the length of a ventilation duct would invalidate the fingerprint approach.

Loose contamination measurements can be made by "taking swabs," i.e., rubbing a piece of filter paper or similar material over the contaminated surface and then taking the paper away for measurement (using a dose rate counter or laboratory analysis).

In situ High Resolution Gamma Spectrometry may be carried out to further investigate any area considered to have significant radiation levels and usually only when it is impracticable to take samples for laboratory analysis. This is typically used to provide nuclide specific measurements for Co-60 and Cs-137 in pipes and drains.

9-5. Sampling and Analysis

In situ methods are suitable for initial surveys, but do not provide comprehensive information about the specific nuclides present. This information can best be obtained by taking samples for laboratory analysis. Depending on

the circumstances, it may be appropriate to carry out a broad range of radiological and chemical analysis on the same set of samples. It is necessary at the outset (based on the known history of the facility and the initial survey) to specify the range of species for which measurements are needed and the required lower limit of detection.

Accurate characterisation requires representative sampling of materials. For example, nonhomogeneous samples (e.g., concrete) require careful sampling and homogenisation to ensure that representative samples are taken for analysis. If contamination is not uniform, but an "averaged" value of activity is required, then some form of systematic sampling (e.g., using a grid, see Figure 9-1) and homogenisation of the samples should be used. Statistical methods may need to be employed to demonstrate that the measured values are representative of the bulk activity. Care must be taken during sampling and sample storage to ensure that the sample radionuclide content is not disturbed. For example, drying of samples may lead to loss of tritiated water, biological degradation of organic samples could result in loss of ^{3}H and ^{14}C, and heating of samples could result in loss of volatile nuclides such as ^{3}H, ^{14}C, ^{35}S, ^{99}Tc, $^{103/106}$Ru, ^{137}Cs, ^{210}Po, etc. Sampling may also disturb secular equilibria within radionuclide decay chains, which can make interpretation of the analytical results more difficult, e.g., isotopes in the natural U and Th decay chains are often analysed by gamma spectrometry. Usually, gamma emissions from daughter radionuclides are used to infer activities of the U and Th parents. If sampling results in a loss of Rn from the sample, the resultant decay chain will not be in secular equilibrium, and measurement of daughter radionuclides, e.g., ^{214}Pb, may not give an accurate representation of the U content of the sample. Similarly, the presence of purified U in a facility may not be detected using gamma spectrometry if the daughter radionuclides have had insufficient time to attain detectable concentrations.

Most radioanalytical techniques for the measurement of alpha or beta emitters require the use of a sample dissolution stage. This is often a simple leach of the sample with a suitable acid such as nitric, hydrochloric, or aqua regia. However, some chemical species (e.g., some forms of Pu, U, Th, and fission product insolubles) will not be completely dissolved in this way, and a more aggressive form of dissolution may be required. Total sample solubilisation, often employing hydrofluoric and/or perchloric acids, is widely used. Alternatively, the sample may be mixed with a flux such as lithium borate or potassium hydrogen fluoride, and the mixture fused at high temperature to produce a melt. The specific approach chosen will depend on the sample matrix and the radionuclides for analysis.

To ensure that all radioactivity within a plant is detected, radiological surveys should be conducted right through the facility. However, certain areas should receive particular attention:

- *Floors* — noting in particular areas of potential spills, e.g., beneath plant;
- *Walls* — where dusts or sprays may have settled;
- *Horizontal surfaces* — such as window sills and the tops of door frames, where dust may have settled;
- *Ceilings* — particularly around ducts and ventilation outlets; and
- *Pipes, tanks, and ducts* — take swabs from inner surfaces where possible.

Look carefully for cracks and hidden penetrations in walls, floors, and ceilings where contamination may have seeped. Take samples of paint from walls and ceilings — they may cover alpha contamination that will not be detectable at the surface. Any liquids in pipes, tanks, and sumps should be sampled; also insulation material. Note unexpected changes in floor levels — a sign that contamination may have been covered up by adding an extra layer of screed. Take up flooring materials such as linoleum or carpets and sample or monitor the underlying floor.

Where active liquids or gases have been stored within vessels and may have diffused into the bulk of the vessel, it may be necessary to section and depth profile the vessel to determine the radionuclide distribution through the vessel thickness. Similar profiling may also be required where a significant thickness of material has been subjected to neutron activation (e.g., concrete reactor bioshields) — the specific activity of the activation nuclides will decrease with distance from the neutron source.

When dealing with low levels of contamination in a building, it is necessary to take account of the natural background radiation. This may affect both the *in situ* dose rate measurements and the radioactivity measured during destructive analysis. Typical natural radionuclides which may be encountered include those from the U and Th decay chains and ^{40}K. The contribution of the natural background radiation will vary according to the geographical area and the facility construction materials, and must be determined on a case-by-case basis.

Air sampling is used to monitor low levels of airborne contamination, e.g., from suspended radioactive particulates. It is used to control worker intake of hazardous materials. Air samplers are instruments that suck a controlled flow of air through a filter paper. The radioactivity collected on the filter paper can subsequently be monitored. Air samplers are typically run for a week at a time and give results in Bq/m^3. Air samplers can also be used to monitor for asbestos and other hazardous dusts such as beryllium oxide.

Gamma spectrometry is used to measure a wide range of gamma emitting radionuclides including the activation products ^{51}Cr, ^{54}Mn, ^{60}Co, ^{65}Zn, and ^{134}Cs, and fission products ^{131}I, ^{241}Am, and ^{137}Cs. Radionuclides in the natural U and Th decay chains can also be determined with this technique.

Gamma spectrometry requires little sample preparation and no separation chemistry. Samples may be homogenised and prepared in a standard geometry for counting. For accurate assessment of the gamma inventory, calibration standards which have been matrix-matched to the sample should be prepared and measured in a similar geometry. If the presence of radionuclides with weak gamma emissions (e.g. ^{55}Fe, ^{129}I, or ^{241}Am) is suspected, the sample geometry should be kept thin to avoid self-shielding effects. Figure 9-2 illustrates a typical gamma camera in action.

A range of alpha-emitting radionuclides may be encountered during decommissioning including:

$$^{224}\text{Ra}, \, ^{226}\text{Ra}, \, ^{226}\text{Th}, \, ^{232}\text{Th}, \, ^{234}\text{U}, \, ^{238}\text{U}, \, ^{237}\text{Np},$$

$$^{239+240}\text{Pu}, \, ^{238}\text{Pu}, \, ^{241}\text{Am}, \, ^{242}\text{Cm}, \, ^{244}\text{Cm}, \, ^{252}\text{Cf}$$

They may be analysed using several techniques depending on the information required. Alpha spectrometry allows identification of the individual isotopes in a sample, based upon resolution and measurement of the energy of alpha particles emitted from a sample. Such measurements may require that the alpha emitters be separated from the bulk sample to produce a thin source with minimal self-shielding. In some, cases, there may be spectral overlap between alpha-emitters of different elements, which may only be overcome by chemical separation of these elements. However, for energy overlaps between

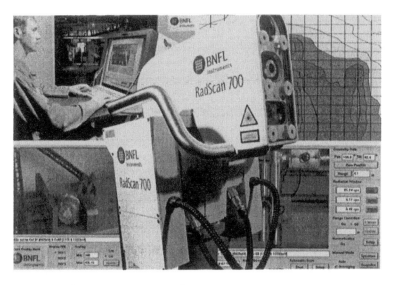

Figure 9-2. BNFL "RadScan 700" Camera in Action.

radioisotopes of the same element, deconvolution is not routinely possible (e.g., ^{239}Pu and ^{240}Pu cannot be readily resolved by alpha spectrometry). The separation requirements of this technique render it too time consuming for routine activity assessment, but it is often used during radionuclide fingerprinting.

Where a waste stream requires routine assessment of alpha activity, Liquid Scintillation Counting (LSC) may be used. It has limited resolving capability, and so should be used in conjunction with other techniques to identify the isotopes being measured. It has a low detection limit, making it useful for environmental monitoring. The technique requires samples capable of dissolution or suspension in a scintillation cocktail, and is thus particularly useful for measuring activity in liquids (e.g., pump oils, waste solutions, environmental waste waters, etc.). By use of careful separation techniques and yield tracers, it is possible to extend the technique to include radioisotopes extracted from bulk solids (e.g., soils and building materials).

Beta-emitters include the following:

- activation products ^{3}H, ^{14}C, ^{35}S, ^{41}Ca, ^{63}Ni,
- fission products ^{89}Sr, ^{90}Sr, ^{99}Tc, ^{129}I, ^{147}Pm, and
- actinide ^{241}Pu.

The spectra of beta-emitting radionuclides show broad ill-defined peaks, so spectrometry is of limited use in determination of the nuclides present. Determination of specific nuclides is normally by chemical separation of the element required, and measurement of the radiochemical activity separated. Measurement may be either by use of solid-state detectors or by Liquid Scintillation Counting

(LSC), which is particularly useful for low-energy beta emitters. Nuclides commonly analysed include tritium and ^{14}C in aqueous samples. By use of separation techniques and yield tracers, it is possible to extend the technique to include radioisotopes extracted from bulk solids. For example, tritium and ^{14}C have been determined in concrete by roasting samples and trapping the evolved water and carbon dioxide for analysis by LSC. Similarly, ^{36}Cl has been determined in reactor graphite using oxidative dissolution to remove the graphite, and analysing the resultant solution by LSC. Activation products ^{41}Ca and ^{63}Ni have been determined in reactor bioshield by dissolution and chemical separation of Ca and Ni from the surrounding matrix. The radioisotopes were then measured by LSC.

9-6. Quality Assurance Requirements

Generally, Quality Assurance (QA) for characterisation is part of the larger decommissioning project QA Program (see Chapter 8). Key aspects relating to radiological characterisation are:

- Personnel — qualifications, experience, and training;
- Procedures;
- Instruments — appropriateness and calibration; and
- Documentation and records.

Figure 9-3 illustrates the monitoring of components prior to sentencing.

An example of the equipment used for weld sampling prior to decommissioning the Trawsfynydd Magnox type reactors (located in North Wales, UK) is described below.

Figure 9-3. Monitoring Components.

Figure 9-4. Some of the Tools Developed for Remote Weld Sampling (Photograph courtesy Schilling).

Two modified, radiation-hardened, Titan 2 manipulator systems with remotely interchangeable tools and custom control systems were supplied to BNFL Magnox Generation. Each manipulator arm is mounted on a remotely operated mobile vehicle for deployment beneath an off-line reactor pressure vessel. The manipulator arms, which can be operated under both tele-operator control and model-based graphical control, deliver much of the equipment necessary for remotely investigating the reactor vessel welds and cutting weld samples with ultra-high pressure water jets. The tools developed include standard items (such as jaws, drills, rotary brushes, snips, and claws) as well as task-specific tools (including a sample retrieval tool, instrument probe, heat gun, suction and inflation tools, and a complex tool for locating and marking weld center-lines). The weld-locating tool contained two lasers, camera, lights, a traversing mechanism, and a grinding tool (see Figure 9-4).

9-7. Characterisation Report

The output of the radiological characterisation will be a report that documents the methods used and the data obtained. This report will be used in planning the next phase of decommissioning. It is important that the information should be as complete and accurate as possible, as it will be required to plan the methods used in dismantling and the handling, storage, and disposal of radioactive wastes. Errors or misunderstandings arising from inadequate recording of the characterisation data could have potentially serious safety implications and/or lead to unnecessary costs later in the project.

The Characterisation report should contain:

- description and operational history of the facility,
- methods used for the characterisation survey,
- instrumentation — types and sensitivities,
- results — radioactive inventory, other hazardous materials, unexpected findings, and
- appendices — references, tables, figures, maps, calibration, and analytical results.

9-8. Reference

1. International Atomic Energy Agency. *Radiological Characterisation of Shutdown Nuclear Reactors for Decommissioning Purposes*, IAEA Technical Report Series No. 389, IAEA, Vienna (1998).

Chapter 10
Decontamination Techniques

10-1. Introduction

Decontamination is a process by which radioactive contamination is removed from a surface, including surfaces that are porous or fissured. Judicious use of decontamination techniques can reduce the radiation levels and/or minimise the volume of radioactive waste produced when a facility is dismantled.

A range of decontamination techniques is available, such as scabbling or pressure jet washing. The choice of technique depends on individual circumstances. This chapter describes the pros and cons of various decontamination techniques which are available, giving examples from a number of successfully completed projects. It also considers instances of novel applications where such practices have not been so successful, and the lessons that have been learned.

10-2. Objectives and Constraints for Decontamination

There are a number of reasons for wanting to decontaminate. First, there is a need to reduce locally the inventory of radioactive material in a facility or an item of equipment, in order to reduce the radiation levels and minimise the potential for a release of radioactivity to the environment. For example, a shielded cell which has been used for postirradiation examination of spent fuel will normally have high levels of radioactivity distributed widely across all the internal surfaces of the cell and on the equipment within it. This will include dust and larger fragments arising from the fuel itself. The radiation levels will typically need to be reduced sufficiently to allow man-access for further dismantling operations. In some cases, the facility will be kept in a safe state, pending final dismantling at a later date, in order to gain the benefit from radioactive decay. However, safety considerations for long-term care and maintenance require loose contamination to be removed as far as practicable, and any remaining radioactivity to be fixed or sealed-up in order to minimise the hazard.

Secondly, there is a need to reduce the quantity of radioactive waste produced. Decontamination can be used to reduce the level of radioactivity on a contaminated surface, so that the contaminated item can be categorised at a lower level (e.g., low-level waste rather than intermediate level). In some cases, it is possible to reduce contamination to the "free release" level (defined in the UK as $<0.4 \, \text{Bq} \, \text{g}^{-1}$), which allows materials to be dispatched for recycling or disposal as nonradioactive waste. The radioactivity removed in the decontamination process is concentrated into a (usually much) smaller volume.

Thirdly, there is a need to complete the final stages of decommissioning, which requires decontamination of buildings prior to demolition and remediation of the site to remove contamination from the foundations and surrounding land. However, remediation of ground contamination is a major topic in its own right and is dealt with separately in Chapters 23–25.

There are a number of constraints involved in decontamination. Decontamination necessarily involves the generation of secondary wastes. The operator must ensure that there is a suitable disposal route available for these secondary wastes, and that their volume is small enough to justify the operation. The use of chemical techniques may create liquid effluents containing materials (such as chelating agents) which could interfere with down-stream processing or could be unacceptable for release to the environment (e.g., heavy metals such as lead).

Decontamination operations may involve exposure of the operators to radiation dose. This needs to be justified in terms of the reduction of dose in subsequent care and maintenance and dismantling operations. Similarly, the financial cost of the decontamination operation needs to be justified in terms of the savings, which will accrue from subsequent care and maintenance, dismantling, and waste disposal. For example, the justification for decontaminating some steelwork to free release level should take account of the cost of doing so (including treatment and disposal of secondary waste) compared with both the scrap value of the steel and the avoided cost of disposal as radioactive waste.

There is also a need to take care that a decontamination operation does not exacerbate the contamination, either by spreading it more widely within the facility, or by converting it into a more intractable form. For example, mechanical abrasion techniques (e.g., scabbling) can create dust which, if it is not trapped in some way, can spread throughout a facility and increase the overall extent of contamination. Alternatively, pressure washing a concrete surface can have the effect of driving the contamination further into the concrete. Decontamination of building fabric by removing contaminated material can, if taken too far, affect the structural integrity of the building. In some cases, where complete walls or structural supports have to be removed, temporary alternative supporting members may need to be inserted.

In some circumstances, it may be difficult to prove conclusively that all of the radioactivity has been removed from a contaminated item. This is particularly true of contaminated equipment such as pumps and motors with complicated internal structures. Painted surfaces can trap alpha activity, which cannot be detected through the paint. Paintwork can be removed, but it may be difficult to remove all of it from crevices. In such circumstances, there may be little benefit in trying to decontaminate the item, as it will still need to be sentenced as radioactive waste because it cannot be shown to satisfy the free release criteria.

The end-point for the decontamination operation should be clearly defined in terms of bulk activity $(Bq\,g^{-1})$, surface activity $(Bq\,cm^{-2})$, and radiation levels (typically $\mu Sv\,hour^{-1}$ at the surface). From a practical consideration, it should be noted that monitoring instruments are generally calibrated to give radiation dose rates, which then need to be interpreted to calculate surface activity or bulk activity. The free release level is defined only in terms of bulk activity. The extent to which the activity levels can be monitored in real time depends on the nature of the radioactivity present. Some forms of activity (alpha and soft beta) are difficult to detect. However, in many circumstances, it is possible to determine a "fingerprint" that allows the total activity to be inferred from measurements of gamma activity.

10-3. Characteristics of Decontamination Techniques

Decontamination techniques can be classified as follows:

- Nonattritive methods of simple cleaning such as swabbing, sweeping, and vacuuming, which leave the substrate surface essentially unchanged;

- Chemical (and electrochemical) treatment to remove a layer of the substrate surface, along with radioactivity. The depth of treatment depends on how far the radioactivity has penetrated beneath the surface; and

- Physical attrition to remove a surface layer, such as the scabbling of concrete or milling the surface of lead bricks.

Most techniques can be applied either *in situ* or to material or components removed to a decontamination facility. The effectiveness of each technique will not be the same in all situations. The method of application must be considered for each technique in the context of the situation in which it is used. The application away from the facility is usually undertaken when the aim is to lower the waste category (for example, to clean the material to allow its free release).

The application to facilities being decommissioned is often somewhat different from the way they are applied to operational facilities. In the latter instance, it is important not to damage the equipment, plant, or facility, whereas when decommissioning a plant the use of aggressive methods is acceptable. An example of this is that the methods used to decontaminate the primary circuit of a water-cooled reactor during operational shutdowns must not affect the long-term integrity of the pressure circuit. When the reactor is being decommissioned, more aggressive chemical reagents can be applied to remove more activity and produce a higher reduction in the radiation levels than can normally be achieved with the chemicals used when operational.

10-3-1. Nonattritive Cleaning

Nonattritive methods remove contamination from a surface without damaging the surface itself. They include simple cleaning techniques universally used in facilities under the heading of good housekeeping, as well as more sophisticated methods such as ultrasonic cleaning. Inside hot-cells, caves, gloveboxes, and any similar facilities, it is good practice to keep the insides physically clean by the application of such methods as given in Table 10-1.

10-3-2. Chemical Decontamination

With chemical decontamination, the aim is to remove the radioactivity which has penetrated into the surface of the contaminated item. This is achieved by the dissolution of a layer of the substrate surface. The radioactive material will either end up dissolved in the chemical with a significant amount of the substrate or, where the radioactivity is not itself soluble in the chemical, it will be suspended in the substrate solution.

Table 10-1. Nonabrasive Methods of Decontamination

Technique	Typical uses
Vacuum cleaning	Applied to clean up in all types of facility. Can be applied using remote equipment. Vacuum cleaner fitted with output filter.
Sweeping/ brushing/ dusting	Conventional process — for large areas. Can be undertaken with remote handling equipment.
Washing	Usually applied where the facility can deal with water. Surfactant can be added.
Swabbing	Picks up particles well. Can use various liquids to wet the swabs.
Scrubbing	For smooth surfaces (but could wash contamination into cracks).
Strippable coating	Good method for sealing in the contamination. Reduces the likelihood of airborne suspension. Extensively used.
Ultrasonic cleaning	Used principally for cleaning smaller components by immersing them in a tank of liquid agitated ultrasonically. Often used on contaminated items removed for repair.
Freon cleaning	For small components intended for reuse which can be put into a special enclosure containing the freon cleaning equipment. No longer acceptable, as freon is not environmentally friendly.
Steam cleaning	Can be more effective than simple washing.

Table 10-2. Chemical and Electrochemical Decontamination Methods

Method of application	Typical uses
Circulation through chemical plant	Applicable to chemical plant where reagents can be readily circulated. Can produce large volume of waste.
Spray reagent	Need a method of collecting the liquid reagent so use is limited. It can be difficult to reach all areas.
Foam reagent	Foam increases the reagent residence time. Foam can be readily collected using wet vacuum cleaning. The foam is then collapsed, thus minimising the amount of reagent.
Gels	Similar to foams in application, but removal involves washing off rather than vacuum removal of the reagent.
Immersion in tank of reagent	Mainly for components and not applicable for *in situ* decontamination. Need to size reduce to fit into tank. Used for lowering the waste category, often to allow free release.
Local use	Special devices developed to apply reagent to a surface locally then possibly wash after the reagent has done its job. Can be used/applied by a programmed robot.

There are many types of chemical in regular use, the most common being simple mineral acids such as nitric acid. Details can be found in the many publications on decontamination. Table 10-2 lists various methods of applying chemical decontamination processes. Addition of a complexing agent such as citric acid helps to solubilise some radionuclides. In some cases, an electrochemical rather than a simple chemical reaction is necessary. The choice of reagent to use will depend on the material being decontaminated and the form of the contamination itself. One variation of chemical decontamination is to use chemicals to remove contaminated layers of paint, thus removing the contamination at the same time.

A MEDOC process is used by SCK-CEN/Framatome to decontaminate metallic items such as pipes, tanks, and heat exchangers from reactor dismantling (BR3 PWRs). It uses Cerium IV as an oxidising agent and it is claimed to sentence nearly all of the treated material as free release with a 95% volume reduction. The spent solution is precipitated, filtered, and encapsulated in asphalt (see Figures 10-1–10-3).

Figure 10-4 shows an electrochemical decontamination head attached to a robot arm as successfully used to decontaminate stainless steel lined remote handling cells at Harwell.

10-3-3. *Physical Attrition*

Chemical methods only work well when the contaminated surface is metallic. For other surfaces, such as concrete or plaster, methods which strip off layers of surface material by physical attrition may be required. Such methods can be applied either *in situ* or in a special facility away from the original location. Attritive methods are often used to decontaminate structural material. This then allows it to be released for unrestricted disposal or recycling, or allows a building to be demolished using conventional demolition. Figure 10-5 illustrates the use of a CO_2 abrasion technique.

The solid CO_2 pellet blasting process can be used to remove contamination in the form of paint or surface coatings or surface layers of soft materials. The process involves entraining dry ice pellets in a propellant air stream. The pellets simultaneously provide the effects of abrasion, thermal shock, and vaporisation to remove the surface coating, without the need for water, abrasive grit media, or chemical solvents, as secondary waste. The process is also capable of removing metal slag. The contaminated material arising from the process is collected in the filters of locally applied ventilation systems.

Figure 10-1. Foam Cleaning Equipment.

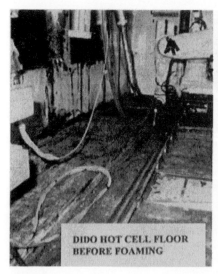

Figure 10-2. Foam Cleaning a Hot Cell.

10-4. Waste Minimisation and Treatment

In the UK, the Nuclear Installations Inspectorate requires operators of a nuclear licensed site to minimise as far as is reasonably practicable the rate of production and total quantity of radioactive waste accumulated on their sites. Decontamination techniques can assist in minimising waste volumes by concentrating the radioactivity, leaving the bulk of the once contaminated material or item in an uncontaminated state, or reducing its waste category to a lower level, thereby making for easier storage or disposal. The extent to which this is possible depends

Figure 10-3. Scabbling.

 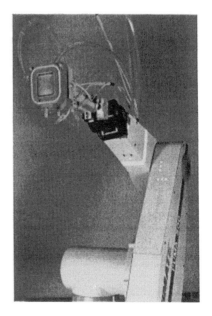

Figure 10-4. Electrochemical Decontamination.

on careful characterisation of the nature and extent of the contamination and the correct choice of decontamination strategy (see Table 10-3).

However, the generation of secondary radioactive waste from decontamination processes is inevitable. The ease of dealing with secondary arisings depends usually on whether it is solid or liquid. If the waste is in a solid form, such as the arisings from scabbling or milling or that collected within a vacuum cleaner, then it can be dealt with as part of the normal solid waste route. Sometimes, some pretreatment may be required.

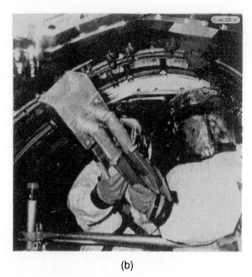

(a)   (b)

Figure 10-5. CO_2 Abrasion (Photographs courtesy ALSTEC). (a) CO_2 pellet blasting nozzle mounted on a hydraulic manipulator cleaning paint off glass. (b) Decontamination of Joint European Torus vacuum vessel with CO_2 pellets.

Table 10-3. Attritive Methods of Decontamination

Technique	Typical use
Scabbling	For concrete — various tools available commercially
Shaving/grinding/ abrasive	For several types of material (concrete, masonry, etc.)
Milling	For metals such as lead bricks
Water jetting (with or without abrasive)	For concrete and other materials. This method might not physically remove the substrate so could be considered in Table 10-1
Jackhammer (and similar devices)	Concrete
Microwaves	Concrete
Explosives	Concrete
Drilling/spalling/ routing	To remove persistent areas of contamination
Sand blasting	To clean the surface – paint removal

Liquid wastes will require treatment in order to either convert them into a solid form or to remove the radioactivity to allow the disposal by a more conventional route. The treatment is often specific and must be considered at all stages. Care must be taken not to produce a liquid waste which, when treated, produces a solid waste which is unacceptable in a future waste repository (because, for example, it will change the chemistry in the repository or effect the integrity of the waste-form).

It is more often the difficulties of dealing with the secondary waste which influences the decision of whether or not to use a particular technique. In the past, the authors have tested a number of sophisticated methods for decontamination, but, as experience has grown, the selection process tends to favor those which have been proven and for which the secondary wastes are most easily treatable.

Melting of metallic components is a potential technique for decontamination and waste volume reduction. In some circumstances, it may be possible to remove the contamination as a slag, allowing the metal to be released for recycling. However, assessments within UKAEA have consistently shown that melting is currently not an economic option, so it has not been used.

10-5. Selecting a Decontamination Technique

In order to determine whether to use a decontamination technique and select the most appropriate one, the following questions need to be addressed:

- What is the nature and extent of the contamination?
- What is the purpose of decontamination? What is the target end-point?
- What processes are available?
- What wastes will be generated?
- Is there a route to deal with these wastes? If so, what is it and is it acceptable?
- How effective is the process — will it satisfy the objective?
- How can the process be applied?

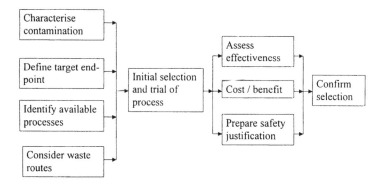

Figure 10-6. Logic Diagram for Selecting a Decontamination Technique.

- Is the process ALARP (as low as reasonably practicable)?
- Does the benefit compare favorably with the cost in both financial and radiation dose terms?
- Can a safety case for its use be produced and approved?

Figure 10-6 illustrates the logic of the decision-making process involved. Before selecting a particular process, consideration should be given to the risk and consequences of failure. UKAEA prepares fall-back strategies to ensure a successful outcome.

10-6. Positive and Negative Experiences from Completed Projects

Table 10-4 summarises positive and negative experience on a selection of projects where decontamination was a significant issue. Useful references [1–8] are given at the end of this chapter. In general, the following are the key lessons which can be learned from this experience.

- Characterise the contamination and plan the work thoroughly. Planning should include a search through past operational records so that accidental contamination can be identified and potential problems anticipated.
- Keep it simple. After examining numerous options for decontaminating lead bricks using chemical techniques, it was decided to opt for a simple technique of shaving off the surfaces of the bricks with a standard planing tool. This has proved very successful.
- Use of liquids to wash down a contaminated building should be avoided unless the floor and other surfaces are impermeable. Decommissioning of the Hermes facility at Harwell was made more difficult by previous efforts to decontaminate by washing, which resulted in contamination penetrating into discontinuities in the floor, walls, and windows.

- Avoid chemicals, such as chelating agents, which can compromise downstream processing. When possible agents for decontaminating the Windscale Advanced Gas Reactor (WAGR) heat exchangers were investigated, nitric acid was chosen partly on the grounds that it would have least impact on the site effluent treatment system. A small amount of citric acid was also allowed. Trials involving the spraying of the acid into one section of a heat exchanger showed that this reagent (applied in this way to avoid producing large volumes of secondary waste) did not achieve the desired decontamination factor. A different disposal strategy, not involving decontamination, was adopted.
- When remote operations are required, test them in a mock-up facility. Before using a TeleRobot for decontaminating a High Activity Handling Cell, a mock-up facility was used to develop specific tooling and train staff [9]. The problems ironed out at this stage would have been much more difficult to overcome if they had been encountered in the active environment.
- Avoid cross-contamination. Extensive use of spray-on strippable coatings can avoid surfaces being re-contaminated once they have been cleaned. This has proved particularly useful in decontaminating plutonium facilities.

10-7. References

1. UKAEA. *Atomic Energy Code of Practice AECP 1085*, Issue 1, UKAEA, Harwell, Oxfordshire, UK, 1998.
2. Bayliss, C. "Practical Applications of UKAEA's Decommissioning & Liabilities Management Toolbox," *Ibc 7th International Conference & Exhibition on Decommissioning Nuclear Facilities*, London (October 2000).
3. Bayliss, C. "Decommissioning – Choosing and Prioritising the Right Options," *Ibc 6th International Conference & Exhibition on Decommissioning Nuclear Facilities*, London (June 1999).

Table 10-4. Summary of UKAEA Experience and Lessons Learnt

Project/Objective	Decontamination technique	Outcome/Lessons learnt
Trials for decontaminating WAGR Heat Exchangers in order to reduce dose levels for dismantling/size reduction operations. Site: Windscale.	Washed with recirculating spray of water, then nitric acid (0.5 M) and citric acid (0.0025 M).	Trials gave a DF ~ 3. In some circumstances, this might be useful, but it was judged too little, so strategy changed. Heat exchangers removed and transported to Drigg LLW repository as single large items.
SGHWR Fuel Pond: removal of sludge and decontamination of pond walls. Site: Winfrith.	Sludge was removed from the pond floor by vacuum cleaning. The pond walls were washed by applying a proprietary surfactant solution following low pressure water jetting, with operatives working off a floating pontoon. Lowering the pond water level as work proceeded gave access to successively lower parts of the pond structure while maintaining shielding and trapping contamination.	Decontamination succeeded in reducing radiation levels to allow free access.
Decontamination of a Plutonium Fuel Manufacturing Facility. Site: Winfrith.	Glovebox ventilation extract system and primary drain lines decontaminated using high pressure water jetting. Building fabric decontaminated using needle guns to remove paint from metal and scabbling to clean concrete surfaces.	Decontamination successful, but progress slow due to difficulty of working in pressurised suit environment. No contamination above free release level was found during building demolition.
High Activity Handling Cell: removal of cell internal equipment and decontamination to allow man-entry for final dismantling. Site: Harwell.	Telerobot (NEATER) used to deploy a variety of tools for size reduction and decontamination of cell internals. Vacuum cleaning and foam washing used for decontamination.	Telerobot reliable and easy to use. Vacuum cleaning picked up fragments of ^{60}Co, which were the main source of background. Foam washing removed surface activity embedded in oil and grease.
Removal of activation and contamination from LIDO concrete bioshield. Site: Harwell.	Radioactivity carefully mapped by core sampling and surface monitoring. Trials funded by CEC on microwave spalling, explosive cutting, and diamond cutting to remove active material.	All methods worked to some extent, but stitch drilling using a diamond-toothed core drill was the simplest and most effective. 5% of the total mass was removed as LLW; the remainder was free release.
Decontamination of a Chemical Engineering Building prior to demolition. Site: Harwell.	Methods used to decontaminate the building fabric include washing, concrete scabbling, and paint removal. Hydraulic platform used to access high level surfaces.	The building was successfully decontaminated and subsequently demolished to time and budget.
Decontamination of Lead Bricks. Site: Harwell.	Following assessment of various options involving chemical and electrochemical techniques, the chosen method was to shave a thin layer off the surface of each brick using a simple planing tool.	Planing method is quick and simple, with minimal secondary arisings. Several hundred tonnes of lead brick have been successfully decontaminated to free release level.

4. International Atomic Energy Agency. *Radiological Characterisation of Shutdown Nuclear Reactors Decommissioning Purposes,* IAEA Technical Report Series No. 389, IAEA, Vienna (1998).

5. UKAEA. "Radioactive Decontamination," *Atomic Energy Code of Practice AECP 1057,* Issue 4, UKAEA, Harwell, Oxfordshire, UK, 1992.

6. Bartholomew, P. "The Decommissioning of the UKAEA's SGHWR Ponds," *The Nuclear Engineer,* 38(1), 23–25 (January 1998).

7. Smith, D. "Decommissioning a Plutonium Fuel Processing Facility to a Green Field Site," *The Nuclear Engineer*, 41(2), 60–65 (February 2000).

8. Abel, E. and C. Hamblin. "Seven Storeys of Decommissioning – The Complete Decommissioning of a Chemical Engineering Building," *Nuclear Decommissioning 98 Conference*, London (December 1998).

9. Inns, A. J. and K. F. Langley. "Decommissioning Three Minor Nuclear Facilities at Harwell to Green Field Sites," *Ibc 5th International Conference and Exhibition on Decommissioning of Nuclear Facilities*, London (February 1997).

Chapter 11
Dismantling Techniques

11-1. Introduction

Nuclear decommissioning invariably involves dismantling of plant and equipment which has some degree of radioactive contamination. Chapter 9 has described how the contamination can be characterised. Chapter 10 has discussed the techniques by which contamination can be removed prior to dismantling. The methods which are available for dismantling are described in this chapter.

The choice of method will depend to a large extent on how successful efforts at decontamination have been, and the nature and extent of any remaining contamination. The presence of radioactivity will require measures to be taken to contain radioactive contamination and shield operators from radiation. Dismantling can lead to the generation of large amounts of dust, and the potential for release of gases and liquids. Hence, it is usually necessary to provide some form of containment around the items to be dismantled. Buildings may be stripped and decontaminated to the point where conventional demolition is possible; otherwise it may be necessary to cocoon the entire building. Workers may need to be provided with personal protective equipment. Alternatively, remote handling methods may be necessary to deploy dismantling tools.

Many nuclear facilities have built-in remote handling equipment, such as cranes, master-slave manipulators, etc., which will have been used during routine operations and maintenance. Provided they are still serviceable, they may be used during the early stages of dismantling operations, for example, for size reduction of in-cell components. However, if the existing equipment has deteriorated or is not robust enough for the tasks to be undertaken, additional remote handling equipment may need to be deployed.

11-2. Cutting Techniques

Dismantling methods generally involve size reduction of the plant and equipment to allow it to be handled and packaged. Dismantling may involve disassembling by undoing bolts. Generally, however, this is too time-consuming.

It is more usual to cut items by a variety of mechanical, thermal, or other methods. Concrete and masonry can be broken down by conventional demolition methods. Table 11-1 summarises the main classes of dismantling techniques, which are described in more detail in the following paragraphs.

11-2-1. Mechanical Cutting

Mechanical cutting involves techniques where mechanical force is used to cut material. Mechanical methods have the advantage of producing relatively easily handled secondary wastes which can be collected for disposal.

Saws

There are many types of saws: reciprocating, circular, band, and wire saws which can be used on almost any scale imaginable from small hand-held hacksaws to very large bandsaws (a form of bandsaw was used to cut the front section off the Russian nuclear submarine "Kursk" as it lay on the bottom of the Barents Sea). Where the material to be cut is particularly hard (such as concrete or masonry), diamond tipped saws can be used.

Shears

Sawing can be slow, particularly in inaccessible positions. An alternative approach is to use shears (of the type used by rescue services to cut the tops off wrecked cars to allow rapid access). These are available in a variety of sizes and are particularly suitable for cutting through metal pipes and structural framework made of girders or I-beams. Shears can be manually actuated or powered pneumatically, hydraulically or electrically. There are three basic types of shears:

- two bladed shears, like scissors, which are suitable for lightweight uses such as small pipework;
- a blade and anvil device, where the blade forces the work-piece against a fixed anvil. This type is suitable for cutting components of larger cross-section and thickness than the scissors-type; and

Table 11-1. Summary of Dismantling Techniques

Dismantling technique	Suitable for	Comments
Mechanical cutting		
Saws	metal, wood, plastics	Available as reciprocating, circular, band, and wire saws
Diamond saws	masonry	Diamond tipped versions of above
Hydraulic shears	metal pipes, ducts, wiring, etc.	Potential problems with high velocity fragments due to elastic rebound
Nibblers	sheet metal and plastic	Avoids recoil and elastic rebound
Drill bits	sheet metal and plastic; concrete	Stitch drilling can cut through large sections of concrete shielding
Milling and routing	metal, sheet plastic	Can strip off surface layers
Hydraulic jack-hammer	concrete, masonry	Suitable for demolition work
Thermal cutting		
Flame cutting	mild steel (not stainless)	Can be used manually or remotely, in air or under water
Plasma arc	all metals	Ditto. Fast, mature technology. Generates aerosol
Thermic lance	steel, concrete	Fast. Large quantity of aerosol — not suitable for highly active materials
Electro-discharge	all metals	Slow, but ideal for small scale applications under water
Laser	all materials	Applications limited by high capital cost
Other methods		
Expansive grout	hard concrete	Proprietary chemicals used to initiate crack formation
High pressure water jets	all materials	Fast. Abrasive water jets used to cut concrete.
Explosive cutting	concrete, masonry, metals	Shaped charges can be very precise, but safety concerns limit uses

• heavy-duty demolition shears, which are used in conjunction with mechanical excavators for cutting structural steel-work and crushing concrete.

Shears have a tendency to produce sizeable projectiles due to elastic rebound as the workpiece fractures, which can be a hazard to operators. This effect can be mitigated by the use of protective screens or, if the circumstances dictate, by applying the force to the shears intermittently using an electronic control device (see Figure 11-1).

Nibblers

A nibbler is a tool that uses a punch-and-die cutting mechanism reciprocating at high speed to cut through sheet material as well as small bore tubing. Nibblers avoid the problem of stored energy accumulating in the workpiece. They can be used in remote handling environments and are considered to be a mature technology.

Grinding

Grinding is a technique which allows a cut to begin in the middle of a plate as well as at the edge. It can allow a good range of movement and approach from different angles. It is not as prone to jamming as some blades, and can produce a deep cut in a number of passes; although realignment to an existing cut can be difficult. Figure 11-2 shows a remote grinding rig deployed at Sellafield, UK on a PaR manipulator for cutting a 2 mm thick plate. It can also cope with welds and brackets.

Drilling, Milling, and Routing

Holes can be drilled in sheet metal as a starting point for a reciprocating saw. Hollow cylindrical drill bits are frequently used to extract core samples for the purpose of chemical or radiological sampling. A series of adjacent holes can also be used drilled to create a continuous cut — a technique known as stitch drilling. Large sections can be cut out of thick concrete shielding using this technique. This was used some years ago at Harwell to remove activated concrete from the LIDO reactor bioshield. More recently, it was used to create an export penetration in the shield wall of a solid waste store (see Figure 11-3).

Milling and routing employs a range of cutting tools in a rotating chuck. These can be used to shave the surface off the workpiece, thereby removing radioactivity, or to

Figure 11-1. A Selection of Mechanical Cutting Saws and Shears.

Figure 11-2. Remotely Mounted Grinding Rig Cutting a Thin Plate.

cut slits. These tools are usually used with sheet metal, but have been successfully used with hard plastic materials in dismantling hot cells at Harwell.

Power chisels and Jackhammers

Pneumatically or hydraulically operated tools with a hammer–chisel end-effector are available in a variety of sizes. They are particularly useful for breaking up

Figure 11-3. Removal of the Export Penetration Plus Following Stitch Drilling.

large pieces of concrete or masonry. Smaller electrically operated breaker tools can be deployed by power

Figure 11-4. Needle Gun Attached to a Hilti™ Breaker Tool with Second PaR™ Manipulator in the Background.

manipulators for more precise remote tasks, such as the removal of concrete and encapsulation grout. Figure 11-4 shows a Hilti™ model TE104 hand-held breaker selected for its high speed and relatively low impact energy mounted in a PaR Systems 3000™ manipulator. Figure 11-4 also shows a standard needle gun attachment mounted in the Hilti™ breaker tool for removal of paint, scale, and concrete surfaces for decontamination purposes.

11-2-2. *Thermal Cutting*

Thermal cutting is generally faster than mechanical cutting, and the equipment tends to be lighter. However, it has the disadvantage of producing aerosols and particulates, which are a potential hazard to workers and the environment and can spread contamination. This means that an efficient system for air filtration is required, preferably as close as possible to the workpiece. They are also a potential fire hazard, particularly if there is inflammable or combustible material nearby.

Flame Cutting

Flame cutting is a well established method which uses a mixture of fuel gas (typically acetylene, propane, or hydrogen) and oxygen to produce a high temperature flame. It is generally used with mild (i.e., carbon) steel. The oxygen in the center of the flame oxidises the metal, which is blown away by the flame to produce the cut. Flame cutting can be used to cut a wide range of steel thicknesses up to about 3 meters. It can be used

under water. The technique is less successful with stainless steel due to the high melting point of chromium oxides produced.

Plasma Arc Cutting

Plasma arc cutting involves the creation of a stream of ionising gas (plasma) by the passage of an electric current between a tungsten electrode and the surface of a conducting metal. The arc causes local melting of the metal, which is blown away by the gas stream. The process is well established and very fast. The cutting heads are lightweight and, therefore, easy to deploy remotely. It can be used in air or under water. The main difficulty is the collection of the copious amount of aerosol generated. This can be done by placing an extraction nozzle close to the cutting head. Large metal tanks are cut up by attaching a temporary extraction system to the tank and cutting slits down the sides of the tank, leaving small sections of uncut metal at the top and bottom of the tank. This enables the aerosol to be contained within the tank and collected efficiently, keeping the integrity of the tank structure until nearly all the cutting has been done. If necessary, the dismantling can be completed by using a slower method which does not generate aerosol, such as sawing. (see Figure 11-5).

Figure 11-5. Plasma Cutting of a Steel Tank.

Thermic Lance

The thermic lance is a method of cutting concrete, steel, cast iron, and other materials. It is ideal for demolition work, where noise or vibration are unacceptable, or where speed is essential particularly on reinforced concrete. The equipment is extremely simple and easy to operate. The lance consists of a steel tube packed with steel rods (aluminum or magnesium are often added to the packing to increase the heat output), where oxygen is passed through, so that when the lance is ignited it becomes a great source of heat, and forms a fluid slag, which flows out of the cavity being cut. The lance is ignited by applying heat to the end of the tube with oxygen-acetylene equipment. The heat generated from the iron/oxygen reaction is sufficient to melt concrete (1800–2500°C). The formation of iron silicate increases the fluidity of the slag produced; therefore, the silicate content of the material has an appreciable effect on the speed of operation and the rate of consumption of packed lance and oxygen. Lances vary in length from about 0.5 to 3 meters and have a range of diameters. It is not recommended for highly activated or contaminated components, as it produces large amounts of aerosol. It has been used in the UK to cut the top bioshield of WAGR.

Electro-Discharge Cutting

Electro-discharge cutting involves the erosion of a metal through the passage of an electric current between an electrode and the metallic substrate. This causes evaporation of the metal substrate, in contrast to thermal methods which melt the metal. The method is slow but suitable underwater applications, particularly for "surgical" operations such as bolt-cutting where some precision is required. Arc-saw cutting is a variation of the technique, using a circular toothless sawblade.

Laser Cutting

Lasers can be used to cut almost any material. They are typically used for precision machining in the manufacturing industry. Applications to nuclear decommissioning have been limited by high capital costs, although R&D trials have been carried out in France and Japan. Laser cutting was used in the Dounreay fast reactor fuel reprocessing plant in the 1980s and early 1990s to cut fuel element wrappers.

11-2-3. Other Methods

Expansive Grout

Expansive grouting is a recognised technique in civil engineering for demolishing concrete structures. The technique involves drilling holes into the concrete and then inserting proprietary chemicals which react together, swelling in volume and, thereby, exerting a splitting force. This is similar to the action of frost on the weathering of rocks. It is recommended for nonradioactive structures. Figure 11-6 shows it being used to split a concrete shielded storage block which had proved very difficult to cut by more conventional methods.

High Pressure Water Jet Cutting

Water jet cutting, with or without abrasive, can be used to cut just about anything. Water pressurised up to 60,000 PSI is forced through a small ruby orifice at more than twice the speed of sound and is directed at the workpiece. Abrasive water jet cutting adds an abrasive, e.g., garnet sand, to the water for cutting hard or thick materials. A water jet without abrasive handles soft or thin materials.

Explosive Cutting

Explosives are widely used for demolition work in the civil engineering industry, although applications on nuclear licensed sites are limited by safety concerns surrounding the use of explosives. Shaped charges can be used to cut pipes and tanks with considerable precision in a controlled manner.

Figure 11-6. The PLUTO Test Reactor External Storage Block Split Using Expansive Grout.

11-3. Remote Handling Techniques

Over the past 20 years, a number of devices have been developed which can be deployed for carrying out dismantling operations remotely. There are many examples of these devices available commercially. The following paragraphs describe specific examples of remote handling devices which have been used successfully in decommissioning projects in the UK.

There are four generic classes of devices for remote deployment of dismantling tools:

- *Master-slave manipulators*, in which an end-effector (e.g., jaws) can replicate the movements of the operator. A variety of tools can be attached to the end-effector (screw-drivers, saws, etc.). Uses are limited to lightweight tools and tasks.
- *Power manipulators*, which use servo systems to amplify the force exerted by the operator. Heavier tools can be used. Lifting capacity depends on the reach required of the operating arm, but can be in the order of 200 kg.
- *Telerobots*, which can perform similar functions to a power manipulator with an operator guiding the robot using a joystick controller, but they can also be programed to repeat the function autonomously.
- *Wheeled or tracked vehicles*, which can deploy manipulators, cameras, and other sensing devices into areas which would not be accessible by humans.

Power Manipulators

ARTISAN™ is a heavy duty hydraulic manipulator designed specifically to meet the needs of a wide variety of demanding remote handling tasks within the nuclear industry. The design of the manipulator arm (see Figure 11-7) is simple and robust to provide a cost effective solution based on modular design principles. An open and accessible arm layout greatly simplifies maintenance and repair activities in contaminated environments. The arm structure is manufactured from stainless steel for ease of decontamination. Sensitive system components are positioned outside the hostile cave environment where they are easily accessible. The manipulator modules can be assembled into a large number of configurations, with varying reach and payloads, from an extensive range of modules and spacers.

Radiation Tolerant NEATER Series Electric Telerobotic Manipulators

NEATER™ Series electric manipulators from RWE Nukem are available for use in a radioactive environment

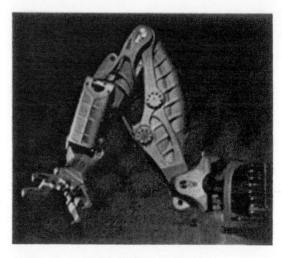

Figure 11-7. ARTISAN™ Heavy Duty Hydraulic Manipulator (Courtesy RWE Nukem).

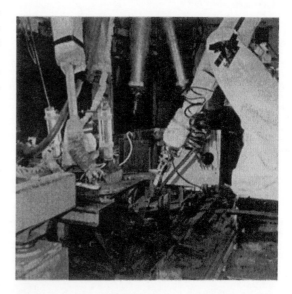

Figure 11-8. NEATER™ Robot in Use in DIDO Test Reactor High Activity Handling Cell (Courtesy RWE Nukem).

with lifting capacities up to 100 kg. The NEATER 600 series and NEATER 800 series robots are constructed out of modules which provide a certain degree of flexibility in specification — configurations can be provided with four, six, or seven rotational axes. They can be installed in both ceiling and floor mounted configurations (see Figures 11-8 and 11-9).

Figure 11-9. Scarab™ Remotely Operated Vehicle (Courtesy RWE Nukem).

NEATER telerobots feature:

- A range of input devices — hand control pendant, keyboard, brake release pendant, twin-joystick, and force reflection joystick.
- Manual tool — change station with quick-release couplings.
- Automatic tool — change station with robotic tool change flange.
- A series of electric and hydraulic tools for decommissioning operations.

Remotely Operated Vehicles (ROV)

Standard crawling or climbing vehicles are offered by firms such as ROV Technologies Inc., e.g., Scarab™ series. These ROVs are radiation tolerant, may be utilised in wet or dry applications, and are controlled using a vehicle operator's control console. They are designed as stable platforms for mobilising a variety of accessories including ultrasound probes, vacuum heads, and orbital welding devices. These vehicles may be wheel or track driven and may be custom sized to meet any project needs.

Gemini Dual-Arm Manipulator System

The Gemini system is a crane-deployable, dual-arm work system for decommissioning. It includes two Schilling Titan 3™ manipulators with remotely interchangeable tools. This system was designed to perform stand-alone remote manipulation tasks in radioactive environments. The first Gemini™ system was delivered to West Valley Nuclear Services in the US for use in a waste vitrification plant. The Titan 3 arms are mounted on a stainless steel, U-shaped, center body that contains an integrated

hydraulic power unit, a high capacity fluid reservoir, and a radiation shielded electronics enclosure. The entire system operates from a single umbilical and can be deployed from an overhead crane or gantry. The system is operated on a Windows™ based personal computer user interface and a pair of replica master arms (see Figure 11-10).

11-4. Radiological Protection During Dismantling

Dismantling nuclear facilities inevitably involves hazards associated with radioactive contamination. Measures need to be taken in order to protect both the workers and the environment, and to minimise the arisings of radioactive waste through the spread of contamination. These measures can be classified as:

- contamination containment, which minimises the spread of radioactivity, and
- personal protective equipment (PPE), which allows workers to enter a contaminated area safely.

It is assumed at this point that, as far as practicable, the radiological inventory has been characterised (Chapter 9), and decontamination works have been carried out (Chapter 10). The extent to which radioactivity will be present during dismantling operations will vary greatly. In some cases, a building may have been stripped and decontaminated to very low levels, so that minimal containment and PPE is required. In other cases, e.g., a reactor, the levels of radiation may be too high and require remote handling equipment, as described in Chapter 10. In between these extremes, there are a wide range of possible scenarios. Please also refer to Chapter 23.

11-4-1. Contamination Containment

For short-term dismantling operations, it has long been standard practice to construct a temporary tent-like enclosure constructed of a tubular metal frame and plastic sheeting. However, this can result in the generation of significant quantities of secondary contaminated waste. In the UK, the introduction of a Modular Containment System (MCS) has significantly improved the robustness of temporary containment enclosures and minimises secondary waste by allowing the materials to be reconfigured and reused on successive projects. The MCS consists of fiberglass reinforced plastic panels which can be bolted together to form a self-supporting enclosure. Figure 11-11 shows an example of the system being used to contain a suite of large fixed alpha-active gloveboxes.

Stippable coatings of an acrylic latex material are applied to the walls and ceiling of the MCS. The coating

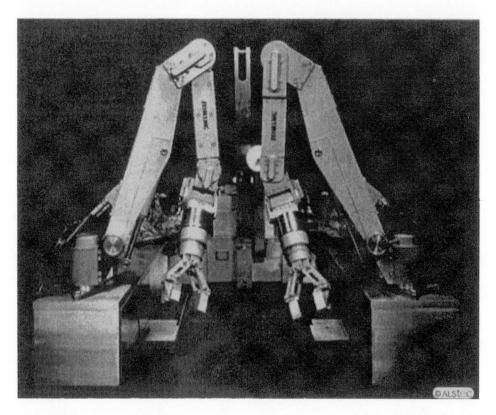

Figure 11-10. Gemini Dual-Arm Titan 3 Manipulator System (Courtesy Schilling).

Figure 11-11. A Modular Containment System (MCS) Constructed Around a Suite of Large Fixed Alpha-Active Shop Window Gloveboxes (Building B220 at Harwell).

becomes contaminated during the dismantling operation. The contamination can be fixed by applying second and successive coats of the latex. When the work is complete (or at appropriate intermediate intervals), the coating can be removed and disposed of, leaving a clean surface ready to take a fresh coat of latex.

The MCS can be fitted with a mobile ventilation/air filtration system and can incorporate airlocks for access by operators.

It is sometimes feasible to carry out dismantling operations under water. This allows the water to act as a low cost form of radiation shielding. To maintain the clarity of the water and minimise the build up of radioactive contamination, it is necessary to have an efficient filtration and purification system. Cooling ponds previously used for fuel storage have sometimes been used to dismantle large items of equipment.

11-4-2. Personal Protective Equipment

There is a range of protective equipment available commercially. The choice depends on the extent of contamination:

• respirators, including positive pressure respirators,

- air hoods,
- pressurised suits,
- armored gloves, and boots, and
- radiological monitoring.

Please refer to Chapter 23, Section 23-8-3 for further information.

11-5. Case Study: WAGR Decommissioning

11-5-1. Introduction

The Windscale Advanced Gas-cooled Reactor (WAGR) was built as a prototype for the UK's commercial advanced gas-cooled power reactor system. Constructed between 1957 and 1961, WAGR achieved full design output in 1963 and operated at an electrical output of 33 MW (E) for 18 years (average load factor of 75%). In 1981, the reactor was shut down after satisfactory completion of all the research and development objectives.

Following shut down, it was decided that the reactor should be decommissioned promptly to Stage 3 as a demonstration project — to show that a reactor core can be decommissioned shortly after shut down and provide a test-bed for development of dismantling and waste handling techniques. Subsequent reviews of the project concluded that the end-point should be redefined as the completion of core and pressure vessel decommissioning, with demolition of the bioshield and containment building deferred.

11-5-2. Decommissioning Plan

The principle technical difficulty associated with the removal of the activated components of the core and pressure vessel was that radiation dose rates of approximately $1 Svh^{-1}$ were anticipated. This high dose rate indicated a need for remote dismantling techniques. However, dose rates from the HotBox and associated components were found to be significantly lower because of the incorporation of a Neutron Shield between the core and the HotBox. Thus, in these areas, dismantling could be achieved by a combination of remote, semi-remote, and manual operations. The principal systems conceived to undertake the remote work comprised the following components:

- A remotely operated machine to deploy tools to dismantle the high dose components;
- A recovery and transport system to remove the dismantled sections;
- A waste route through which to move the components, sort them, take assay measurements, and pack them in suitable containers;

Figure 11-12. The Remote Dismantling Machine (RDM).

- A conditioning plant where the waste is treated for disposal or storage;
- A storage/disposal container; and
- An interim storage facility for ILW boxes.

The Remote Dismantling Machine (RDM)

The Remote Dismantling Machine (RDM) (Figure 11-12) consists of two handling systems deployed beneath a turntable mounted at the reactor operating floor level. First, an extendable mast from which a remotely controlled manipulator is suspended and, secondly, a series of suspended crane rails enabling a 3 tonne hoist to travel across the reactor vault into the adjacent cells. Operators are shielded from radiation by a lead shot filled shield floor within the turntable construction. To minimise dose uptake during RDM construction, a temporary shield floor was built over the exposed surface of the HotBox. The contract to design and build the RDM was let in 1986, and the completed system was installed over the reactor in 1993 after extensive testing.

The Waste Route

The waste route (Figure 11-13) was constructed through two of the heat exchanger bioshields to gain benefit from their shielding concrete. To achieve this, the heat

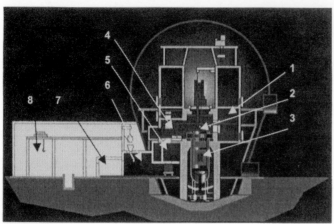

1 Maintenance Cell
2 Reactor Vault
3 Core and Pressure Vessel
4 Sentencing Cell
5 Upper Loading Cell
6 Lower Loading Cell and Concrete Filling Cell
7 Transfer Cell
8 Export Facility

Figure 11-13. The Waste Route.

exchangers were first raised by 12 meters to make the space available. Diamond drilling techniques were used to create the openings into the reactor vault providing access for the 3 te (te — metric tonnes) hoist transport system integrated with the RDM.

The waste is moved laterally from the reactor to the waste packaging plant for characterisation and encapsulation. Starting from the reactor end, the sentencing cell is encountered first, where the waste is placed in box furniture, either racks or baskets. Located immediately below this, the upper loading cell provides a relatively low background environment within which to make γ-dose rate measurements on the waste. A communicating trap door allows the box furniture containing the waste to pass through and be loaded into a WAGR box standing on a trolley in the lower loading cell. The 8 tonne capacity hoist mounted in the hoist room above the sentencing cell is used for this operation.

The Waste Packaging Plant

In the Waste Packaging Building (Figure 11-14), all waste removed from the reactor vault is placed in WAGR concrete waste boxes and encapsulated in a cementitious grout. The anticipated high radiation dose rates dictate that the process has to be remotely undertaken. The Waste Packaging Building, therefore, comprises two shielded cells, the Lower Loading Cell and the Concreting Cell.

Having loaded the box in the Lower Loading Cell with the waste/box furniture, the containment doors are opened and the container on its trolley is driven through to the Concreting Cell. In-fill grout, mixed in a purpose-built grout and concrete plant, is pumped into the container

Figure 11-14. Waste Packaging Plant.

to take up all the voidage. A reinforced concrete lid is cast on the box to complete the container. The shield doors leading to the transfer station are then opened and the trolley driven through to allow the container to be lifted by the 60 tonne building crane and after a detailed radiological survey to check for surface contamination, placed into the curing and weighing pit.

LLW boxes are transported by road to the LLW Repository at Drigg for disposal, whilst ILW boxes are sent to the WAGR ILW Box store for temporary storage awaiting the provision of a national ILW Repository.

The WAGR Waste Box

The container adopted at WAGR (Figure 11-15) for the storage/disposal of LLW and ILW comprises a rectangular reinforced concrete box 2.4 × 2.2 × 2.2 meters with top entry. The enclosing walls of the container provide both structural integrity and radiation shielding of the contents, whilst the dimensions are chosen to accommodate WAGR thermal shield plates and graphite blocks without cutting. The box is fitted with twistlock corner castings top and bottom to enable lifting, stacking, and restraint during transport, and is designed and tested to meet the integrity requirements of an industrial package Type 2 (IP-2). Please refer to Chapter 22, Transport.

The WAGR ILW Box Store

The WAGR waste packages will subsequently be stored in a purpose-built store situated a short distance from the waste encapsulation plant. A ventilation system is incorporated to protect the operators from the truck's exhaust fumes, during box handling operations. There is no requirement for the building to be heated; thus, the temperature and humidity levels within the store are not controlled, but the conditions are monitored.

11-5-3. The Dismantling Campaigns

The reactor is being dismantled in a series of 10 campaigns (Table 11-2); each associated with a particular core component as follows.

Campaign 2 — Operational Waste

Operational waste generally consisted of cylindrical items that formed part of the fuel stringer and removable items from reactor operation. These items were removed as part of the defueling operation and the LLW fraction disposed of to Drigg. The parts of these items classified as ILW were size reduced, fitted with lifting pintels, and returned to the fuel channels to await the decommissioning of the reactor.

Removal of all items was undertaken using the 3 tonne hoist (Figure 11-15), and a lifting grab designed to engage with the pintels fitted to each waste item. Each box contained furniture to hold 110 items of operational waste and, in all 770 items, were removed from the reactor.

Table 11-2. WAGR Dismantling Campaigns

Campaign 1	Preliminary operations — controlled manual activity to prepare the top of the Hot Box for remote operations
Campaign 2	Removal of Operational Waste from the fuel channels
Campaign 3	Dismantling of the Hot Box
Campaign 4	Removal of the Loop Tubes
Campaign 5	Dismantling of the Neutron Shield
Campaign 6	Removal of the Graphite Core and Steel Restraint structure
Campaign 7	Dismantling of the Thermal Shield
Campaign 8	Size reduction and removal of the Lower Structures
Campaign 9	Size reduction and removal of the Pressure Vessel and Insulation
Campaign 10	Size reduction and removal of the Outer Ventilation Membrane and experimental thermal columns. Also, the final clean out of the reactor bioshield

Some of the waste items had significant activity and, thus, high-density boxes were used.

The total dose uptake for the operations team was 3.72-man mSv, with highest individual dose being 0.53 mSv.

Campaign 3 — Hotbox

The hotbox was the gas manifold used to divert the hot coolant gas into the heat exchangers. It was a short flat-ended cylindrical pressure vessel, fabricated from carbon steel. It was approximately 5 m diameter and 1 m high, effectively in the shape of a large pillbox. Internally, the hotbox was lined with insulating material, comprising multi-layers of alternate, dimpled/plain stainless steel foil (Refrasil), made up to around 19 mm thickness on the underside of the top plate and 38 mm on the bottom plate. The side wall of the box has this insulation at approximately 25 mm thickness. The hotbox contained 253 stainless steel fuel element guide tubes and 100 carbon steel stay tubes. The hotbox weighed 31 tonnes.

Industrial plasma arc cutting was adopted to undertake the size reduction, as it proved the most adaptable of the potential systems, with a narrow kerf producing least particulate. The hotbox was dismantled in a series of mini campaigns using 40–200 amp plasma torches deployed both by remote rigs and used manually.

Efficient plasma arc cutting relies on the cutting head being maintained at a constant offset from the subject. The deployment tool used to remove the Upper Refueling Tubes (URTs) attached to the top of the hotbox was designed to stand on three legs over the tube with

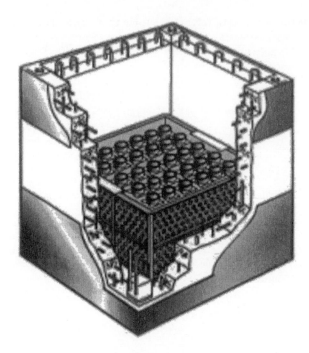

Figure 11-15. The WAGR Waste Box.

the torch suspended within. By vertical and radial movements, the torch was intended to cut between the tube flange and the hotbox top plate. In operation, there was great difficulty maintaining the torch offset and cutting at the correct point, resulting in many failed cuts and damaged torches. Sixty cuts were achieved in 2 months, with an accrued dose of 8-man mSv. After due safety consideration, manned access was adopted to undo the bolts with power wrenches. The remaining 129 URTs were removed in 4 days for a dose of 9-man mSv. It was found that, by removing the more radioactive components first, it was possible to remove the sidewalls by controlled manual intervention.

Despite the additional manual intervention, the total dose of 55.6-man mSv came within the dose budget of 65.5-man mSv, with the highest individual dose being 2.7 mSv. The waste was packed in 14 normal density boxes and as LLW was despatched to Drigg for disposal. The campaign took 13 months.

Campaign 4 — Loop Tubes

There are six loop tubes. These were the six experimental fuel channels used to undertake fuel experiments. All six loops were constructed from work-hardened stainless steel. The loop tubes were installed in the core for the lifetime of the reactor and had become highly

activated, potentially giving a dose rate of 120 Sv/hr from the central sections. To avoid spreading fragments of such active material around the reactor, size reduction using an hydraulic shear was adopted in preference to flame cutting or sawing. To minimise the risk of the tube becoming trapped in the shear blades, and to make the cutting process more efficient, the loop tubes were filled with high-density cement grout.

The campaign was very successful, taking only 3.5 months, with a significant proportion being grout-curing time. Although the equipment could be installed totally remotely and was used for the first installation, manned intervention was used to make the six service line connections to the equipment. This activity accrued little additional dose, but reduced time and ensured that no damage was caused to the plugs and sockets by using the manipulator.

The campaign was completed with a total of 8.3-man mSv, within the budget of 15.7. The highest individual dose was 1.0 mSv.

Campaign 5 — Neutron Shield

The neutron shield is effectively in two major parts, referred to as the inner and outer neutron shield, respectively. The inner neutron shield (INS) contains the reactor upper fuel channel sections, and consists of stainless steel guide tubes surrounded by graphite blocks. The outer neutron shield (ONS) is free from channel sections and is effectively a number of solid blocks of graphite (in the main) surrounding the inner neutron shield. The neutron shield contained nearly 2300 components weighing over 80 tonnes.

The neutron shield was removed in a series of 11 mini campaigns completed in April 2002. Most of it was consigned as LLW and only those sections of graphite containing stainless steel guide tubes were disposed of as ILW. Ninety tonnes of graphite and steel were removed and packaged in 32 WAGR boxes (22 LLW, 10 ILW). The total dose uptake was 17-man mSv compared with the dose budget of 43-man mSv.

Campaign 6 — Graphite Core and Restraint Structure

The graphite core consists of 200 tonnes of graphite blocks in eight layers, each comprising 253 fuel channel blocks surrounded by a graphite reflector forming a flat cylinder approximately 5 m in diameter and 800 mm deep. The layers are each restrained by a tensioned steel beam slotted into grooves around the top circumference of the reflector. The WAGR core was heavily instrumented and the graphite blocks were interlaced with many thermocouple wires and flux scanning tubes.

Many of the tools used in the removal of the neutron shield are used for the core removal: ball grabs; drilling tool, manipulator fitted with various tools to remove thermocouples and flux scanning tubes.

At the time of writing, dose rates within the reactor vault have increased by at least two orders of magnitude, as the graphite core has been removed. In areas previously accessible for tool changes, the rate has become 30–40 mSv/hr, whilst the dose rate at contact with the exposed core components is ∼500 mSv/hr. In consequence, manned access is no longer permitted. This is having a significant affect on the dose accrual; it is now predicted that the whole campaign will be completed with a total dose of <10-man mSv compared with the dose budget of 35-man mSv.

Despite the difficulties, progress has been excellent, with the first three layers removed within 4 months of starting the campaign leading to the expectation that the program duration of 18 months can be reduced to 11 or 12 months.

Campaigns 7, 8, 9, and 10

As this book is published, the development of the tooling and methodologies for these future campaigns is currently in progress and proceeding well.

11-5-4. Future Strategy

A series of studies is being carried out to review the options for the facility after the current phase of decommissioning has been completed. Current strategy, driven by the tritium activation of the core concrete bioshield, is to defer dismantling until 2040. A range of other options is being considered including (i) immediate demolition of the whole facility and (ii) dismantling the building whilst cocooning the bioshield for later removal.

Chapter 12
Site Environmental Restoration Program Management

12-1. Introduction

It is necessary to have a rigorous process which integrates technical, safety, security, value for money, and regulatory aspects of the proposed project works. This enables the case for release of funds to be made such that the decommissioning, waste management, or environmental remediation may be carried out.

This chapter describes a process by which a Management Company responsible for the remediation of a number of redundant nuclear sites and/or facilities may derive, from its core values, policies, mission statements, and goals as set out in its Corporate Plan, its site remediation program.

12-2. The Framework for Environmental Restoration Program Management

Individual projects stem from the overall program of work determined for a facility, a site or for the total decommissioning management company. The program, in turn, is derived from the mission, corporate objectives, strategy, culture, and policies of the company. These are normally set out in the company's "Corporate Plan," which is a formal forward looking statement of where the company is heading, how it intends to get there, and what measures it will use to demonstrate progress along the way.

Individual "Site Strategies and Plans" are derived from the principles set out in the Corporate Plan. However, such plans necessitate a prior evaluation of what exactly the liabilities that need to be decommissioned actually are. These may be described in a "Definitive List of Liabilities." The costed outline programs to completion and profiled project costs over a manageable future period (typically 4–10 years for major works) to liquidate the liabilities then also need to be determined. In this way, an estimate of the total costs for liquidating the nuclear liabilities from the current status to a recognised end point

is formulated in a rigorous manner. The "Liabilities Estimate" for the company is a key figure which appears in the company accounts and must, therefore, be fully auditable. The derivation of the forward decommissioning program in this manner is shown in Figure 12-1.

Therefore, in essence, there are four main elements to a company's decommissioning and waste management program management.

- Determining long-term what has to be done and laying down the policies, strategies, and priorities — "program formulation."
- Putting together the portfolio of tasks that at any one time can be done within the funding and other resources available — "developing the program with plans."
- Monitoring and controlling progress and spend to be able to accommodate variances from the budgeted schedule and make best use of available funds and other resources.
- Reporting to the fund holders (company Board, etc.) as the 'client'; both on the stewardship of funds and on the overall progress with the program.

Examples of typical Corporate Plans can be found on the World Wide Web [1]. BNFL's arrangements for the management of its nuclear liabilities are described by Warner [2] and the methodologies used by UKAEA for its strategic planning by Bayliss [3].

12-3. The Strategic Plan

12-3-1. Introduction

Figure 12-2 further elaborates the iterative nature of the planning process used to derive the decommissioning management company's strategic plan. An understanding of the forward program costs involved in moving forward from the "Definitive List" of liabilities to the individual "Site Strategies and Plans," the individual

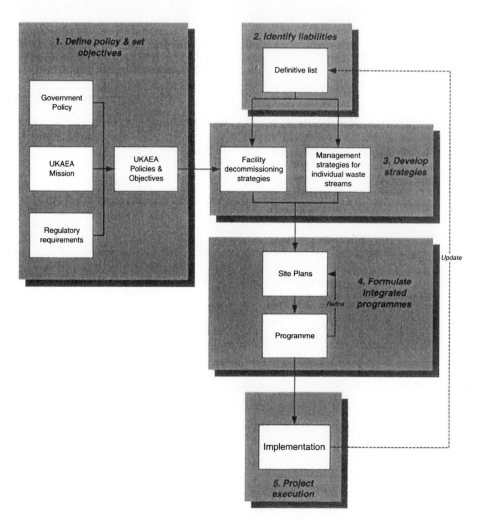

Figure 12-1. The UKAEA Planning Process for the Derivation of a Project Portfolio to meet the Decommissioning and Waste Management and Environmental Restoration Company Mission.

projects involved and the resulting view on the "Liabilities Estimate" requires detailed and rigorous planning.

12-3-2. A Strategic Planning System

A typical Strategic Planning System (SPS) software used for this process needs to be interactive and take into account:

- The timescales of decommissioning of all facilities as dictated by safety or technical considerations;
- The logical linkages which dictate when some facilities can be decommissioned. For example, decommissioning may need to be delayed until a waste route has been established or until there is no longer a need for a plant to perform a service for other operating facilities;

- The complexity of the interrelationships when waste management facilities are called upon to deal with the wastes from several decommissioning works. Essentially, the unit costs of waste treatment are dependent upon the decommissioning strategy. Waste plant throughputs will depend upon volume and timing of decommissioning waste arisings; and
- Infrastructure costs (such as personnel, finance, property, security policing, etc.).

At a particular site, there will be direct and indirect costs linked to the decommissioning strategy in a complex manner. For instance, certain elements of the infrastructure cost will depend upon the total decommissioning work at a given point in time. Other components will depend upon the amount of property occupied and the services required directly by the facilities involved.

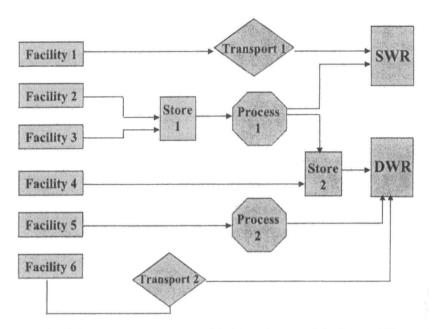

Figure 12-2. Example of the Strategic Features Modeled in a Decommissioning and Waste Management Strategic Planning System (DWR, Deep Waste Repository; SWR, Shallow Waste Repository).

Security costs will depend upon plant categorisation and whether safeguarded materials are present. Early decommissioning may be justified if such high security costs are reduced after shipping fuels and high category wastes off the site. Similarly high cost fire services may be reduced once, say, large quantities of radioactive sodium are neutralised and disposed of. Note: SPS is the name given to UKAEA's Strategic Planning System software tool.

Figure 12-2 illustrates the strategic features that such a planning system would model. The waste model requires cost data (capital costs, refurbishment during lifetime costs, decommissioning, storage, processing, transport, and disposal operating costs, etc.) for each of the types of process detailed in Figure 12-2. Associated with these elements of waste management strategy, there are also a series of constraints which have to be taken into account in the overall modeling. Processing plants have throughput rates, and transport operations have annual limitations. Such constraints may well alter over time as new plant comes on stream.

Typical Strategic Planning System (SPS) [3] software output associated with the modeling of hypothetical buffer storage facility (necessary for interim storage between decommissioning waste arisings and despatch to a disposal facility) is shown in Figure 12-3.

Such modeling is also an important help when gauging the totality of a nuclear site's decommissioning liabilities. Figure 12-4 illustrates the SPS modeling of infrastructure, decommissioning and waste management costs against

a particular decommissioning strategy for a site. Such software allows the storage of all information about the decommissioning strategies in a way that is easily accessible and with an auditable rigor such that the liabilities estimates so derived may be placed in the decommissioning Company accounts.

12-3-3. *Managing the Care and Maintenance Process*

Chapter 6 describes the advantages and disadvantages from inserting periods of Care and Maintenance (C&M) into the decommissioning program for a particular facility. A systematic analysis of the existing facility is required so as to make the case for continuous decommissioning or decommissioning interspersed with periods of care and maintenance. Only in this way will all costs be taken into account (including the high infrastructure costs associated with a dormant nuclear facility) so as to ensure that appropriate safety standards are maintained in a cost efficient manner. Such an approach will:

- Screen and define C&M options;
- Establish a baseline facility status at hand-over for C&M;
- Identify and assess bounding options for hazard reduction;
- Develop a program of preparatory work for hazard reduction;

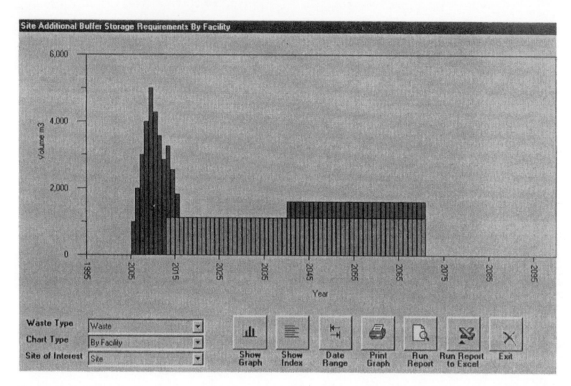

Figure 12-3. Waste Buffer Storage Requirements as Modeled in a Strategic Planning System (SPS) Software Site Decommissioning Analysis.

- Identify and prioritise systems required during C&M;
- Assess the facility environmental monitoring system requirements;
- Rationalise the C&M Examination, Inspection, Maintenance & Testing (EIMT — see Chapter 18) activities;
- Consider methods to minimise waste during the C&M period;
- Assess the resources required during the C&M period; and
- Help identify project risks.

This then forms the basis for developing the Care and Maintenance plan for the facility. UKAEA have developed a CARe and Maintenance Electronic Notebook (CARMEN) [3] using database software so as to apply the necessary auditable rigor to the process.

12-3-4. *Program Risk Management*

A risk may be defined as:

- "Real or potential events which reduce the likelihood of achieving business objectives. Or, put another way, uncertainty as to the benefits. The term includes both the potential for gain and exposure to loss," [4] and

- "Exposure to the possibility of economic or financial loss or gain, physical damage or injury, or delay, as a consequence of the uncertainty associated with pursuing a particular course of action" [5] or, more succinctly, as

- "Likelihood" (probability of occurrence or frequency) × "Impact" (consequence) of an identified threat.

At the company-wide level these may include:

- *Inter-site risks.* These would arise because of interdependencies between sites, e.g., between a decommissioning project at one site and a waste management project or operation at another.
- Risks between sites and third party operators or projects common to more than one site, e.g., risks arising from transport of wastes from several sites to one disposal facility.
- Other company-wide risks including:

  - A failure of corporate services;
  - Inadequate corporate resources;
  - Inadequate finance;
  - General problems concerning interactions with third parties;
  - New legislation or regulations not previously envisaged;

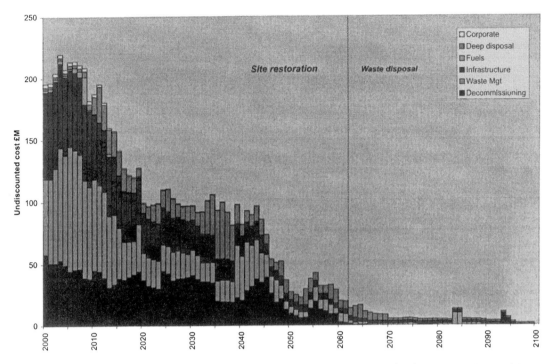

Figure 12-4. Overall Decommissioning Plan (Hypothetical) and Associated Infrastructure, Waste Management, and Decommissioning Costs over the Lifetime of the Site as Modeled using SPS.

- Problems arising from national economic or infrastructure conditions, including inadequate supply of trained personnel, increases in real wage rates, and changes in discount rates;
- External events such as nuclear incidents;
- Any systematic optimism or pessimism among project staff about the speed of which projects may be implemented, e.g., because of some overall resource or systems problem which cannot be individually recognised;
- Inability to recruit and/or (re)train a suitably skilled workforce;
- New on-site discoveries such as leaking drains or large areas of underground contamination; and
- Lack of competent contractors.

Program Risk Assessment and Management (PRAM) [3] is required to provide a formal control of such risks and be embedded in the decommissioning management company's overall program management as a continuous process. In the UK, this control needs to be compliant with the recommendations of the Turnbull Report published by the Institute of Chartered Accounts in 1999 [6]. Program risk management is in addition to, and at a higher level than, the more normal project risk management processes. Some key project risks may, however, be sufficiently

significant to form part of the top risks managed by the decommissioning company. PRAM is intended to ensure that decommissioning and waste management processes and activities within the forward strategic plans have properly identified risks, risk logs prepared, risks assessed, managed and reviewed under a rigorous process with named personnel responsible for each risk so identified.

Following a formal interview procedure and in the absence of precise information about the likelihood and impact of a particular risk, managers may wish to develop a view of the importance of the risk by reference to Tables 12-1 and 12-2. Program risks should be reviewed by the company's management quarterly, with an annual assessment of significant risks prepared for Board scrutiny.

Each identified risk may be assigned to a box of this matrix, depending on the size of its likelihood and impacts on the costs and schedule. The use of four probability and four impact assessments in the standard risk calibration scheme gives 16 possible categories of risks. This degree of separation is, however, likely to be unrealistic given the subjective nature of the assessments which have to be made. Hence, risk category scores may be reduced to the six shown in Table 12-2.

Table 12-1. Program Risk Ranking

Likelihood (probability of occurrence)	Description	Range	Impact on (Discounted) Program Cost (Range)	Impact on Program Schedule (Range)
Very high	Very likely though not certain to occur	>80%	>£100M	>10 years
High	More likely than not to occur	51–80%	£11–£100M	5–10 years
Medium	Less likely than not to occur	20–50%	£1–£10M	1–4 years
Low	Unlikely to occur but not impossible	>20%	<£1M	<1 year

Table 12-2. Site-Wide/Decommissioning Company-Wide Program Risk Impact Matrix

Impact	Likelihood			
	Very high	High	Medium	Low
Very high	6	5	3	2
High	5	4	3	2
Medium	3	3	2	1
Low	2	2	1	1

12-3-5. *Program and Project Prioritisation*

Program prioritisation is described in Chapter 14. Program prioritisation is largely about senior management gaining an understanding of where the priorities lie, based upon company values (such as safety, environmental acceptability, value for money, and public acceptability). Application of the process offers management guidance about which projects within the overall program should be accelerated or slowed up should funding constraints require such action. At the boundaries of an overall suite of projects within the forward decommissioning and site restoration program, prioritisation may also help management decide which projects fall within those to be sanctioned within a financial year.

This is not to be confused with the prioritisation of individual activities within a project plan or project prioritisation within an integrated decommissioning site plan. Here, other drivers at a lower level take precedence. In particular, the concept of "critical path" activities and Program Evaluation and Review Techniques (PERT) will concentrate the mind of project managers. Typical drivers for a site integrated decommissioning suite of project works include:

- Regulatory requirements and Government Policy;
- Safety and security considerations;
- Environmental considerations;
- Delaying work to permit radioactive decay;

- Availability of waste disposal routes and treatment/storage facilities;
- Availability of staff with specialist knowledge of the plant;
- Confidence with technology;
- Interactions with other facilities;
- Site-specific infrastructure costs which are related to the presence, or absence, of other facilities;
- Specific planning consents which might include a requirement for early clearance of a building; and
- Need to reuse the building or land area.

12-4. The Integrated Site Restoration Plan

The development of a comprehensive site environmental restoration program for dealing with existing and future radioactive wastes on a redundant nuclear site requires waste management and decommissioning strategies to be fully integrated. A strategic planning system, as described in Section 12-3-2, may be used at the highest planning level. Conventional project planning tools may be used at the lowest project level to help plan, monitor, and control individual project works. However, there is a middle program level of integration of the individual projects that is required to assist in the management of the decommissioning site program. A typical integrated decommissioning program will address:

- The overview of the site's restoration process including a definition of end points and full integration of the parts making up the whole;
- The decommissioning plan which outlines the work necessary to demolish and dismantle the various facilities (perhaps on a zone by zone basis);
- The radioactive waste management plan which describes the strategies, waste routes, and facilities for dealing with existing and future waste arisings as the decommissioning advances;
- The estates and utilities plan which addresses the long-term site infrastructure (i.e., facilities, services, contractor's accommodation, and lay-down areas, etc.) as required to support the restoration of the site;

- A nuclear fuels inventory and management plan which addresses the management of all fuels on the site; and
- A contaminated ground decommissioning and restoration plan which outlines the approach to restoration of contaminated ground, both radiological and nonradiological.

Successful integrated plans, as for example used on the Hanford and Rocky Flats decommissioning programs in the USA, contain logic-linked programs using conventional project Work Breakdown Structure (WBS) formats. The WBS is a graphic portrayal of the overall site environmental restoration plan, exploding it, in a level-by-level fashion, down to the degree of detail needed for effective planning and control. It must include all deliverables required to deliver the fully restored site. The advantage of the WBS over other methods (which simply list all the items, or possibly just put them in bar chart format) stems from the hierarchical, structured approach and the ability to visualise the total site restoration program in terms of all its major and minor elements. The WBS breaks the overall program down into a series of sub-projects, all focused on, and aligned to, the business of the site's restoration (as opposed to treating the site as an operational facility on which some decommissioning will be carried out if funds allow). Once the WBS has been prepared; then an aligned Cost Breakdown Structure (CBS) and Organisational Breakdown Structure (OBS) may be generated. In this way, all work packages, costs, and human resources are assigned to the business of the site restoration management Company's goal — that of decommissioning the site. All work phases (operational, decommissioning, care and maintenance/surveillance, and postrestoration) are dealt with in this manner.

12-5. Making the Case for a Project to Proceed

There are many steps to go through to make the case for the particular decommissioning, waste management, or environmental restoration project. The ground remediation around a liquid effluent treatment plant at Harwell during 1989 involved some "39 steps" from proposal to implementation (see Table 12-3).

Chapter 13 describes the analytical methods used to make a financial assessment of individual decommissioning project works so as to secure the necessary funds. This chapter concentrates on the overall process, framework, and program management within which such assessments are made. The general principles to be followed are described below, together with a typical project sanction case and a case-study for consideration by the reader. It should be recognised that the extent of the work necessary

to make the case for a project may well involve large sums of money in itself. As such, securing funding for these early initiation and definition stages in the project life cycle may be seen as a separate project. All the rigors of a formal project management process should be brought to bear so as to keep a tight control on the costs and schedule involved.

12-6. The Project Sanction Process

12-6-1. *Introduction*

A sanction is an approval by an authoritative body for the expenditure of funds. A project sanction case needs to reassure the sanctioning authority (the company Board, the Chief Executive Officer (CEO), a particular Director or Group Leader depending upon the delegated authority for expenditure involved) that:

- The work is necessary;
- It is consistent with company policies and with approved plans;
- A proper assessment has been undertaken;
- The work program has a clearly defined implementation plan against which progress can be measured;
- The proposed program is achievable and represents the best option for meeting the requirement taking account of safety, environmental, security, and value for money criteria; and
- Funding is available and will be provided in the most appropriate way, taking account of different funding options (for example Private Financial Investment (PFI) and Public Private Partnership (PPP) initiatives; see Chapter 13, Section 13-7).

To achieve this, a sanction case needs to be prepared and show:

- How the proposal fits into the company's forward strategy and its priority;
- That all sensible options have been properly assessed with an appropriate level of safety, technical, and financial appraisal;
- That risks have been adequately addressed;
- That the necessary funds and other resources are allowed for in the current plans;
- What will be delivered and when; and
- If there are any residual uncertainties, e.g., over the exact work to be done of the achievable end point.

To achieve this, it is normal for several papers at different levels of detail to be produced and assessed by different committees or groups within the company. Nuclear decommissioning projects often cost millions of pounds for which only Directors have the authority to

Table 12-3. The "39 Steps"

1	Level 1 Studies	A high level assessment of the most likely methodology and option for doing the work.
2	Level 2 Studies	An assessment of various options, costings, and sensitivity to risks.
3	Feasibility Studies	Practical assessment of the preferred options.
4	Option Studies	Further analysis of options.
5 & 6	Justification of Timing Studies	Including Best Practical Means (BPM) and Best Practical Environmental Option (BPEO).
7	Sanction Paper	Justified case for release of funds for the work.
8	Program Directory	Inclusion of work within overall portfolio of projects within the company program.
9	OJEC Article	Preparation of advertisement of forthcoming work in the European Journal. OJEC — *Official Journal of the European Community.*
10	Contractor Prequalification	Preliminary assessment of contractors interested in bidding for the project works.
11	Project Plan	Program for the works including production of all key paperwork submissions to meet Regulatory requirements (conventional health and safety, Environment Agency, Nuclear Installations Inspectorate, planning authorities, etc.).
12	Preliminary Categorisation	Assessment of radiological hazard and discharges, plant categorisation, etc.
13	Safety Case	Production of safety case.
14	Peer Review of Safety Case	Independent review and incorporation of feedback.
15	Environmental Impact Assessment	
16	Technical Specification	For inclusion in project works tender documentation.
17	Pretender Health, Safety & Environment Plan	For inclusion in project works tender documentation.
18	Contract Document	
19	Invitation to Tender	To prospective contractors identified in steps 9 and 10 above.
20 & 21	Tender Assessment Report	Tender evaluation (Technical, Financial, & Safety).
22	Contractor's Health & Safety Plan	From selected contractor following competitive tender process.
23	Notification to Health & Safety Executive	
24	Cost Estimate	And feedback following tender process into company's database.
25	Handover Document	Pass possession of site to contractor.
26	HAZOP Study Report	
27	Method Statements	For agreement between Client/Contractor (18).
28	Risk Assessments	For agreement between Client/Contractor (18).
29	Excavation Permits	General permit to work processes.
30	Safety System of Work	Working method arrangements.
31	Waste Form Specifications	For Low Level Waste (LLW) to meet BNFL's Drigg disposal facility waste receipt criteria.
32	Activity Assessment Justification	
33	Summary of the Waste Fingerprint	
34	QA Sub-Program for Disposal to Drigg	
35	District Council Applications	
36	Environmental Surveys/Support Procedures	
37	QA Plans	
38	QA Program	
39	Authority to Operate	Approvals.

allow the works to proceed. At Director level, sanction papers will be more closely examined in areas associated with how the proposal fits in with strategy, on the soundness of the safety, technical and financial appraisals, on risks and how they will be managed, on funding issues, and generally how the project will achieve its objectives to time and cost. Only those technical arguments crucial to the recommendation would normally be examined at this stage. Appendices to the paper may, of course, be used to cover earlier more detailed analysis.

12-6-2. *Typical Sanction Paper Structure*

Papers should be concise with appendices used for backup information. The arguments for the case should be sustained independently of the appendices, since not all

Directors will read the back-up material. Overly long analysis should be avoided and abbreviation should be used sparingly, and only after definition. Jargon should not be used.

A recommended structure is:

(1) *Objective.* A brief description of why something needs to be done, how this fits in with company strategy, and the objective to be achieved. Mention should be made of any related previous sanctions or submissions. Where the proposal is related to an item on the company's definitive list of liabilities, the Definitive List reference should be given. (Any historical background felt to be necessary should be given as an Appendix.)

(2) *Recommendation.* An unambiguous statement of what is being recommended to the sanctioning authority. This may also need to include a recommendation on how subsequent stages in the sanctioning process are to be followed.

(3) *Options.* The paper should demonstrate that all plausible options have been considered. These may relate to differences in timing or technical approach. "Do nothing" and "delay for a year" should always be considered. It will often be helpful to list these in tabular form in the main text (if a limited number), indicating those which can be rejected on technical, safety, or licence compliance grounds. Brief statements (one or two sentences) justifying rejection should be given in the text, but care should be taken not to dismiss options too lightly. Normally, there should be an Appendix containing a table listing all the options, characterising each, and summarising its advantages and disadvantages. Arguments for rejection should be made clearly in this. It may sometimes be helpful to supplement this table with more detail on each option, perhaps included as additional appendices.

- All options that cannot be ruled out on technical, safety, or licence compliance criteria should be subjected to more detailed appraisal. A Risk Assessment should be carried out, identifying the threats each option is exposed to, the likelihood of these occurring, and the impact they would have. Except for the smallest projects, there should be an Appendix on the risks with a ranked Risk Assessment for each option. The text should also, in a few sentences, comment on the risks attached to option. The depth to which the Risk Assessment is taken will depend upon the nature and size of proposal.

- Each option, that has not been rejected on technical, safety, or licence compliance grounds, should be financially appraised.

- A sensitivity analysis should be carried out to see how far the conclusions of the financial appraisal would be affected by different assumptions on costs or timings. What percentage change in estimate would change the balance?

- The text should summarise the outcome with a table showing for each option best estimates of the discounted and undiscounted cost to completion. Ranges should be given, taking account of both the risk assessment and uncertainties over the cost estimates. The main risks/uncertainties that determine the range in each case should be commented on. Refer to Chapter 13 for a discussion on discount rates.

- For many nuclear decommissioning and waste management safe environmental remediation projects, the financial appraisal alone will seldom determine the choice of option, but will be taken into consideration with other factors. Such factors may include the scope for further reducing hazards, environmental considerations, making use of worker skills and facilities while they are still available, dose uptake, interaction with other facilities including waste routes, the extent to which an option may help to implement wider aspects of site strategy, etc. For projects in other areas (such as property development not involving nuclear facilities) there may be other, nonfinancial considerations. The text should comment on these and justify the preferred option.

This section should end with a clear statement of the option chosen. Care should be taken to assess all options objectively. Proposals which use the Option section to argue solely in favor of the Recommendation are likely to be criticised. The author should ask himself whether he has covered all the options that should be considered, and what would have to change to alter the preferred option. Is the choice clearly supportable?

(4) *Implementation.*

(a) *The Proposal.* A brief description of the work proposed, in sufficient detail (but no more) to understand what the project will comprise, what its end point will be, when is its target completion date and sanction completion date, how its main costs will be incurred, and (if appropriate) any future review or decision points. For very large projects, it may be appropriate to seek sanction for the release of funds only up to specific, defined stages in the project, with further release of funds dependent on review by the sanctioning authority. (Any Gantt charts may be presented as an Appendix.)

Where there are a number of individually separate items that can be grouped together into a conglomerate project (such as separate plant improvement schemes or building demolitions), careful consideration should be given to the extent of the proposal for which sanction is to be sought. In general, the presumption should be in favor of aggregating related work where this forms part of a program to achieve a common objective. Where not all of the work has been planned to the same extent and costs for some are less certain, the sanctioning authority may be asked to give qualified approval for the whole, with funds released in tranches at a lower level of delegated sanction against appropriate justification for each tranche.

(b) *Intermediate Deliverables.* Set out, preferably in a short table, what is expected to be achieved by when. Milestones should be set for key stages. These will be the intermediate achievements needed to secure the final objective of the sanction paper.

(c) *Risk Management.* How will the risks already identified be managed? Which can be transferred to other parties, which can be mitigated, and which remain to be accepted? Would there be serious consequences, safety or financial, if the project slipped much beyond its planned completion date? Since the ability to complete the project successfully to time and cost depends on how risks will be managed, the sanctioning authority will pay particular attention to how this will be done. It is not enough merely to list the risks without describing the intended management response. If there are low-probability high consequence risks which, if they occurred, would take the project beyond its sanction, these should be explicitly discussed and the implications for any resanction stated.

(d) *Contract Strategy* (see Chapter 13). How will the work be implemented? How will contractors be used? What will be the contractual arrangements? This must indicate that company policy on the use and control of contractors has been taken into account and, in particular, that the use and selection of contractors has been followed in deciding the extent, if any, to which it is appropriate to use contractors. Explain how the work will be packaged and what the contractual arrangements will be. Where appropriate, explain why alternative contractual arrangements have been rejected.

(e) *Safety Approvals.* What approvals or clearances are needed? What is the timescale for achieving these?

(f) *Other (if necessary).* If the project involves the generation of radioactive wastes, how will these be managed? Confirmation should be given that no problems/bottlenecks are expected, or if they are an explanation should be given of how they will be resolved. Similarly, any issues relating to discharges should be discussed, as also any significant interaction with or dependence on other projects.

(g) *Project Management.* Describe the project management structure, usually in the form of a chart, which can be presented as an Appendix. Responsibilities should be clearly defined, and evidence of having prepared a Work Breakdown Structure (WBS), together with the associated Organisation Breakdown Structure (OBS) and Cost Breakdown Structure (CBS) should be demonstrated.

(5) *Resources Required for the Project.*

(a) *Costs.* Include a table showing the make up of the sanctionable costs (vertically) against years (horizontally), giving best estimate figures and totals for each year. Any existing sanctioned expenditure (e.g., at the Project Initiation stage) should be included in the totals. The vertical breakdown should be enough to enable the sanctioning authority to see how the costs are made up and the timing of when these will be incurred. The total of these columns will be the Project Estimate.

The basis for these estimates should be given so that the sanctioning authority can judge their quality.

A further risk provision should be added (allocated across the years) to bring these to the Approval Estimate required for sanction. The justification for this additional risk provision should be given, where appropriate, with reference to the Risk Assessment.

For long duration projects, the figures should be presented in constant money terms, and this should be made explicit. Otherwise, cash of the year figures should be used, with the assumptions on inflation made clear.

Note that sanctionable costs will include only those elements of waste management costs that are "new money" and which would not be incurred if the project was not going ahead. If in doubt, advice should be sought on the figures to use.

(b) *Funding.* Compare the best estimate cash of the year figures with the provisions already made in the most recent Site Strategy & Plans. Funding from any other sources should be indicated.

Any mismatch between funding requirements in a year and the funding source should be discussed with an explanation of how it will be resolved.

(c) *Other Resource Requirements.* Identify other key resource requirements, such as project management resources, and confirm that these will be available as required.

(d) *Priority of Project* (see Chapter 14). The ranking of the project within Site (and where appropriate) company priority lists should be indicated.

(e) *Control of Contingencies.* Normally funds only up to the level of the Project (or Central) Estimate will be released to the Project Manager, the balance between this and the sanctioned Approval Estimate (which includes contingencies) being held as a contingency at a more senior level. The paper should say who will hold this contingency and the arrangements for releasing it. For projects requiring Director or CEO sanction, the difference between the Approval Estimate and the Project Estimate will normally be held by the relevant Director.

(6) *Public Relations Aspects.* Where a project is likely to evoke public interest, the paper should discuss how it is intended to handle the public relations aspects, whether through press releases, the local liaison committee, or other means, and if there are likely to be contentious issues, how these will be handled. Where a project is unlikely to evoke public interest, this should be said.

(7) *Conclusion.* It will often be helpful to pull the main points together into a concise concluding statement, summarising the key points. This would be the place to reiterate any qualifications that relate to the sanction, for example that it may not cover certain risks, or that there is a requirement to come back to the sanctioning authority for review at a certain hold point.

An example of a proforma sanction case cover sheet based upon the principles outlined above is shown in Figure 12-5.

12-7. Principles for Carrying out Financial Appraisals

General modern financial appraisal and analysis techniques are described in Chapter 13. When applied to nuclear decommissioning, waste management, and site environmental restoration projects, they should be applied in the following context:

(i) A common end point should be defined for all options to be appraised. (e.g., waste to be treated, packaged, and transported to an interim store or disposal site on the same basis for all options under consideration).

(ii) All costs that will be incurred directly or indirectly as a result of following the option should be identified. Similarly, any savings that will result (e.g., to infrastructure) should be identified.

(iii) The costs should be expressed in constant money values and discounted to a common base at the appropriate discount rate.

(iv) The costs for each option should be best estimates. Where appropriate, a contingency should be added to the base estimate to give this level of confidence.

(v) Care should be taken to avoid including any fixed costs that will not be affected by whether or not the work proceeds. Waste tariff costs, for example, usually contain large fixed costs and should not be used for assessing alternative decommissioning options. However, any variable element in the waste costings should be included. The assumptions used should be made explicit in notes.

(vi) Similarly, depreciation and interest payments should not be included.

(vii) Sunk costs should be ignored. Sunk costs are those that have been expended up to the point where a decision on whether to go ahead with the project or not may be made. The viability of the project itself, once initiated, should not have to carry these initial exploratory works which should be budgeted for separately.

(viii) Allowance should be made for any real cost increase over time. For example, it may cost more to do a task in the future if the future workforce is unfamiliar with the type of work and has to be trained. Care and maintenance costs may increase as buildings become older, etc. The assumptions should be made explicit.

(ix) Risks should be factored in through a formal risk assessment.

(x) Sensitivity analysis should be carried out to test the conclusions to different assumptions about costs and timescales. Where closely balanced, the effect of using a different discount rate should be tested.

(xi) The costs to be evaluated should be the costs of the project, not the costs falling to any one particular funding source where these may be abated by contributions from other parties.

Appendix 2 shows a sanction case study of repacking site $\times$ legacy intermediate level wastes.

Synopsis of Sanction Case

TITLE OR OTHER REFERENCE:

SANCTIONING AUTHORITY:

PROJECT SPONSOR: PROJECT MANAGER:

DESCRIPTION OF WORK FOR WHICH SANCTION IS SOUGHT

APPROVAL ESTIMATE   £
APPROVAL ESTIMATE SCHEDULE COMPLETION DATE

PROJECT ESTIMATE (funds allocated to Project Manager)   £

PROJECT ESTIMATE SCHEDULE COMPLETION DATE (schedule for which Project
Manager is responsible for achieving)

Checklist of Issues covered in Submission (give references to paragraph, table or appendix)

• Full description of work proposed

    . . . .

• Project Core Team

    . . . .

• Listing of all options considered

    . . . .

• Technical evaluation of options

    . . . .

• Financial appraisal and sensitivity analysis of all viable options

    . . . .

• Source and provision of funding

    . . . .

• Consideration of alternative sources of funding

    . . . .

• Prioritisation

    . . . .

• Decommissioning, WM, safety, and environment issues

    . . . .

• Public Relations (issues and strategies)

    . . . .

• Risks and risk management

    . . . .

• Safety, planning, and environment approvals

    . . . .

• Contract strategy and Project Management

    . . . .

• "Approval" and "Project" Estimate Cost and Schedule requirements

    . . . .

Deliverables   . . . .

Figure 12-5. An Example of a Sanction Case Cover Sheet.

12-8. References

1. www.ukaea.co.uk and http://managementsystem.ukaea.org.uk.
2. Warner, D. R. T. "The Management of Nuclear Liabilities," *7th International Conference on Decommissioning of Nuclear Facilities*, London (October 2000).
3. Bayliss, C. R. "Practical Application of the UKAEA's Decommissioning and Liabilities Management Toolbox," *Ibc 7th International Conference on Decommissioning of Nuclear Facilities*, London (October 2000).
4. Stock, Copnell and Wicks. *KPMG Review – Internal Control: A Practical Guide*, Gee Publishing Ltd., London, UK, ISBN 0860896618 (October 1999).
5. Cooper, D. and C. Chapman. *Risk Analysis for Large Projects*, John Wiley & Sons, ISBN 0-471-91247-6 (1987).
6. *The Turnbull Report*, Institute of Chartered Accountants in England and Wales (ICAEW) – September 1999, London, UK.

Chapter 13
Project Investment Appraisal and Contract Strategy

13-1. Introduction

Some decommissioning tasks initially start with the closure of an operating facility which is relatively straightforward to plan, monitor, and accomplish (for example, a modern Materials Test Reactor) to time, cost, and quality (including taking into account safety, security, and environmental factors). Other facilities (such as older engineered ILW interim storage wet silos or shaft ILW geological disposal repositories) require a considerable amount of new waste treatment plant and waste storage facility construction to provide the required waste routes before decommissioning can take place. In all cases, the reduction in nuclear liabilities involves expenditure. Money is needed for:

- planning, designing, and building a given facility,
- operating and maintaining it,
- refurbishment,
- decommissioning,
- waste management (operations and new build), and
- waste storage and disposal (operations and new build).

It is totally wrong to think of decommissioning as merely a demolition job. The direct costs associated with decommissioning, which generates nuclear and conventional waste, include waste treatment, packaging, storage, and transport of wastes to a recognised end-point, as well as on-going plant maintenance and all the requirements to be compliant with legal and regulatory requirements.

A break-down of expenditure on care and maintenance, operations, ongoing, and new projects for a decommissioning Company might typically be as shown in Figure 13-1. Whatever is done throughout the project life cycle (as shown in Figure 13-2) can, therefore, be expressed in monetary terms. These terms provide a common yardstick for establishing the financial commitment during the different project phases. Much of the responsibility for decommissioning historic nuclear liabilities arising from early research programs lies with the

public sector (you and me as the taxpayer). The work requires Government funding and, as such, Government is particularly interested in knowing that the case for the expenditure is well founded, that the money is being well spent, and that cash flow and "in year" spend are all under tight management control.

Project evaluation in purely financial terms is generally insufficient to convince an investment house or decision-maker on the merit of a project. Not all issues can be converted into hard cash terms. The merit of the overall case involves economic as well as financial considerations. In the economic analysis, costs and benefits are all converted to money terms on a common basis. Such an economic appraisal is sometimes referred to as a cost-benefit analysis. Classic economic cost-benefit cases include the 20th Century Victoria Underground Railway construction in London and how best to deal with foot and mouth disease in cattle.

This chapter will confine itself to financial project appraisal and also consider the most appropriate contract strategies for decommissioning projects.

13-2. Capital Investment

The aim of a decommissioning or waste management project is to spend money *now* on capital goods in the expectation of the project works efficiently contributing to the safe and secure remediation of the nuclear facility or nuclear materials involved at a later date. In other words, *investment* now in the hope of reducing the nuclear liabilities in the future. More normally, a capital investment is made in the hope or expectation of making future profit from the revenue streams that the end project works produce. The investment or expenditure may be for:

(i) replacement of equipment,
(ii) expansion of productive capacity,
(iii) provision of new production facilities,
(iv) new build (waste plants, stores, etc.),

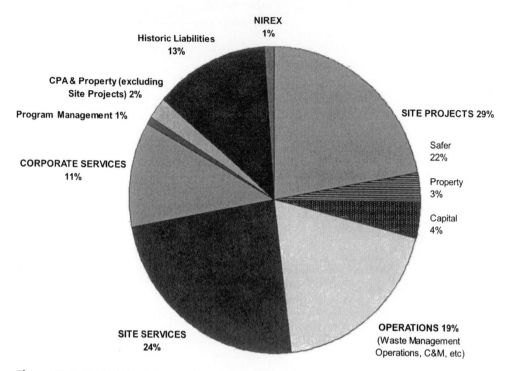

Figure 13-1. Typical Breakdown of Expenditure for a Decommissioning Management Company.

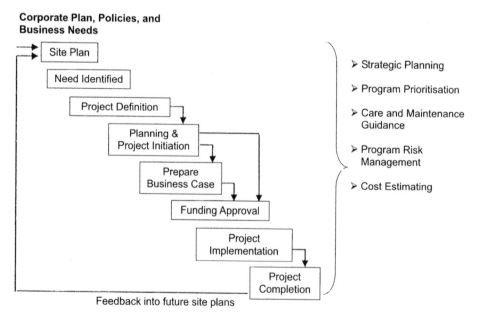

Figure 13-2. Nuclear Decommissioning Project Life-Cycle.

(v) plant upgrades, and

(vi) decommissioning.

13-3. Project Identification

In normal commercial business, the problem is to find "good" projects by imagination, creativity, and alertness in order to spot the investment opportunity... coupled with a bit of luck. Such projects are difficult to find for the entrepreneurs, banks, and money lenders. In the case of Government aid agency projects, it is particularly difficult to find "good" projects which will route the benefits from the investment into helping the communities for which the project was intended. In the case of nuclear decommissioning projects, the issue is more to do with making the most efficient use of scarce investment resources so as to meet Government decommissioning and waste management policy and Regulatory obligations. Apart from possible revenue gained from the exploitation of intellectual property accumulated from decommissioning experience or revenue from the remediated land, the cash flow is normally all outgoing. The decommissioning firm and/or Government still has to consider the appropriateness of the individual project option in comparison with other methods or options for doing the work, including the "do nothing" option. First of all, lets consider conventional investment projects.

A project which increases plant capacity or throughput (an expansion of current facilities) may increase net income. Investment to modernise existing plant or to bring about operating efficiencies (cost reduction) may reduce costs and, thereby, improve profitability (see Figure 13-3).

When assessing projects, it is often necessary to look at both the financial *and* the economic costs and benefits. Indeed, nuclear projects require justification on financial and economic grounds coupled with an assessment of the Best Practical Environmental Option (BPEO) within the context of Tolerability of Risk (ToR) principles.

13-4. Appraisal Methods

13-4-1. *Rate of Return*

Consider the two projects A and B in Table 13-1. To understand if Project A is a better investment opportunity than Project B, consider:

(i) *Cash inflows.* Project B has a higher total cash inflow (24 vs. 15);

(ii) *Total net profit.* Project B has a higher total net profit over 3 years (12 vs. 9);

(iii) *Average annual profit (or return).* Project B has a higher average annual return (4 vs. 3); and

(iv) *Rate of return on investment.* Project A has a higher rate of return on investment (50 vs. 33%).

The *average* annual rate of return on investment is a simple, easy to understand, and a generally good project appraisal methodology. It tells about the profitability

Table 13-1. Annual Rate of Return — Project A vs. Project B

	Project A ×£'000,000	Project B ×£'000,000
Investment, End Of Year EOY 0	−6	−12
Cash Inflows EOY 1	+3	+7
EOY 2	+4	+8
EOY 3	+8	+9
Total Cash Inflows	+15	+24
Total Net Profit	+9	+12
Average Annual Profit	+3	+4
Average Annual Rate of Return $= \frac{\text{Average Annual Profit}}{\text{Initial Investment}}$	3/6 = 50%	4/12 = 33%

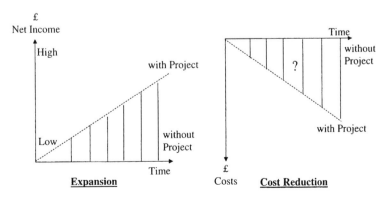

Figure 13-3. Expansion or Cost Reduction Investment Projects.

Table 13-2. Annual Rate of Return — Project C vs. Project D

	Project C ×£'000,000	Project D ×£'000,000
Investment, End Of Year EOY 0	−6	−6
Cash Inflows EOY 1	+1	+6
EOY 2	+2	+2
EOY 3	+6	+1
Total Cash Inflows	+9	+9
Total Net Profit	+3	+3
Average Annual Profit	+1	+1
Average Annual Rate of Return $= \dfrac{\text{Average Annual Profit}}{\text{Initial Investment}}$	1/6 = 16.7%	1/6 = 16.7%

of a capital project over the lifetime under consideration; which in this case is only 3 years. It tells nothing about the timing of the cash streams that flow from the project. The averaging process eliminates such relevant information about timing. In this particular case, it does not take into account the fact that Project A has a rapidly increasing year-on-year cash inflow, whereas the returns from Project B are relatively static. Further, the investment required for Project B is twice that required for Project A.

Consider projects C and D in Table 13-2, where the investment required for each project is the same. In this example, the timing of the relative magnitudes of the annual income streams is in reverse order, but the total income over the 3 year project appraisal period is the same. The average annual rate of return on investment for each project is now also the same. However, it is obviously sensible to get the return sooner than later. Under project D, the extra £5,000,000 income received at the end of year 1 could be taken out of the project and usefully invested for 2 years on the capital markets to yield a further positive return. Thus, taking into account what is known as the "opportunity cost of capital," Project D is superior.

13-4-2. *Payback*

This simple appraisal methodology allows for the timing of returns. Payback indicates how many years it will take before the original amount invested in a capital project is "paid back," i.e., this is the time before cumulative returns exceed the initial investment. Perhaps one may consider that the shorter the return period the better. This may be because the "value" of money is now considered, at this

Table 13-3. Cash Flows for Project E and Project F

		Project E (£'000,000)	Project F (£'000,000)
Investment	Year 0	−6	−6
Cash inflows	Year 1	+3	+8
	Year 2	+4	+4
	Year 3	+8	+3

point in time, better than money at some time in the future because of:

- inflation,
- giving up right to spend immediately, and
- risk from delay.

Consider projects E and F in Table 13-3. The "payback period" for Project E can be easily calculated. £3,000,000 is 'repaid' in year 1, and £4,000,000 in Year 2. Thus, Project E's payback period is 1¾ years (1 year at £3,000,000 + ¾ of a year at £4,000,000 equals the initial investment of £6,000,000).

Similarly, Project F's payback period can be calculated at 9 months. Only ¾ of the £8,000,000 cash inflow in Year 1 is needed to recover the initial investment of £6,000,000 (assuming that cash is received evenly throughout the year). These results can be shown graphically, as in Figure 13-4.

The payback method has one clear advantage over the average rate of return on investment: it does take timing into account.

Project F's payback period is 9 months, and project E's is 1¾ years. However, one has not set a payback period target by which different projects are judged. In other words, one has not considered the maximum payback period acceptable or upon what criterion such judgments are based as being important.

The payback method ignores cash receipts accepted after payback. This could be vital when comparing project viability, especially if one project has a rapidly increasing cash stream over time. The payback method is, therefore, a measure of risk, but *not* of profitability. It may be considered as a rough screening device for assessing which projects to invest in.

13-4-3. *Time Value of Money*

Discounted Cash Flow (DCF) techniques may be adopted for project appraisal using the methods of Net Present Value (NPV) and Internal Rate of Return (IRR). With the advent of the spreadsheet and the personal computer, these calculations are relatively easy to perform. However, it must be remembered that projected cash streams from an initial investment in a project are merely estimates.

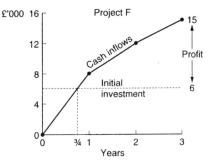

Figure 13-4. Payback Periods for Project E and Project F.

The use of a computer does not, in itself, bring greater surety to the future profitability of the investment.

Consider £100.00 invested at 10% over 3 years:

EOY 0   £100.00
EOY 1   £100 × 1.1 = £110.00
EOY 2   £110 × 1.1 = £121.00
EOY 3   £121 × 1.1 = £133.10

Thus, the "future" value of £100 is now £133.10 at the end of 3 years. So, the present value of this investment (without considering inflation) is £133.10. So, as long as inflation is less than 10% per annum, the investment, in real terms, will be worth more (likely to be able to purchase more) at the end of the 3 year investment period than it was at the beginning. Discounted cash flow investment appraisal techniques take into account the time value of money by allowing for such effects.

13-4-4. *Discounted Cash Flow*

Net Present Value (NPV)

Consider the case where a decommissioning Company wishes to invest £10,000 now in Project G, with the following anticipated returns:

EOY 0 outlay = −£10,000 (initial investment)
EOY 1 return = +£3000
EOY 2 return = +£4000
EOY 3 return = +£5000

The Company would have to consider whether to invest in Project G (with all the inherent risks) or simply bank the investment money (where the investment would be relatively safe) at an interest rate of, say, 10% pa.

The Net Present Value (NPV) evaluation method compares cash receipts and payments expected to result from the capital project investment — discounts expected cash flows to present values — i.e., to end of year 0 in money terms — using a given discount rate.

Discounted Cash Flow (DCF)

There is a need from the earlier analysis to forecast both the *amounts* and the *timing* of revenue streams. The question is does the PV (present value) of the project's discounted cash flows exceed the cash investment involved? Note that no method of analysis gives a precise answer and indeed the interpretation of the answer, especially when used in comparison with other investment opportunities, is where the real investment appraisal skill lies. The analyst has to consider if the timing is correct, if the opportunity cost of capital is set at the right level and met, and if there are important nonfinancial aspects (cost benefit analysis, economics, etc.) to be considered.

Internal Rate of Return (IRR)

The method of analysis is really the same, but the question asked in this case is "what discount rate reduces the NPV of the project to zero?" In general, the higher the Internal Rate of Return (IRR) the better.

13-5. Project Investment Examples

13-5-1. *NPV Example*

A (nonreturnable) investment now of £10,000 in Project G is expected to produce £3,000 cash at the end of year 1, £4,000 at the end of year 2, and £5,000 at the end of year 3. Assuming that money could otherwise be invested (e.g., with a bank) to earn 10% a year, should the company invest in Project G or not? The analysis is shown in Table 13-4.

13-5-2. *IRR Example*

In Project G, a 10% discount rate produced an NPV of −£210. It is known that a zero discount rate would produce an NPV of +£2,000. (This is reached by simply

Table 13-4. Project G — Net Present Value (NPV) Analysis

End of Year (EOY)	Cash Flows (£)		Discount factor (at 10%)	"Present" EOY 0 Value (£)	(£)
0	−10,000	$[\div(1.10)^0] \times$	1.000 =	−10,000	−10,000
1	+3,000	$[\div(1.10)^1] \times$	0.909 =	+2,727	
2	+4,000	$[\div(1.10)^2] \times$	0.827 =	+3,308 $\Big\} =$	+9,790
3	+5,000	$[\div(1.10)^3] \times$	0.751 =	+3,755	
				Net Present Value =	−210

Table 13-5. Internal Rate of Return (IRR) of Project G

End of Year	Cash Flows (£)		Discount factor (at 9%)	Present Value (£)	(£)
0	−10,000		1.000 =	−10,000 =	−10,000
1	+3,000	$[\div(1.09)^1] = \times$	0.917 =	+2,751	
2	+4,000	$[\div(1.09)^2] = \times$	0.842 =	+3,368 $\Big\} =$	+9,979
3	+5,000	$[\div(1.09)^3] = \times$	0.772 =	+3,860	
				Net Present Value =	−21

adding up the undiscounted cash flows: −£10,000 + £3,000 + £4,000 + £5,000 = +£2,000.) Therefore, since the sign changes, the internal rate of return (IRR) — which has to produce an NPV of zero — must lie between 0 and 10%. And, since −£210 is much closer to zero than is +£2,000, the IRR will lie closer to 10% than to 0%.

Using a 'trial and error' method of finding the internal rate of return, one could first try a discount rate of 9%, as shown in Table 13-5.

The net present value of −£21 is close enough to zero; so in practice one would reckon the internal rate of return as being (just under) 9% a year. Since the required rate of return should certainly be superior to that of a safe bank deposit investment at an interest rate of 10%, Project G's 'internal' rate of return is not high enough to justify investing in it.

In fact, the net present value of project G could be plotted for a whole range of different discount rates. More complex applications of DCF on a spreadsheet allow sensitivity analysis to be carried out. This involves changing the basic assumptions and seeing how the NPV alters as a result (see Figure 13-5). For example, the future cash flows resulting from the project are definitely going to be subject to uncertainty, optimism, and pessimism. The "what if" question may be applied by varying the cash flows, tax, operating costs, capital investment, etc., and seeing how sensitive the overall project is to each of these changes.

Three issues in particular are worth noting.

(1) At a 0% discount rate, the NPV is +£2,000. This can be found simply by deducting the (undiscounted) cash outflow from the (undiscounted) total of the cash inflows.

(2) Using a 10% discount rate, the NPV is −£210. This is the figure found earlier, when using 10% as the 'opportunity cost' criterion rate.

(3) The net present value is zero at a discount rate of 8.9%. This is the 'precise' internal rate of return.

13-5-3. NPV vs. IRR

Most financial analysts prefer expressing their project investment appraisal analysis in terms of IRR rather than NPV. Back in the 1980s, the use of computers for financial appraisal was far less widespread. Out of a study on 150 of the largest UK firms in retailing and manufacturing, relatively few used Discounted Cash Flow (DCF) techniques, as shown in Table 13-6. The situation has now very much altered, with payback and annual rate of return being used as a quick initial check before a more detailed DCF analysis is undertaken.

With the introduction of the Personal Computer since 1980 (and on a Microsoft Excel spreadsheet, Future Value (FV) and Present Value (PV) functions are available by the click of a button), more and more firms are now using DCF and, if not careful, making mistakes along the way. Undoubtedly, many cash flow forecasts are subject to wide margins of error. However, this hardly justifies using a theoretically incorrect method or project appraisal. (It does, perhaps, call for special focus on cash flows in the early years of a project.)

Some experts believe that the process of estimating the future cash flows arising from a project is the most valuable part of the appraisal procedure; hence, that it may not matter too much which "appraisal method" is actually used. (A similar argument is sometimes used in favor of a

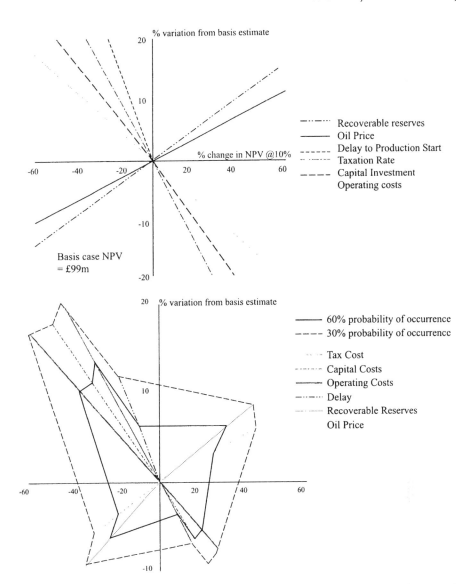

Figure 13-5. Sensitivity Analysis — The Affect of Different Situations on the Viability of a Capital Project.

"decision tree" analysis, where the precise "probabilities" employed are extremely uncertain as a rule.) In any case, it would be naïve to suppose that only technical financial considerations are relevant in deciding on the commitment of funds which may help shape the whole future of an enterprise. Strategic considerations may be equally, or more, important for large projects.

13-5-4. *Project X, Other Problems, and Discussion*

This section includes some examples for the reader to work through so as to demonstrate the application of the financial appraisal techniques discussed earlier in this chapter. Appendix 2 is an example of the application of such techniques to a real decommissioning example.

Consider "Project X", which is expected to produce the following cash flows:

End of year 0   −£900 (Initial Investment)
End of year 1   +£400
End of year 2   +£400
End of year 3   +£400

(1) Assume £900 is held on a bank account until the end of year 3. How much will it accumulate to by the end of year 3 if the bank pays interest at 10% at the end of each year?

Table 13-6. Project Appraisal Methods Used by 150 Large UK Firms, 1980

Primary[a] method (%)		One method (%)		Two methods (%)		Three methods (%)		Four methods (%)	
a	32	a	11	ab	12	abc	10	abcd	10
b	32	c	8	ac	12	abd	6		
c	41	d	4	ad	6	acd	10		
d	17		3	bc	3	bcd	1		
				bd	1				
				cd	3				
	122[a]		26		37		27		10

[a]This equates to more than 100% because some methods ranked 'equal first'.

a = payback, b = average rate of return, c = IRR, d = NPV.

(2) Assume the company invests £900 in the project and receives £400 per year as above. How much will the £400 per year accumulate to if the bank pays 10% interest at the end of each year?

(3) Should the company invest in Project X? How much better or worse off would it be at the end of year 3 by investing in this project?

(4) Suppose the company did not have £900 available but borrowed this from the bank. Interest of 10% was to be charged at the end of the year. The £400 per annum would be used to repay the bank. How much would the company owe or be in credit by the end of this period? Should the company invest in Project X?

(5) Compute the net present value of Project X using tables. Assume a 10% discount rate. Why is your answer different to that in question 3? Reconcile the two answers.

(6) What is the maximum amount the company could invest now in Project X and not end up worse off? (Assume the inflows of £400 per annum remain unchanged.)

(7) What is the maximum rate of interest a company could pay for a loan to finance Project X and still break even on the project?

(8) What is the minimum equal amount which the company could receive per annum and break even on the project? Assume a 10% discount rate.

Other Problems

(9) Compute the Internal Rate of Return for the following cash flows:

0   −£1000
1   +£300
2   +£500
3   +£800

(10) What is the net present value of a constant stream of cash flow of £2,678 per year starting at the end of year 1 and finishing at the end of year 20, assuming the interest rate to be 12%?

(11) You wish to borrow £15,000 from your building society. The interest rate is 12% per year and repayments are required at the end of each year for the next 20 years. How much will your repayments be each year (ignore tax)?

(12) Suppose the purchase of new equipment involves the following incremental cash flows:

Year 0   −£1,000
Year 1   +£420
Year 2   −£48
Year 3   −£48
Year 4   +£52

What annual level of net savings for a 4 year period beginning in year 1 ending in year 4 is required to justify the project? Assume a 12% discount rate (ignore tax).

(13) Using the data in question 12, calculate the before tax savings necessary to justify the purchase of the equipment. Assume the tax rate is 52%, and tax is paid after a 1 year delay. Assume also that the cash flows shown above from year 1 to year 4 are not subject to tax.

One has seen how discounted cash flows may be used to help evaluate projects. For engineers, the maths is not complex, and often it is the project staff that have to provide the raw data used in such computations.

(i) One must not assume that a spreadsheet and DCF gives a "right" answer. Both the time taken to build the project and the period over which returns take place is a subject for judgment.

(ii) Further complications arise from tax and risks. These can be catered for in the analysis.

(iii) One has to ask oneself if DCF is applicable to projects where:

- timescales are long, and
- there is no "return on investment" that can be put in strict money terms.

13-6. Modern Contract Strategy in the Nuclear Industry

13-6-1. Introduction

The selection of a contract strategy for a project is a key decision which will have a major impact on the project's outcome. Any project incorporates a degree of risk which, once initiated, may be countered by effective change control, producing a revised clear scope and work definition, rescheduling both the program and the cash flows, together with application of insurances, payment bonds, advance payments, and retentions as appropriate. The type of contract employed to complete the project works should take into account risk identification and then the appropriate apportionment of the risks with those (Client or Contractor) who are best able to manage them.

It is immature to expect a project to have anything other than a compromise of emphasis between time, cost, and quality (safety, environmental, and public relations) issues. Some projects will require more emphasis on the completion date than on cost, etc. Figure 13-6 illustrates this and describes the different types of contract for placing risks with either Client or Contractor. Consider, for example, where you would place the emphasis when formulating the most appropriate type of contract for the following projects:

  (i) A nuclear power station construction project;
  (ii) An overseas electricity distribution aid contract;
  (iii) Repair to a section of city Metro tunnel rail track;
  (iv) A new City motorway road bypass construction;
  (v) Production of a new waste plant Pre Commissioning Safety Report (PCmSR);
  (vi) Introduction of a new document management system into a Company;
 (vii) Preparation of the safety case for an Intermediate Level Waste (ILW) deep waste repository;
(viii) Repair of a failed electrical cable feeding a fuel cycle area on a nuclear licensed site;
  (ix) A nuclear fuels accountancy audit; and
  (x) Efficiency improvements (turn-around) of a National Health Service/Nationalised Railway or Nationalised Nuclear Decommissioning Company.

This Section describes a methodology for considering how best to apply these principles to nuclear decommissioning works.

13-6-2. *Modern Contract Selection Appropriate to Nuclear Decommissioning*

Regardless of the contract strategy adopted, the client and nuclear site licence holder must always have, and be able to demonstrate to have, ultimate control and "day-to-day" control of all activities. The contract strategy begins with the Client's business model and objectives before moving to the individual project's objectives and requirements. Each project is then examined in terms of its complexity and the need for Client involvement to give a "first cut" guide on appropriate contract formation. The main contract models are described in relation to such an analysis, together with requirements for managing risk, pricing, and the market.

Specific objectives for the Nuclear Decommissioning Management Company (Client organisation) will be the successful outcome of the project, the predictability of costs, and provision of demonstrable value for money within the context of a safe and secure environment. Assuming the project implementation work is to be competitively tendered by the Client organisation to a variety of Contractors, then plotting the project's complexity against the need for Client involvement is helpful, and illustrated in Figure 13-7 [1].

Project complexity issues include:

• Performance requirements and associated constraints;
• Level of technical challenge (i.e., novelty);
• Development work required;
• Scope for innovation;
• Scope for cost reduction;
• Requirement for multiple specialisations;
• Opportunities for Private Finance Initiatives (PFI) or Private/Public Partnerships (PPP); and
• Confidence in the technology.

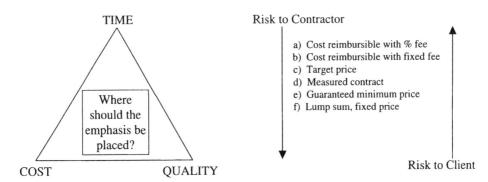

Figure 13-6. What Are the Driving Factors on a Particular Contract and Where Should the Project Risks Best be Placed?

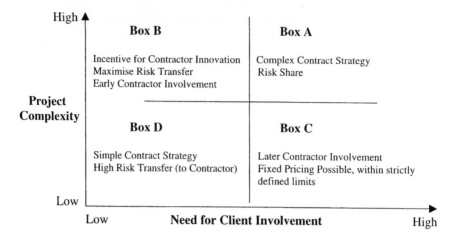

Figure 13-7. Characterisation Matrix Relating Project Complexity with Client Involvement.

Table 13-7. Applicable Contract Strategies to Suit the Figure 13.7 Project Characterisation Matrix

Application	Applicable Contract Strategy
Box A	Alliances
Box B	Consortium
Box C	Traditional
Box D	Prime Contractor

Client involvement issues include:

- Strategic importance (of the project);
- Risk of failure;
- Stakeholder interest;
- Regulator interest;
- Proximity to existing nuclear facilities; and
- Safety issues in implementation.

From such a plot, the most applicable contract strategy for a particular application may be determined. UKAEA experience suggests the strategies in Table 13-7. The main advantages and disadvantages of these contracting strategies are described in Table 13-8.

13-6-3. *Types of Contract*

Within the nuclear decommissioning market, it is recognised that projects have a history of not always being completed to time and budget. This may be due to:

- Ill-defined deliverables;
- Poorly executed risk management;
- Genuine lack of knowledge about the facilities being decommissioned (see Chapter 10);
- Regulatory interactions; and
- Poor project management/performance by either or both Client and Contractor.

Often, there is only an adversarial contractual framework within which to develop disputes. This may not be appropriate and is certainly not efficient when applied to nuclear decommissioning where uncertainties about the works are a genuine risk.

Standard forms of contract are available for use on projects. The important issue at the outset is to carefully consider the most appropriate form for the works being undertaken.

- *Cost reimbursible with % fee* — The contractor agrees to carry out the work for whatever it actually costs him to complete it (as substantiated by receipts, time-sheets, etc.) and then charges this amount plus a percentage fee based upon these costs. The disadvantage, or risk to the Client, is that the contractor may not keep his costs under tight control. There may be no particular incentive for the contractor to keep his costs down, since he will receive a larger fee the longer and more expensive he makes the job. Such conditions of contract may be necessary for research work, where only a few contractors have the capability and the outcome may not be known for certain. Client and contractor must be in complete harmony, working for the same goal, for this type of contract to be considered.
- *Cost reimbursible with fixed fee* — This form of contract puts a limit on the costs by imposing a fixed fee upon the contractor. Often, this form of contract is used by engineering design consultants. Normally, the reputation of the consultant is at stake and abuse of such conditions, therefore, unlikely with reputable firms.
- *Target Price* — The contractor agrees to perform the works within a given cost ceiling and/or time frame. If the contractor manages to complete the works within budget or time frame, then a bonus is paid. This type of contract has been very successful for relatively conventional engineering projects, where completion to

Table 13-8. Advantages and Disadvantages of Different Applicable Contract Strategies

For	Against
Alliances (Box A in Figure 13-7)	
• The Client can demonstrate overall control through direct works contracts and chairing the Alliance Board. • Competition can be used for the appointment.	• Experience of this type of contracting is limited in the nuclear sector. • It requires a high level of commercial capability to set up and arrange.
• Successful application of alliancing is claimed to have achieved significant schedule and cost savings. • There is a strong incentive on contractors to work together on all aspects of the project, including safety, and innovate to reduce costs. • The use of an integrating contractor removes much of the managerial load from the Client, provided that the Client and Site Licence holder is able to demonstrate overall day-to-day control of activities, and has an adequate contingency plan in the event of failure of the Contractor.	• Selection criteria have to include "soft" issues which can be difficult to define and assess objectively.
Consortium (Box B in Figure 13-7)	
• Brings a mutual interest between key contractors, with incentive for innovation and cost reduction. • Opportunity to transfer risk to the main contractor and on to the other members.	• At least two layers of subcontracting. • Weak overlap of Client and Consortium objectives unless strongly incentivised. • A Consortium formed for the purpose of successful bidding does not always work well for project implementation.
Traditional (Box C in Figure 13-7)	
• Maximises competition. • Permits maximum use of in-house/contractor assets and resources. • Gives demonstrable Client control over the project.	• Conflicting objectives of contractor and Client leading to claims. • High degree of Client involvement in managing the project. • Client may require large in-house design team under his direct control. • Client often carries most of the design work responsibility. • Poor record of delivery to schedule and cost. • Can lead to compartmentalisation of projects, to the detriment of overall value for money.
Prime Contractor (Box D in Figure 13-7)	
• Transferred risk rests clearly with the prime contractor (providing the specification is sound). • Clear responsibility and accountability for all aspects of the project. • Relationships are relatively straightforward.	• Tends to lead to long subcontractor lines. • Weak overlap of Client and Contractor objectives (though incentivisation can help to provide a common interest). • A Consortium formed for the purpose of successful bidding does not always work well for project implementation. • Tends to be weak overall commitment to the project from key subcontractors. • Innovation and cost reduction are only likely if effectively incentivised. • Project management organisations in this role can filter information flows between Client and the "doers," thus reducing Client control and influence.

a set schedule is required by the Client and the incentive of a large bonus has driven such works to a successful conclusion by the contractor.

• *Measured Contract* — A bill of quantities is prepared to describe the works in great detail. Rates are attached to each item of work, and the contractor is paid according to the amount of work performed. This is applicable to conventional well-defined works. If the works scope definition is insufficient, there is a risk to the contractor that he will not have adequately priced the bill of quantities (B of Q) at the tender stage and he will have little recourse if he did not fully understand the extent of the work at the outset. Interim payments are made to the contractor on a regular basis as the work proceeds so as to assist him in maintaining a relatively low level of working capital. Variations to the estimated quantities in the original tender document B of Q invariably occur in practice, leading to possible friction between Client and contractor. If these increases or decreases materially affect the overall intent of the original contract works (often judged by whether the overall contract has altered by more than $\pm 15\%$), then the rates originally quoted may also have to be varied. By this mechanism, the risk to Client and contractor is kept within manageable bounds.

- *Guaranteed Minimum Price* — The Client and contractor agree to a guaranteed minimum price for completion of the works. This may then be varied should the scope of the works change during the contract period. A guaranteed minimum price reduces the risk to the Client, but increases it for the contractor. This type of contract requires good definition and a minimum amount of interference and change requests by the Client during the contract period.
- *Lump Sum, Fixed Price* — The Client and contractor agree a fixed price for carrying out the work. The risk here is greatest to the contractor, since unforeseen circumstances (e.g., the discovery of more contamination than originally anticipated when decommissioning a facility) would alter the cost of the work considerably. The Client has effectively placed the risks involved with unforeseen circumstances with the contractor with this type of contract. Of course, the contractor will price the works accordingly to cover the perceived risks involved with a large contingency to cover any lack of definition. It is important with this type of contract that the Client does not impose any significant changes to the scope or definition of the work during the contract period. If the Client does this, then the contractor will be able to correctly claim for the extra costs. This type of contract is most applicable to well-defined conventional works.

One such Standard Form of Contract of particular interest for application to nuclear decommissioning works is the New Engineering Contract (NEC). This is a family of contract forms including the Engineering and Construction Contract (ECC) for major new build projects, and the Professional Services Contract (PSC) for consultancy. The benefits of this contract which has been developed since the late 1980s are:

- a move to milestone payments,
- suitability to partnering and target cost contract works,
- applicability for design, build, and operate contract works,

- allowance for multi-disciplinary construction project working,
- a framework to allow movement away from more conventional confrontational contract forms (ICE 6th Conditions, etc.), and
- inclusion of positive steps to encourage the avoidance of the adversarial disputes that have occurred on many large construction projects.

Payments under these terms and conditions of contract are made against activity schedules, linked to a program. This ensures that the contractor has thoroughly researched his program and it allows both parties to understand their cash flow constraints. It attempts to ensure that the project is program-driven — payment only being made against completed activities. In addition, this form of contract insists upon timely agreement of the value of compensation events that arise during the works. Claims are not, therefore, allowed to languish until the overall works are completed. They are agreed as they arise, thus allowing the contractor due compensation in good time and avoidance of having to fund a large working capital account.

13-7. Alternative Sources of Funds

13-7-1. *Introduction*

Since the nuclear legacy is largely a result of nuclear research carried out by different Governments in the mid-to-late 20th Century, the cost of the clean-up programs falls largely to these Governments. However, an optimal decommissioning program may well require "bulges" in expenditure in the program to pay for the capital costs of constructing the necessary waste treatment and handling facilities and stores. Nuclear decommissioning involves uncertainties, and Governments look more for smoothed and sure demands on funding and also may wish to share the risks of such large projects with the private sector. In addition, at any point in time, Government may have many alternative pressures on the public purse for such expenditure (for example, for use on schools and hospitals), and these pressures may be seen to outweigh the benefits to the community from those derived from a particular site environmental restoration program. One such alternative approach to funding nuclear decommissioning work is a Private Finance Initiative (PFI), which is a contractual structure used generally for delivering asset-based services to the public sector.

13-7-2. *What is PFI?*

The structure is a development of limited resource project finance which has been used since the 19th

Century — most notably for the private sector development of railways in the UK and US, and more recently for power and process plants worldwide. It is a form of partnership between the public and private sectors (Public Private Partnership or PPP), which has been successfully applied to the provision of hospitals, schools, prisons, government accommodation, light rail schemes, and roads. Instead of the standard approach to asset delivery, the specification for a PFI project is based on the required outputs rather than the defined inputs. This allows the private sector greater flexibility to deliver innovative solutions to public sector requirements, thereby delivering, in theory at least, improved value for money.

Figure 13-8 illustrates the most applicable area of contract strategy, which matches with the PFI concept. PFI pushes the contract model away from detailed specifications (where the contractor carries out the detailed instructions of the client) towards performance specifications — where the contractor takes on much more risk and has much more freedom to innovate. In this respect, it is a model closely related to outsourcing — a model under which an organisation identifies its "noncore" activities and then contracts with another company to provide them, rather than continue to provide them itself. Outsourcing is successful when the outsourcing provider:

- Can create economies of scale by rolling several contracts together and, for example, by reducing procurement costs;
- Has access to specialist expertise that the client organisation does not have "in house"; and
- Is able to manage the functions more effectively because of superior market knowledge, processes, relationships with suppliers, or similar as a result of the work being the outsourcing provider's core activity.

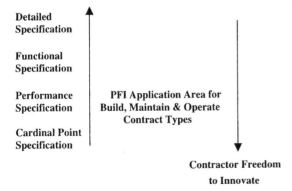

Level of Detail and Client Risk

Detailed Specification

Functional Specification

Performance Specification — PFI Application Area for Build, Maintain & Operate Contract Types

Cardinal Point Specification

Contractor Freedom to Innovate

Figure 13-8. Area of Application Within a Spectrum of Different Contract Strategies for Private Finance Initiative (PFI) Funding Arrangements.

Standard PFI contracts have a life of some 25–30 years during which the asset is to be delivered and maintained. Commonly, the contract will also provide for the delivery of ancillary services throughout that period, thereby allowing the public sector to concentrate on its "core" activities. However, in the nuclear context, there is a clear obligation on the site licence holder to be in "day-to-day" control of the site. Pushing the contracted out services boundaries for operations management or nuclear facility maintenance too far has caused Regulatory intervention in the UK.

The contractor will secure funding for the project from the private sector through a mixture of debt and equity. Payment to the contractor from the public sector will be based upon maintaining availability of the facility or delivery of the service to the standards specified in the output orientated contract specifications. Crucially, security for the debt raised by the contractor is derived from the income streams set out in the contract rather than from the physical assets themselves. The banks, therefore, require:

- Strong covenants from the public sector with long term surety; and
- Low risk that the contractor can meet the output specifications set out in the contract.

This second point imposes an additional discipline on the contractor, in as much as PFI aligns the interest of the public sector, and the banks, in so far as they both require the contractor to deliver the specified outputs, and who will, therefore, both take a strong interest in the conduct and management of the project.

13-7-3. *Fixed Price/Risk Premium and Value for Money*

The private sector should not be required to assume risks over which it has no control or cannot hope to mitigate. However, value for money for the public sector client comes partially from risk transfer to the contractor and the ability to allow the contractor to innovate. If these factors more than offset the extra cost of commercial debt and equity finance, then the PFI scheme may be seen to offer value for money to the taxpayer. The degree of risk transfer and the scope for innovation are, therefore, crucial factors. It is not just about transfer of risk, since the private sector will undoubtedly demand a risk premium in order to enter into a long-term contract on a fixed price or output driven basis. Rather, the key to value for money is whether the private sector is better able to manage a given risk and, therefore, whether the aggregate effect of the risk transfer premium, ability to innovate, and cost of funding produce a cheaper alternative than a sole public sector funding solution.

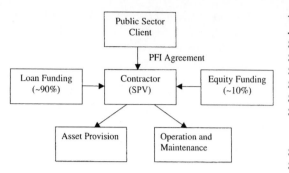

Figure 13-9. A Typical PFI Contractual Model.
Note: SPV — Special Purpose Vehicle, a legal entity set up for the sole purpose of conducting the contract to be funded under the Private Finance Initiative (PFI) scheme.

Table 13-9. Staged Process to a PFI Contract

Stage 1	Establish Business Need
Stage 2	Appraise the Options
Stage 3	Business Case for Reference Project
Stage 4	Developing the Team
Stage 5	Deciding Tactics
Stage 6	Invite Expressions of Interest from Contracting Organisations or Alliances (In Europe Publish in *Official Journal of European Community*).
Stage 7	Prequalification of Bidders
Stage 8	Selection of the Shortlist
Stage 9	Refine the Appraisal
Stage 10	The Invitation to Negotiate
Stage 11	Receipt and Evaluation of Bids
Stage 12	Selection of Preferred Bidder(s) and Final Evaluation
Stage 13	Contract Award and Financial Arrangements Closed
Stage 14	Contract Management

13-7-4. *Technical Viability and PFI Project Set-Up Costs*

A major constraint on the application of PFI is that commercial lenders will require a high degree of certainty as to the ability of the private sector to deliver the output specification. At its simplest level, this is because lenders want to be repaid with interest. At a secondary level, if those involved fail to deliver the project, then lenders will need to be able to exercise step-in rights and appoint others to complete the works. A difficult issue in a highly regulated industry. Higher risk of delivery or constraints will adversely affect loan margins and, hence, value for money.

Borrowing money from the "market" (commercial banks) to fund such works is inevitably more expensive than seeking funding from Government. In addition, the costs of setting up PFI projects is significant. With respect to projects in new sectors without a history of success and no standard contract forms, then fees for legal, financial, and technical advice will be large and perhaps as high as £3–£5M for both the public and private sector partners involved. The scale of the project has little bearing on the set-up costs to be incurred. As a result, PFI funding mechanisms have generally been used on larger projects with over, say, £50M of capital value, which can absorb the expense involved. A typical funding model is illustrated in Figure 13-9.

13-7-5. *The Staged Approach to PFI*

UK Treasury guidance refers to a staged approach to PFI, as shown in Table 13-9. Such a process may well take 18 months to 2 years to navigate to a successful conclusion.

In summary, therefore, PFI is applicable to:

- larger (typically £50M plus) contracts,
- more self-contained contracts and, therefore, those with a greater degree of operational control and fewer interactions,
- ideally those with a large construction element so as to benefit from capital tax allowances, and
- contracts that may be characterised as being largely based upon proven technology, particularly if the first of a kind in the nuclear sector.

13-8. References

1. Nicol, R. D. *ImechE Conference, Nuclear Decom 2001, Contract Strategy Selection*, UKAEA, Dounreay, C596/043/2001 (2001).
2. Bayliss, C. R. and Bundy, M. Ibc 8th International Conference on Decommissioning of Nuclear Facilities, The Development and Implementation of Modern Project and Engineering Management Processes for Nuclear Decommissioning, November 2002.
3. Bayliss, C. R. Nuclear Congress 2003 (18/19 June 2003), Manchester, UK. The UKAEA Approach to Alliancing, BNES, London.

Chapter 14

Hazard Reduction and Project Prioritisation

14-1. Introduction

A hazard is defined as the "intrinsic property or disposition of anything to cause harm." Risk is correctly defined as the product of probability × consequence and in this respect also "the chance that someone or something that is valued will be adversely affected in a stipulated way by a hazard." In practice, therefore, the hazard associated with redundant nuclear facilities on a site together with the layers of technical and procedural safety precautions taken are equivalent to the residual risk of the facility:

Hazard less Safeguards = Residual Risk.

This chapter describes a methodology for understanding the hazards on a site and why hazard reduction may be used as one driver for a site's decommissioning program.

This chapter also describes a methodology for understanding the relative importance of the different projects within an overall integrated site decommissioning program. The application of this prioritisation process assists management in making tough decisions as to which projects should take precedence when seeking to fund an overall program of work or what projects may need to be deferred in a particular year because of budget constraints.

14-2. Understanding Risk and Dose

Whether people are prepared to accept a risk has a lot to do with whether they feel that they have control over the actions that they are taking or whether circumstances that involve a degree of risk are being imposed upon them. Table 14-1 illustrates risks associated with everyday occurrences.

Note that $20\,\mathrm{mSv\,y^{-1}}$ is used by ICRP in setting limits, but not for use as a target dose. 20 mSv per year over a 50 year lifetime gives an accrued dose of 1 Sv corresponding to a risk of 1 in 20 (5×10^{-2}). A very simplistic, but not accurate, risk:dose relationship might, therefore, assume 5% per Sv. See also Chapters 2 and 18.

Table 14-1. Relative Average Risks Associated With a Variety of Everyday Events

Event	Risk
Aircraft accidents (passenger journeys)	1 in 125,000,000 (8×10^{-9}) of death
Rail travel accidents (passenger journeys)	1 in 43,000,000 (2.3×10^{-8}) of death
Lighting strike (UK average annual)	1 in 18,700,000 (5.3×10^{-8}) of death
Rail travel accidents (passenger journeys)	1 in 1,533,000 (6.5×10^{-7}) of injury
Road accidents (km travelled)	1 in 1,432,000 ($\sim 7 \times 10^{-7}$) of injury
0.02 mSv per year to a worker over 50 years	1 in 1,000,000 (1×10^{-6}) of death
Surgical operation anesthesia (operations)	1 in 185,000 (5.4×10^{-6}) of death
Natural radon-induced lung cancer (UK annual)	1 in 29,000 (3.4×10^{-5}) of death
Construction (UK industry, annual)	1 in 17,000 (5.8×10^{-5}) of death
Road accidents (UK all forms, average annual)	1 in 16,800 ($\sim 6 \times 10^{-5}$) of death
Burn/scald in the home (UK average annual)	1 in 610 (1.6×10^{-3}) of injury
Smoking 10 cigarettes a day (lifetime)	1 in 200 (5×10^{-3}) of death

Source: [1], pp. 70 and 71.

About 156,000 people in the UK are exposed to man-made radiation and the associated possible risk to their health as a result of their work. Most workers receive doses less than $5\,\mathrm{mSv\,y^{-1}}$ from natural and man-made radiation sources as a result of their work. Average annual doses within the UK nuclear industry are less than $1\,\mathrm{mSv\,y^{-1}}$, radiation workers receive only some 0.5 mSv $\mathrm{y^{-1}}$ and medical radiation workers some $0.1\,\mathrm{mSv\,y^{-1}}$. These exposures should be seen against a background of a general fall in exposure levels, with average annual doses falling by half in the UK in the period 1987–1991. By far the greatest exposure to the UK population is from

Table 14-2. Radiation Exposure to the UK Population

Radiation source	Percentage contribution to average population radiation dose
(a) *Natural Sources*	
Radon gas from the ground	50%[a]
Gamma rays from the ground and buildings	14%[b]
Food and drink	11.5%[c]
Cosmic rays	10%[d]
(b) *Artificial (man-made) sources*	
Medical	14%[e]
Occupational	0.3%[f]
Fallout	0.2%[g]
Products	<0.1%[h]
Nuclear discharges	<0.1%[i]

[a]The yearly average annual dose from radon is 1.3 mSv in the range 0.3–100 mSv. Radon is a gas given off from uranium bearing rocks such as granite and is part of the natural decay chain from U-238 to Pb-206. Up to 6% of the annual incidence of lung cancers in the UK (2000–3000 cases per year) are believed to be initiated by the radioactive decay products of radon.

[b]The dose depends on the local rocks, soils, and building materials. The yearly average dose is 0.35 mSv within a range of 0.1–1 mSv.

[c]The yearly average annual dose from diet is 0.3 mSv in the range 0.1–1 mSv.

[d]The yearly average annual dose at ground level is 0.26 mSv in the range 0.2–0.3 mSv. Some 24,000 UK aircrew receive on average an annual dose of 2 mSv from cosmic rays which readily penetrate the fuselage of aircraft.

[e]The dose from a lower spine X-ray diagnostic is typically 2 mSv in a range from ~1 to ~6 mSv. Medical radiation is the largest source of man-made radiation to the public. The average diagnostic dose is some 0.37 mSv.

[f]The largest group of occupationally exposed workers (some 50,000) are those whose work place is in radon prone areas. They have an annual average dose of ~5 mSv.

[g]Average annual doses in the UK from radioactive fallout are from 5 μSv to 15 μSv in high rainfall areas. An increase in 1986 was due to the Chernobyl nuclear reactor accident.

[h]Consumer products typically include smoke detectors, luminous watches, natural radioactivity from gas mantles, etc. The average dose is 0.4 μSv with a range up to 100 μSv.

[i]Average annual doses to the public from weapons tests have declined from 140 μSv in the early 1960s to ~ 5μSv now.

natural radiation sources arising from, for example, living in high radon areas such as in Devon and Derbyshire, resulting in an average yearly dose in the UK population of some 2.6 mSv y^{-1}. Table 14-2 describes the make up of radiation exposure in the UK [2].

Nuclear regulation in the UK drives risks to be As Low As Reasonably Practicable (ALARP — see Chapter 18) such that risks below 1 in 1,000,000 (1×10^{-6}) are considered to be broadly acceptable. Certainly, for events leading to risks less than this, then there is no requirement for detailed working to demonstrate that the risk is as low as reasonably practicable. It is, however, necessary to maintain assurance that the risk stays at or below this

Table 14-3. UK Dose Limits for the Public and Occupational Workers and Industry Constraints for the Design of New Plant

	Public	Occupational Worker
Dose Limit	1 mSv y^{-1}	20 mSv y^{-1}
Dose constraint (or target)		10 mSv y^{-1} (new plant design)

Figures from [3] and ICRP literature — see Chapter 2, Section 2-5. Prior to 1999, the public and worker dose limits were 5 and 50 mSv, respectively. Constraint figures are not legislative but set by the Industry as design targets for new plant.

level. This risk is equivalent to the risk of dying in a fire or gas explosion or being electrocuted at home. The risks associated with nuclear plants are maintained at extremely low levels by layers of protection (from the design through to the maintenance regime and the procedures involved in plant operation). Typical dose limits and dose constraints set by the industry in the UK for the design of new plant are as shown in Table 14-3.

The relationship between dose and risk is a complex subject and takes into account the concept of the critical group. For these purposes, a dose uptake of 20 mSv y^{-1} for 50 years roughly equates to a risk of some 5% or 5×10^{-2}.

It is a harsh but realistic fact of life that there must be due consideration to the proportionality in terms of cost of the benefit arising from precautionary measures put in place to reduce risks. However, it is not an equitable playing field with some public services being prepared to pay more for infrastructure involved in risk reduction and life saving measures than others. For example, the costs of road works to save a life (removing bends in roads, installation of traffic lights, etc., to reduce accident black spots) is less than that applied to rail transport improvements and far less than that applied to the nuclear industry in practice. HSE guidance suggests that £1–£2M should be invested to save a life from cancer arising from the associated work activities. From Table 14-1, this would equate to an additional expenditure of £10^5 Sv^{-1}. When one looks at the expenditure on nuclear waste operation plants to reduce discharges, it can be seen that the expenditure in the UK nuclear industry is more than an order of magnitude greater than this being driven by political issues and public perception.

If the driver for nuclear decommissioning was based upon risk alone, then there would be little incentive for progress. Indeed, the very action of decommissioning produces doses, however small, to workers and creates risk in itself. The driver for decommissioning redundant nuclear plant "as soon as reasonably practicable taking all factors into account" is, therefore, better based upon hazard reduction and the requirement to show "systematic and progressive reduction in hazards."

14-3. Hazard Reduction

14-3-1. *Why is Hazard Reduction Important?*

The UK Health and Safety Executive (HSE) requires that hazards be identified, the risks they give rise to are assessed and appropriate control measures introduced to address the risks. Government Policy, Cm 2919 requires:

- Progressive and systematic reduction of hazards; and
- Decommissioning as soon as reasonably practicable, taking all factors into account.

Risk-based analysis is still embedded in HSE/NII's safety assessment principles, but there is a clear move towards hazard as opposed to risk reduction when assessing decommissioning (as opposed to plant operation) programs. In an operating plant, risk can be balanced against the benefits to society of the operation. The principles of ALARP (As Low As Reasonably Practicable) and ToR (Tolerability of Risk) are, here, obviously applicable. Redundant facility decommissioning has no clear benefit against which to balance the risk, which leads to a focus on progressive hazard reduction (see Figure 14-1).

14-3-2. *How are Hazards Reduced?*

The first requirement, in line with Cm 2919, is always to complete, as soon as practicable, the removal of fuel

and loose radioactivity which has the potential to spread contamination. Projects to do this are, therefore, scheduled to take place as soon as the facility is taken out of service. Once the initial work has been done, the next stage, whether to continue with the decommissioning of a facility or keep it under care and maintenance until a later period, depends upon a number of considerations:

- The physical condition of the structure of the facilities and services required for decommissioning (such as cranes).
- Will these deteriorate and require refurbishment, or even replacement, if decommissioning is delayed?
- Does delay reduce the cumulative dose uptake (because of radioactive decay—see Chapters 3 and 6), or increase it because of the longer period of exposure during the care and maintenance regime?
- Are there advantages in using staff knowledgeable of the plant, which would be lost as these staff retire or are dispersed?
- How best would the work fit in with other work being done and the availability of waste routes?

14-3-3. *What Methods May be Used to Gauge Hazard Reduction?*

A hazard-based approach to the benefits derived from a particular program of decommissioning works cannot be usefully based upon source terms alone. These will only

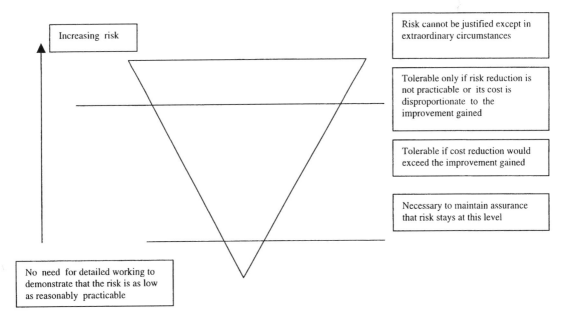

Increasing risk

Risk cannot be justified except in extraordinary circumstances

Tolerable only if risk reduction is not practicable or its cost is disproportionate to the improvement gained

Tolerable if cost reduction would exceed the improvement gained

Necessary to maintain assurance that risk stays at this level

No need for detailed working to demonstrate that the risk is as low as reasonably practicable

Figure 14-1. Levels of Risk and ALARP "As Low As Reasonably Practicable."

Table 14-4. Categorisation Criteria for Radiological or Toxic Hazards used by UKAEA

Category	Fundamental definition	Interpretation for radiation	Interpretation for chemicals
1/A	Off-site hazard	> or = 5 mSv off-site	Dangerous dose off-site
2/B	On-site hazard	> or = 5 mSv on-site, or > or = 0.05 mSv off-site, or > or = 50 mSv in-building	Dangerous dose on-site, outside the building housing the activity
3/C	In-building hazard	> or = 5 mSv in-building	Dangerous dose confined to the building housing the activity
4/D	Hazard confined to local work area	> or = 5 mSv in a building	

reduce through radioactive decay or removal of material. A modified hazard-based approach can:

- consider using an integrated measure such as "categorisation" of buildings,
- be largely hazard-based, but will include a transport term so it will have a risk element,
- be built upon existing, well understood foundations rather than inventing a totally new measure,
- take into account both chemical and radiotoxicity, and
- consider the immobilisation of the activity.

This approach should be able to demonstrate a progressive reduction in hazard potential. Therefore, such a method could be based upon a measure of hazard potential of categorised activities. The categorisation should apply to all activities that could give rise to a radiation or toxic chemical hazard, either directly or indirectly, and should consider both the source term and relative mobility of the hazard. The category should reflect whether the radiological and chemotoxic hazards associated with the specified activity extend off-site, are contained on-site, or are only local to the facility. See Table 14-4 and refer to Chapter 18, Table 18-6, where the UN/IAEA/OECD — NEA International Nuclear Event Scale (INES) is described.

With knowledge of the number of facilities or activities present on a site at any one time, and their relevant hazard category, it is possible to calculate a total "hazard index" for the site. To arrive at a numerical score, each category should be allocated a number of points, e.g., category 1 (highest), 10 hazard points, category 2, 3 hazard points, etc.

It should then be possible to construct a plot of the projected reduction in hazard index with decommissioning over time. It must be recognised that the hazard index of a site may increase in periods where construction and operation of new plant and storage facilities precedes the complete decommissioning and demolition of the original building or facility. However, this should be reflected in the predicted hazard index plot for the site. Figure 14-2 illustrates the typical output from such an analysis and plots the reduction in hazard index, as a site is decommissioned over time in accordance with the integrated site decommissioning program.

14-4. Project Prioritisation

14-4-1. Why Does One Need to Prioritise Projects?

A decommissioning program should be based upon what needs to be done by a licensee to implement long-term plans and manage its responsibilities in a way which is safe, environmentally sound, economic, and publicly acceptable. Funds and other resources (e.g., manpower) may not be available as required to match the optimum program. Therefore, work must be prioritised in order to best understand where scarce resources may be allocated over the decommissioning period.

The prioritisation process will help decide:

- which projects to bring forward;
- which projects to postpone when funds or other resources are not available to do everything in the plan; and
- which work to reschedule when urgent new projects are introduced into the program.

Figure 14-3 illustrates schematically the stages in planning and prioritising a decommissioning project.

The strategic review process, which is an integral part of planning, should consider the strategic, technical, and logistical issues of a project. This should ensure that the project is consistent with the overall organisation's plan. Preliminary option studies should then be carried out and, when the preliminary project proposals are prepared, the prioritisation process should be applied to determine the prioritised score for the project. It is important that the process does not become mechanistic and that common sense is applied. Once major project elements of

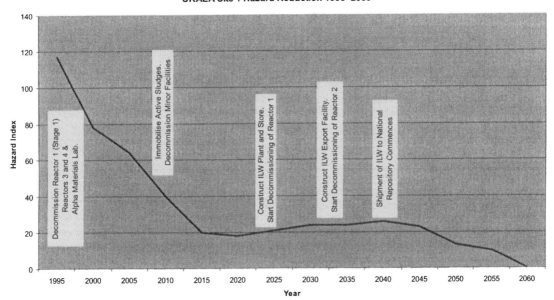

Figure 14-2. Typical Plot of Projected Hazard Index Against Time in Accordance with Proposed Site Integrated Decommissioning Program.

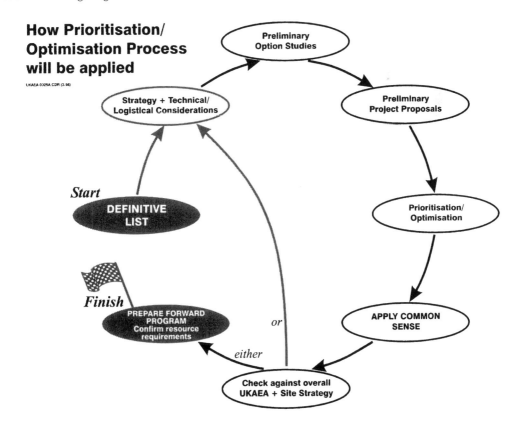

Figure 14-3. Application of the Prioritisation/Optimisation Process.

the integrated plan have commenced, then prioritisation becomes more an issue of conventional planning, including critical path analysis, since it may not be economic to abandon contracts once initiated.

14-4-2. *A Prioritisation Methodology*

A methodology, which allows work to be prioritised by allocating a score for each project, may usefully be developed and will include:

(a) Rigorous investigation of the validity of the project and its drivers.
(b) A multiattribute decision analysis for prioritisation.

Principal stakeholder/interest groups (see, for example, Table 14-5) should be identified and interviewed to determine the decision drivers and priorities which they apply when considering the priorities to be applied to activities. From this, the key decision drivers on which the project priorities should be assessed can be identified. These are likely to include:

- Safety and Security;
- Environmental Responsibility;
- Value for Money; and
- Public Acceptability.

The process is based upon a questionnaire and rigorous interview of program/project managers who have a detailed knowledge of how strongly the project is affected by each decommissioning issue. If possible, the interviews should be carried out by the same person (or by a few trained personnel). The scores of similar projects should be reviewed to ensure that any differences in scores reflect real differences in priority and are not due to inconsistency in scoring methods.

Table 14-5. Typical Stakeholder Set (Whose Opinions may be Valuable when Assessing a Decommissioning Program of Work)

(a) Nuclear Installations Inspectorate Environmental Protection Agency Scottish Environmental Protection Agency	Regulators
(b) Government	Policy makers and often fund providers
(c) Local Authorities	Including planning permissions
(d) Local Population	
(e) Workforce	
(f) Contractors	
(g) Environmental pressure groups	
(h) Press/Media	

14-4-3. *The Model*

The model can be run using database software, e.g., Microsoft Access, and should record the interview details and carry out the calculations necessary to produce a scored and prioritised list of projects. The key features of this model will include:

- The principal decision drivers.
- Subheadings and test questions (see Table 14-6) under each driver so as to assist in ascribing a score. The reasons why each score has been assigned are recorded in the database (see Figure 14-4).
- Projects scored in a consistent manner and a co-ordinator may be appointed to ensure that this happens.
- Weightings applied to the principal decision drivers to arrive at a total score for each project.
- The database calculates the weighted total for each project and generates a ranked list of projects.
- A sensitivity analysis will be carried out to investigate the dependence of the ranked list of changes in the choice of scores or weightings.
- The output of the prioritisation process is a list of projects, ranked in order of priority, with an auditable record of the process by which the scores and the ranked listing are produced.
- The scores will be regularly reviewed as the status of the project may change. The reasons for any amendments to the scores are recorded in the interview record.

14-5. Case Studies

14-5-1. *Hazard Reduction Over Time on Site X*

Colin, the Company Planning Director, looked across his desk at Roy and Jane, thinking how rosy cheeked and young they both looked. They were both brainy, had a "bit of go," and didn't mind hard work. Yes, he would give them a go at this and hope to hell they made a good job of it. The "men from the Ministry" were demanding a performance measure from his Company (a nuclear site licence holder) that would be used to gauge the accuracy of decommissioning planning and forecasting. Colin had sat with George, Roy and Jane's line manager, only the day before, and they had jointly turned up what Government Policy [4] actually said about decommissioning: "the systematic and progressive reduction of *hazards*." "Funny," George had said, "I always thought that we could justify our forward program as based upon Tolerability of Risk."

"If we did that," Colin said, "we would never do any decommissioning. The risk term for our facilities is already so low ($< 10^{-6}$) and, on risk terms alone, there would be no point in pulling down a reactor.

Table 14-6. Test Questions for Prioritisation Interviews

Driver	Sanctionable Project Category	Points	Operational Category	Points
Safety and Security	To respond to regulatory action	10	To respond to regulatory action	10
	To avoid anticipated regulatory action	7.5	To avoid anticipated regulatory action	7.5
	To remedy significant plant/facility deficiency	5	To remedy significant plant/facility deficiency	5
	To improve safety/security performance or reduce hazard	2.5	To improve safety/security performance or reduce hazard	2.5
	No significant safety/security effect	0	No significant safety/security effect	0
Environmental Factors	To respond to regulatory action	10	To respond to regulatory action	10
	To avoid anticipated regulatory action	7.5	To avoid anticipated regulatory action	7.5
	To remedy significant plant/facility deficiency	5	To remedy significant plant/facility deficiency	5
	To improve environmental performance or reduce hazard	2.5	To improve environmental performance or reduce hazard	2.5
	No significant environmental considerations	0	No significant environmental considerations	0
Value for Money: Savings/Cost Ratio	Ratio of 6% discounted additional costs is >1.250	8	Increase efficiency	10
	Where the ratio is likely to be between 1.125 and 1.250	6	Reduce costs	7.5
	Where the ratio is likely to be between 1 and 1.125	4	Maintain plants/facilities in sound functional condition	5
	No savings benefit	0	No VFM consideration	0
Potential Program Impact	Delay brings knock-on effect of > 5 times amount saved	8		
	Delay brings knock-on effect of > 2 < 5 times amount saved	6		
	Delay brings knock-on effect of < 2 times amount saved	4		
	No potential impact on program	0		
Impact on Contractors	Significant damage to contractorisation drive	4		
	Some damage to contractorisation drive	2		
	No effect on contractors	0		
	Divide total VFM points by 2			
Public Perception	To respond to adverse publicity	10	To respond to adverse publicity	10
	To avoid anticipated adverse publicity	7.5	To avoid anticipated adverse publicity	7.5
	To enhance company achievement	5	To enhance company achievement	5
	To publicise company achievement	2.5	To publicise company achievement	2.5
	No effect on public perception	0	No effect on public perception	0
	Total maximum points	**40**		**40**
Weighting Used to Test Spreadsheet				
Safety and Security	0.35			
Environmental Factors	0.30			
Value for Money	0.20			
Public Perception	0.15			

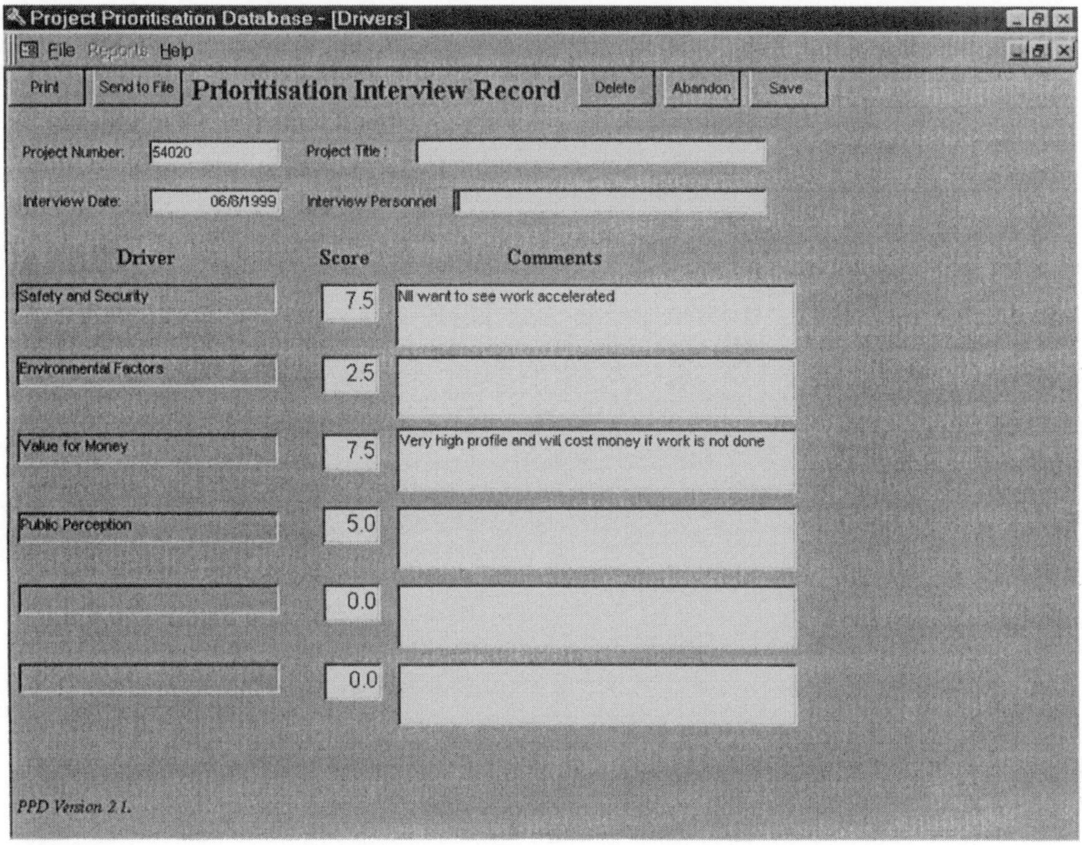

Figure 14-4. Prioritisation Interview Record.

The redundant reactor already has containment second to none and in this respect is a perfectly good waste store. Early decommissioning would expose workers to unnecessary dose uptake (however small). Putting all the reactor's activity into lots of smaller boxes and then into an expensive large purpose built (essentially replacement) store pending a final solution on waste disposal seems a nonsense. After POCO — and all things being equal on safety-related aspects — it is only if the costs of earlier decommissioning give a better value for money case than the costs of ongoing periods of care and maintenance would early decommissioning be worthwhile."

"Well, first of all what are the final solutions?" said George. "We no longer have the likelihood of an ILW disposal facility until 2040 (a planning assumption following the demise of the 1997 Nirex Rock Characterisation Laboratory near Sellafield, UK), and an HLW facility is even further away. Taking this into account, some form of interim storage is inevitable, given the push to get on with decommissioning based on hazard reduction by the Regulators."

"OK then George, lets look at the facilities that we have, what we have got readily available so as to categorise them, and then match the demolition of these facilities against the forward decommissioning program thereby generating a hazard reduction profile. If its OK with you, I'll get Roy and Jane involved with this one, if you come up with the base data for them to work with." George went away and produced a listing of the site facilities and a methodology for categorising them.

Given the following data for Site X, which gives 20 main facilities of different categorisations and their proposed Stages 1, 2 and 3 decommissioning dates, together with the guidance document, Roy and Jane were set the task of generating a hazard index over time profile for the site.

(1) Generate your own profile on graph paper.
(2) How would you/could you use such a profile to generate a Performance Measure for the Company's decommissioning effectiveness?
(3) Is there any benefit from subdividing the categorisation within a given facility?

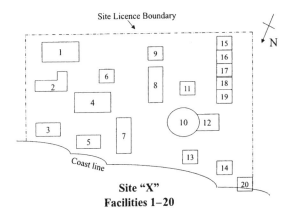

Site "X"
Facilities 1–20

Figure 14-5. Case Study—Plan of Site X Showing Facilities 1–20 (as defined in Table 14-7).

A site plan and listing of the main Site X facilities are shown in Figure 14-5 and Table 14-7 respectively.

Site X Case Study — Suggested Solution

The fist step is to group the facilities:

- *Facilities and plants to be decommissioned*
  10 Shut down reactor
  12 Heat exchangers and electrical generating equipment
  13 Sea-water pump house
  16 Fuel fabrication plant (operational)
  17 Uranium recovery plant
  18 Fuel reprocessing plant
  19 PIE facility

- *"Enabling" facilities (all operational)*
  1 Analytical laboratories
  3 Beta gamma active workshop
  5 Equipment test workshop (beta gamma active)
  6 Active laundry facility
  8 Laundry and safety equipment store
  9 Office accommodation
  15 Change room complex/barrier change, etc.
- *Waste treatment plants and stores (all operational)*
  2 LLW waste store for packaged beta gamma waste
  4 ILW waste alpha beta gamma store for packaged waste
  7 High active liquor storage from reprocessing
  11 Fuel product store
  14 Low level liquid effluent treatment plant
  20 Flask compound

The second step is to understand the plant categorisations (Cat 1, 2, 3 & 0) and the sequence of the decommissioning events. Use Table 14-4 to categorise the facilities. The proposed categorisations and the anticipated decommissioning sequence is as suggested below:

10 *Shut down reactor* — Cat 1 for 3 years until all fuel is removed and then Cat 2 until stage 3 decommissioning is complete in ~40 years time.
12 *Heat exchangers and electrical alternator* — Cat 3 plant. Strip out in next 5 years then 0.
13 *SW pump house* — Cat 3 plant. Demolish in next 5 years then 0.
3 *β γ active workshop* — Cat 2 plant. Needed to support decommissioning for next 40 years then 0.
5 *Equipment test workshop* — Cat 2 plant. Needed to support decommissioning for next 40 years then 0.

Table 14-7. Case Study—Table of the Site X Main Facilities

No.	Site Facility (brief description)	Categorisation	No.	Site Facility (brief description)	Categorisation
1	Suite of analytical laboratories (active)		11	Fuel product store	
2	LLW waste store for packaged $\beta\gamma$ wastes		12	Heat exchangers and electrical generating equipment	
3	Beta Gamma active workshop		13	Sea-water pump house (under C&M)	
4	ILW waste $\alpha\beta\gamma$ store for packaged wastes		14	Low level liquid effluent treatment plant (operational)	
5	Equipment test $\beta\gamma$ workshop (active)		15	Change room complex/barrier controls, etc. (operational)	
6	Active laundry facility		16	Fuel fabrication plant (operational)	
7	High active liquor (HALs) storage from reprocessing		17	Uranium recovery plant	
8	Laundry and safety equipment store		18	Fuel reprocessing plant	
9	Office accommodation		19	PIE facility	
10	Shut down reactor		20	Flask compound	

Simple example — no new enabling plants, no new stores to be built, operated and decommissioned during the forward decommissioning program.
PIE — Post Irradiation Examination; C&M — Care and Maintenance.

16 *Fuel fabrication plant* — Cat 2 plant. Currently operational. Decommission over next 3 years to Cat 3 and then wait for 3 years, then to 0.

17 *Uranium recovery plant* — Cat 1 plant. Needed to support 16 and 18 for next 8 years then to Cat 2 and 3 quickly as POCO and decommissioning is carried out. All done by year 12.

18 *Fuel reprocessing plant* — Cat 1 plant for next 7 years until all spent fuel is reprocessed. Then Cat 2 for 5 years until POCO is carried out. Then Cat 3 for next 10 years and then to Cat 0 as decommissioned and demolished.

19 *PIE facility* — Cat 2 plant. Will remain Cat 2 for 3–5 years then goes quickly to Cat 3 as fuels moved to HLW store. Goes to Cat 0 after further 5–6 years as decommissioned and demolished.

15 *Change room complex* — needed for next 40–50 years to support facilities and waste plants. Cat 3 plant during this period.

6 *Active Laundry facility* — needed for next 40–50 years to support facilities and waste plants. Cat 3 plant during this period.

1 *Analytical labs* — needed (in part) for whole decommissioning period for sampling of liquors/plant environmental samples, etc. Currently, Cat 2, and remains at this level for next ~40 years then goes to Cat 3 and holds this until year 50.

8 *Laundry and safety equipment store* — Cat 3 and will remain until year 48.

9 *Office accommodation* — Cat 0 and required for full 50 year decommissioning period.

14 *Low Level Liquid Effluent Treatment Plant (LLLETP)* — Cat 3 and will be used for full 50 years.

2 *LLW store* — Cat 3 and will remain in use for at least 50 years.

4 *ILW store* — Cat 2 and will remain in use for at least 50 years.

7 *High Active Liquors (HALs) tanks* — Cat 1. Becomes Cat 2 as HALs is immobilised and packaged and remains as Cat 2 over 50 year period.

11 *Fuel product store* — Cat 1 until fuel is sent off-site (say, year 20). Then decommissioned and goes to 0 in 3 years.

20 *Flask compound* — Cat 3 and will be for 40 years as flasks necessary to support waste plant operations.

The third step is to plot the decommissioning activities over time against overall reducing hazard on graph paper.

14-5-2. *"My Project Is More Important Than Yours" : A Case for Project Prioritisation*

Colin, the planning Director of a leading nuclear site licence holder, was getting increasingly frustrated at the Quarterly Progress Meeting, held on Site X, of which he was the Chairman. "Why can't you just all stop bickering? Surely you must appreciate that putting new footpath kerb stones around the local reprocessing plant cannot, by any stretch of the imagination, be more important than getting on with the highly active coolant removal from the 'Pliny' fast reactor?" Silence fell. Then, the General Maintenance Manager ventured to remind the Chairman that the "men from the Ministry," including the Minister himself, were coming to site next month and he wanted it all "ship shape" for the visit. Further, that the contract for the works was part of an existing "draw down" maintenance contract and without the work he would seriously underspend his budget. George, the "Pliny" Decommissioning Project Manager, said, "Look Colin, I've got the Regulator on my back and they are threatening an 'Improvement Notice' unless I get on with these works. I've got deadlines to meet and . . . ."

"Hold it," said Colin, "lets get this analysis onto a proper footing, which at least stands up to a degree of auditable and logical scrutiny. What are the priorities here? Bring me your completed project prioritisation proformas and overall scores and I'll settle the question this afternoon. Now what is the next item on the Agenda?"

The data needed to assess the case is given below. Judge for yourselves how the proforma interviews would have scored the works and fill in the blank forms.

(1) Why is it important for a trained and, if possible, common interviewer to be present when the strengths of these cases are discussed?

(2) Where would these two projects typically fit into the project listing attached?

(3) If both these projects had already commenced, would this alter your views on what should take precedence and why?

(4) Within the context of an integrated decommissioning program, would other issues, such as an analysis of critical path activities, also require consideration?

Table 14-8 shows a Site X Project Priority Listing, and Table 14-9 shows data for assessing projects — weightings and point scores to be used during project prioritisation interviews.

14-6. References

1. HSE. *Reducing Risks, Protecting People*, HSE Publications, HM Stationery Office, London, UK (2002).

2. NRPB. *Radiation Doses, Maps and Magnitudes*, NRPB Publication, Chilton, Oxfordshire, UK.

3. Stationery Office. *Work with Ionising Radiation, Approved Code of Practice and Guidance to the Ionising Radiation Regulations 1999 (IRR 99)*, HM Stationery Office, London, UK (1999).

4. Stationery Office. *Review of Radioactive Waste Management Policy*, Cm 2919, HM Stationery Office, London, UK (1995).

Table 14-8. Project Priority Listing

No.	Scores	Project no.	Project title	Site	Annual spend £k	Cumulative spend £k
1	9.2	54,653	Replace/upgrade/maintenance of existing equipment (>£50 k<£250 k)	Dounreay	2.534	2,534
2	9.2	55,971	Decomm/C&M of pulse column rig glovebox in D2670	Dounreay	756	3,290
3	8.6	54,401	Oxide Treatment Facility	Dounreay	0	3,290
4	8.4	59,726	Electrical Distribution Reinforcement	Dounreay	350	3,640
5	8.1	54,410	Fuel Stabilisation	Dounreay	250	3,890
6	8.0	55,615	Modifications to the Liquid Effluent Pit Complex (D1211)	Dounreay	0	3,890
7	8.0	54,175	New Sludge Treatment Plant & Mods to Ultrafilter Plant (moved out to 03/04)	Dounreay	0	3,890
8	7.8	55,620	FCA Ventilation Improvements	Dounreay	1,225	5,115
9	7.8	55,614	Dounreay Particle Investigation	Dounreay	2,020	7,135
10	7.8	54,321	Hydrogeological Investigations	Dounreay	100	7,235
11	7.8	55,624	D1209 Vent Duct Early Clean Out	Dounreay	500	7,735
12	7.7	55,070	DFR Breeder Fuel Removal & Disposal	Dounreay	3,400	11,135
13	7.7	54,331	SILW Site Clearance	Dounreay	892	12,027
14	7.6	55,055	Plant Upgrades	Dounreay	1,400	13,427
15	7.5	59,785	Site Safety Support & Licence Compliance	Dounreay	6,639	20,066
16	7.5	59,780	Dounreay Senior Management	Dounreay	809	20,875
17	7.5	59,790	Radiation Protection Services	Dounreay	6,033	26,908
18	7.3	54,324	D1225 ILW Shaft Retrieval — Shaft Isolation	Dounreay	702	27,610
19	7.3	59,722	Dounreay Joint Control Centre	Dounreay	700	28,310
20	7.2	54,613	Minor Stack Gaseous Discharge	Dounreay	590	28,900
21	7.2	55,800	Decommissioning/C&M of sodium rigs (now in decommissioning)	Dounreay	0	28,900
22	7.2	55,025	Bulk NaK Removal	Dounreay	2,700	31,600
23	7.1	59,781	Dounreay Project Management	Dounreay	4,080	35,680
24	7.1	54,032	DCP Store Import/Export Facility	Dounreay	500	36,180
25	7.0	54,790	Liquid ILW & HLW Management	Dounreay	948	37,128
26	7.0	54,603	D1200/D1215/D1310 Laboratories Facility Availability	Dounreay	3,741	40,869
27	7.0	59,795	Site Engineering Maintenance & Design Resources	Dounreay	15,361	56,230
28	7.0	54,750	Solid LLW Management	Dounreay	798	57,028
29	7.0	54,780	Liquid LLW & Gaseous Wastes	Dounreay	775	57,803
30	7.0	54,900	Solid LLW Management	Dounreay	1,542	59,345
31	7.0	54,999	WMG Facilities, Services & General Management	Dounreay	1,337	60,682
32	7.0	54,759	Provision of Decontamination Facility (D2900)	Dounreay	151	60,833
33	7.0	55,700	WMG Redundant Facilities C&M/POCO	Dounreay	0	60,833
34	7.0	55,710	Redundant Facilities C&M/Minor Decommissioning	Dounreay	982	61,815
35	7.0	55,950	Decomm/C&M of Nuclear Laboratories Complex D1200	Dounreay	589	62,404
36	7.0	54,333	Dounreay RAM Transport — Operations	Dounreay	386	62,790
37	7.0	54,330	Waste Treatment Plant	Dounreay	1,086	63,876
38	7.0	55,958	POCO & Minor Operations in D2001 (Exc WPC)	Dounreay	280	64,156
39	6.8	54,323	Shaft Waste Retrieval	Dounreay	1,035	65,191
40	6.6	55,803	C&M/Decommissioning of Supernoah	Dounreay	200	65,391
41	6.6	55,953	Decommissioning of Lab 33 in Nuclear Laboratories Complex D1200	Dounreay	881	66,272
42	6.6	59,750	Dounreay Site Infrastructure — Operations	Dounreay	9,101	75,373
43	6.5	55,622	D9867 Ventilation Upgrade	Dounreay	400	75,773

Table 14-9. Data for Assessing Projects : Drivers and Points Used to Test Spreadsheet

Driver		Sanctionable project category	Points
Safety and Security		To respond to regulatory action	10
		To avoid anticipated regulatory action	7.5
		To remedy significant plant/facility deficiency	5
		To improve safety/security performance or reduce hazard	2.5
		No significant safety/security effect	0
Environmental Factors		To respond to regulatory action	10
		To avoid anticipated regulatory action	7.5
		To remedy significant plant/facility deficiency	5
		To improve environmental performance or reduce hazard	2.5
		No significant environmental considerations	0
Value for Money	Savings/Cost ratio	Delay would lead to likely cost increase of 25% of amount saved by delay	8
		Delay would lead to likely cost increase of between 12.5%–25% of amount saved by delay	6
		Delay would lead to likely cost increase of up to 12.5% of amount saved by delay	4
		No savings benefit	0
	Potential Program Impact	Delay brings knock-on effect of > 5 times amount saved	8
		Delay brings knock-on effect of > 2 < 5 times amount saved	6
		Delay brings knock-on effect of < 2 times amount saved	4
		No potential impact on program	0
	Impact on Contractors	Delay likely to have major effect on contractors	4
		Delay likely to have some effect on contractors	2
		No effect on our contractors	0
		Divide total VFM points by 2	
Public Perception		To respond to adverse publicity	10
		To avoid anticipated adverse publicity	7.5
		To enhance company achievement	5
		To publicise company achievement	2.5
		No effect on public perception	0
		Total maximum points	40

Weightings Used: Safety and Security 0.35, Environmental Factors 0.30, Value for Money 0.20, Public Perception 0.15.

Chapter 15
Decommissioning Cost Estimating

15-1. Introduction

It is necessary to be able to gauge the costs of the decommissioning task from the earliest stages of the project (project definition, planning, and project initiation phases of the project lifecycle) through the implementation, commissioning, and operations works. Generally, as work proceeds, the project out-turn costs become more definitive, as indicated in Figure 15-1. Time spent planning and defining the works at the outset reduces the risk of exposure and provides a greater surety of the likely final project cost. Only by having cost estimates of suitable accuracy can those responsible for investing money in the works be satisfied that value for money is being achieved. This chapter describes conventional and parametric cost estimating processes as applicable to decommissioning and waste management works.

15-2. Conventional Cost Estimating

Contracts based upon bills of quantities or schedules of rates are commonly encountered in conventional civil engineering and building services works. Such works have often been defined by the Client or Engineer who has drawn up a detailed design, broken it down into its different tasks, materials and items of work required, and produced a schedule which potential contractors price during the tender stage. The production of bills of quantities, therefore, help fulfill the following purposes:

- To assist in defining the works in detail;
- To enable the tenderer to price an enquiry rapidly and accurately;
- To help analyse the works so that no items or procedures are left unpriced or sources of cost or expense to the Client and Contractor are omitted;
- To help group parts of the works into separate entities as required to suit the Client's or Engineer's cost control systems;
- To establish for each item a unit rate of charge which can be used for calculating both the estimated and actual

cost of the works performed. In addition, to assist in the calculation of variations to the originally anticipated works;
- To help identify and isolate all "once-off" or overhead charges (site offices, management supervision, insurances, etc.), so that they may be built into the overall cost of the works;
- To make financial provision for nominated subcontracts;
- To make financial provision for additions or variations to the works which have not yet been fully defined, but for which acceptable rates may be derived from main bill of quantities items (spares holdings, additional drawings, manuals, etc.); and
- To assist in the tender analysis process. (However, bear in mind that the tender prices themselves as entered into bills of quantities say nothing about the methods, skills, organisation, and financial strength of the contracting company.)

Bills of quantities (B of Qs) may, therefore, be considered as a "conventional" or "bottom up" approach to cost estimating. Their use requires a detailed understanding of the totality of the works to be performed if the summation of the items in the B of Qs is to reflect the final out-turn cost of the works. Typical B of Qs may look like those shown in Table 15-1, where time and materials costs are derived from:

$$\text{Cost} = \sum \text{unit price} \times \text{quantity}$$

In cases where a separation of labor costs is required then the derivation has to also take into account both the applicable time to do the work and the associated productivity.

$$\text{Labour cost} = \sum \text{labour rate} \times \text{unit hours} \times \text{number of operations}$$

Such estimating and contracting terms require good definition without which numerous disputes between Client and Contractor can arise. In the cable installation example shown in Table 15-1, there is no mention of how the cable is to be routed as it nears the equipment

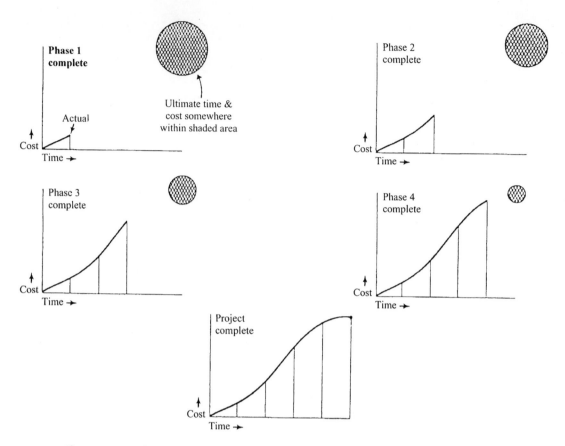

Figure 15-1. Relative Uncertainty of Out-turn Time and Cost During the Project Lifecycle.

onto which it is to be terminated at each end of the cable run. Is the contractor expected to route the cable through ducts into and out of a building? Similarly is the cable to be mounted on cable tray within the buildings where the equipment is housed? If these works are to be included so as to complete the job, then what rates are applicable for running cable through ducts? Who supplies and installs the cable tray? What are the quantities to be installed in this manner and what rates apply for this work?

If such works going on in a nuclear decommissioning environment are now considered, then it is necessary to also take into account any decontamination required before new cable can be routed through the building. Further, one would need to know about the radiological environment within which the cable was to be installed, so the complexity of the work and the need for any special protective clothing could be judged. It is unlikely that this level of detail will be available at the early project stages. The necessary definition and derisking required to get to this level of detail may well be of a magnitude to represent a project in itself. Hence, the uncertainties at the initial stages in the project lifecycle, as indicated in Figure 15-1.

For the estimation of the total financial liability for all the decommissioning and waste management activities, costs must be estimated for tasks which may well have to be undertaken several decades ahead, some on a 10 year timescale, and others which are already underway or shortly to be implemented.

For those projects which are imminent, detailed estimates are required of the likely costs of planning and implementation before invitations to tender for the work can be issued and contracts let. If definition is good and the work of a conventional nature, then estimates based upon Bills of Quantities may well be applicable. For the estimation of those projects which make up a 10 year forward program then a reliable, but not so detailed approach is required.

Four principal steps are involved in the production of any detailed cost estimate, as described below:

Step 1 — Quality First such that estimates

- are prepared by suitably qualified and experienced personnel;
- hold good provenance;
- are consistent;

Table 15-1. Typical Bill of Quantities (B of Q) for the installation of a cable in a trench

Item	Description	Unit	Quantity	Unit rate	Total
1	Excavate 600 mm wide × 1100 mm deep cable trench in ground including reinstatement	Linear meter	255 lin. m	£10 / lin. m	£2,550
2	Excavate 600 mm wide × 1100 mm deep cable trench through asphalt including reinstatement	Linear meter	1680 lin. m	£15 / lin. m	£25,200
3	Supply and install cable tiles	No.	6,450	£0.80 / tile	£5,160
4	Supply and install sand surround to cables	cu. m	350 cu. m	£3 / cu. m	£1,050
5	Supply and install 4c 185 mm^2 XLPE SWA XLPE copper conductor 600/1000 V cable in trench	Linear meter	1,950 lin. m	£19 / lin. m	£37,050
6	Straight joint 4c 185 mm^2 XLPE SWA XLPE copper conductor 600/1000 V cable	No.	3	£256 / joint	£768
7	Terminate 4c 185 mm^2 XLPE SWA XLPE copper conductor 600/1000 V cable onto equipment provided by others. Rate to include all termination equipment and materials	No.	2	£328 / per termination	£656
8	Pressure test, resistance test, and continuity (phasing) test cable installation	Unit	1	£2,650 / per full testing to engineer's specifications	£2,650
				Total	£75,084

- are produced in a rigorous well managed way; and
- have an audit trail to show that the estimating process adopts best practice and includes appropriate supervisory, checking, and approval processes.

Step 2 — Estimate Scope and Assumptions

- purpose;
- level of detail to be adopted;
- scope (especially the end point conditions); and
- project, site, and corporate assumptions.

Step 3 — Information Gathering

- sources of information;
- dedicated documentation;
- drawings and facility plant data;
- safety cases, Health Protection surveys, and Incident Reports;
- interviews with current or former employees; and
- facility visit(s).

Step 4 — Estimate Production

- selection of items from the coded database using standardised cost listings;
- application of project-specific factors (task complexity, radiological conditions, etc.);
- use of derived specific "Norm" values associated with database costs and project specific factors when appropriate; and
- consideration of "Areas" or "Stages" of the works.

15-3. Standardised Cost Listings

For the estimation of costs for longer-term projects, sound judgments based upon experience of past and current projects rather than expensive analysis based on only very outline proposals gives a more fit for purpose solution. In particular, it is useful to have a standardised list of cost items and cost item definitions for decommissioning projects. This allows a database to be built up and rational comparisons made between estimates or actual out-turn costs for the different items of decommissioning work. The NEA/OECD [1] has issued such a standardised listing, which is summarised in Table 15-2. Item 4 in Table 15-2 has specifically brought out the need to consider the costs involved preparing nuclear facilities for long periods of care and maintenance.

In addition to the cost elements listed in Table 15-2, attention must also be made to the peripheral areas, which can add enormously to the overall decommissioning budget requirements, as listed in Table 15-3.

The quality or level of confidence that can be expected from a cost estimate is a function of the quantity and quality of the project-specific data and the accuracy of the decommissioning cost databases used by contractor or client firms. Estimates are produced at various stages in the lifecycle of a decommissioning project. These may be related to a "Level" approach adopted for their study, planning and implementation with

Table 15-2. Cost Elements for Decommissioning Projects

(1) Predecommissioning	(1.1) Decommissioning Planning
	(1.2) Authorisations
	(1.3) Radiological Surveys
	(1.4) Hazardous Materials Surveys
	(1.5) Contractor Selection
(2) Facility Shutdown Activities (POCO)	(2.1) Plant Shutdown Inspection
	(2.2) Removal of Fuel and/or Nuclear Materials
	(2.3) Drainage of all Redundant Systems
	(2.4) Sampling for Radiological Characterisation
	(2.5) Removal of Systems Fluids
	(2.6) Decontamination for Dose Reduction
	(2.7) Removal of Wastes
	(2.8) Removal of Combustible Materials
	(2.9) Removal of Spent Resins
	(2.10) Isolation of Power
	(2.11) Asset Recovery (sale or transfer)
(3) Procurement of Equipment and Material	(3.1) Dismantling Equipment
	(3.2) Equipment for Decontamination
	(3.3) Radiation Protection and Health Physics Equipment
	(3.4) Security and Maintenance Equipment for Long-Term Care and Maintenance
(4) Preparation for Long-Term Care and Maintenance (if required)	(4.1) Sampling and Radiological Characterisation
	(4.2) Dismantling and Transfer of Contaminated Equipment and Material
	(4.3) Reconfiguration of Site Infrastructure
	(4.4) Facility Hardening and Isolation
(5) Dismantling Activities	(5.1) Decontamination of Areas and Equipment
	(5.2) Drainage and Decontamination of Spent Fuel Storage Pond
	(5.3) Radiological Characterisation for Dismantling
	(5.4) Preparation of Temporary Waste Storage Area
	(5.5) Design, Procurement, and Testing of Special Tools and Equipment
	(5.6) Dismantling Operations on Reactor Pressure Vessels and/or other Major Items
	(5.7) Removal of Shielding Structures
	(5.8) Removal of Containment Structures
	(5.9) Removal and Disposal of Asbestos
	(5.10) Building Decontamination
	(5.11) Fuel Radiological Survey
	(5.12) Decontamination of Materials for Recycling
	(5.13) Asset Recovery
	(5.14) Personnel Training
(6) Waste Processing, Storage, and Disposal	(6.1) Safety Documentation and Procedures
	(6.2) Feasibility Studies
	(6.3) Permits for Storage, Transport, and Disposal
	(6.4) Processing of Liquid Wastes
	(6.5) Disposal of Operational Wastes
	(6.6) Packaging and Storage of Decommissioning Wastes
	(6.7) Disposal and Transport of Radioactive Wastes for Disposal
	(6.8) Disposal of Nonradioactive Wastes
(7) Site Restoration	(7.1) Demolition or Restoration of Buildings
	(7.2) Decontamination and/or removal of Below Ground Structures
	(7.3) Final (Independent) Survey and Delicensing

Table 15-3. Additional Cost Elements (Not to be Forgotten When Building Up a Total Decommissioning Cost Estimate)

(8) Site Security	(8.1) Site Security Operations and Surveillance
	(8.2) Inspection and Maintenance of Buildings and Systems in Operation
	(8.3) Site Upkeep
	(8.4) Energy and Water
	(8.5) Periodic Radiation and Environmental Surveys/Monitoring
(9) Project Management, Engineering, and Site Support	(9.1) Mobilisation and Preparatory Work
	(9.2) Project Management and Engineering Services
	(9.3) Public Relations
	(9.4) Support Services
	(9.5) Health and Safety
	(9.6) Demobilisation
(10) Research and Development	(10.1) Research and Development of Decontamination, Radiation Measurement, and Dismantling Processes, Tools and Equipment
	(10.2) Simulation of Complicated Work, Models, and Mock-Ups
(11) Fuel and Nuclear Materials	(11.1) Transfer of Fuel or Nuclear Materials from Facility or from Temporary Storage to Intermediate Storage
	(11.2) Intermediate Storage
	(11.3) Dismantling/Disposal of Temporary Storage Facility
	(11.4) Preparation of Transfer of Fuel or Nuclear Material/Waste from Intermediate Storage to
	(11.5) Final Disposal (if such a route exists)
	(11.6) Dismantling/Disposal of Intermediate Storage Facility
(12) Other Costs	(12.1) Owner Costs
	(12.2) General, Overall (not Specific) Consulting Costs
	(12.3) General, Overall (not Specific) Regulatory Fees, Inspections, Certifications, Reviews, etc.
	(12.4) Taxes
	(12.5) Insurances
	(12.6) Overheads and General Administration
	(12.7) Contingency
	(12.8) Interest on Borrowed Money
	(12.9) Asset Recovery

Level 1: Project Definition and Preliminary Planning.
Level 2: Project Initiation and Planning.
Level 3: Prepare and Approve the Business Case.
Level 4: Project Implementation.

Estimates will increase in accuracy as the degree of uncertainty is reduced. The use of such an approach is illustrated in Table 15-4.

15-4. Parametric Cost Estimating

Rather than itemise all aspects of the works in considerable detail, a "parametric" or "top down" estimate uses validated relationships between cost, schedule, and measurable attributes of systems, hardware, and software. Such techniques are widely used in industry, both in the UK and US, especially in the defense sector. In a parametric cost estimate, the cost of an item is usually related to some easily determined parameter, e.g., length of pipework, drain or ducting, tonnage of structural steel or concrete, etc. Relationships are determined for different degrees of radioactive contamination which, in turn, reflect the different decommissioning productivity levels likely to be achieved. For example, working in a full pressurised suit is likely to be less productive than using lower levels of personal protective equipment.

A PaRametrIc Cost Estimating system (PRICE) has been developed by UKAEA for decommissioning works using a computer database of past decommissioning and waste management experience. The PRICE system requires that a task or project is first described in terms of an hierarchical or work breakdown structure (WBS). At the lowest tier in the hierarchy are the "Components," and it is at this easily determined parameter tier that costs are attributed. The database typically holds some 40 such standard components for selection by the user.

Table 15-4. The Level Approach to Decommissioning Cost Estimating

Input		Estimate Status		Purpose
	Level 1			
Basic data:				Project Definition.
		Preliminary Estimate		
Scope & Strategy	$\longrightarrow$	Initial study	$\longrightarrow$	For preliminary assessment, prepared with very little data, other than the size and capacity of plant.
Facility Data				
	Level 2			
Improved data:		Intermediate Estimate		Project Initiation.
		Detailed study and		
Scope and Strategy	$\longrightarrow$	optioneering	$\longrightarrow$	More detailed, with better definition of major work packages but still broad brush often using parametric data based on past experience.
Facility Data				
	Level 3			
Detailed data:	$\longrightarrow$	Sanction Estimate	$\longrightarrow$	Business Case and initial sanction approval. Increasing degree of engineering input: preliminary design data and drawings. Scope and Strategy likely to be available.
Facility Data				
	Level 4			
Detailed data:	$\longrightarrow$	Implementation control and possible fixed price tendering	$\longrightarrow$	Detailed specifications, drawings, and bills of quantities may be produced and prices sought from suppliers and contractors.
Facility Data				
Tender Information				
Implementation				

Having selected a component, it is necessary to apply two factors that impact on cost: these are "Task Type" and "Complexity." There is a choice of Task Types, covering most Components, which indicate the level of radiological protection that will be required when undertaking the task:

M — minimum protection (see Figure 9-3).
C — complex contact handleable (see Figure 7-5).
R — remote (see Figure 7-3).

For all Components, there are five levels of Complexity to choose from, so as to locate the appropriate "Norm" values of the costs involved. This is generally a function of the physical size of the item being dealt with. Components may also be described as "User Defined" items. In such cases, the system knowledge base is not used and the estimator is able to input a lump sum allowance for that Component.

The performance of such an estimating system needs to be regularly benchmarked and checked for accuracy. At the highest level, a comparison of the estimate with the tender price and the project out-turn costs is the final arbiter for the system. This is now sufficiently accurate to be adopted as standard practice within UKAEA. Such comparisons require close attention to detail and, in particular, the strict application of change controls so as to gain an understanding of any work scope growth during the project.

At a more detailed level, such a cost estimating system is benchmarked against a set of metrics so as to provide a "health check" against which the estimate may be assessed. These metrics include:

• Component/Value Profile;
• Task Type Allocation (value and percentage);
• Work Package Group Usage;
• Decommissioning Stage Breakdown; and
• Knowledge Base vs. User Defined Cost Usage.

The system is such that, at any stage in the estimate, an analysis may be automatically generated to give instant feedback to the estimator on the above metrics.

The PRICE system in itself does not give a formal consideration of the impacts of risk. As a matter of process risk identification and some estimation of risk allowance is, however, considered for Levels 1 and 2, included in some Level 2 estimates, and included in *all* Level 3 and 4 cost predictions at typically the 50 and 90 percentile confidence levels. This will often comprise of a formal qualitative and quantitative risk assessment exercise and may well entail the use of additional software packages such as @Risk^{TM} or Predicte!^{TM}.

Figure 15-2 gives examples of the use of the parametric cost estimating model by comparing the PRICE estimates for different facilities with the tendered prices received

| Facility | Site | Cost Estimate (Decommissioning) | | Tender | Out-turn | Deviation | Deviation |
		PRICE	Level 1			PRICE - Out-turn	Level 1 - Out-turn
B47	Harwell	£415k	£400k	£153k	£290k	30%	28%
B336.17	Harwell	£329k	£297k	£150k	£280k	15%	6%
A52	Winfrith	£698k	£2,038k	£600k	£929k	-33%	54%
A59	Winfrith	£24,847k		£26,307k			
					Mean Deviation :-	3%	29%

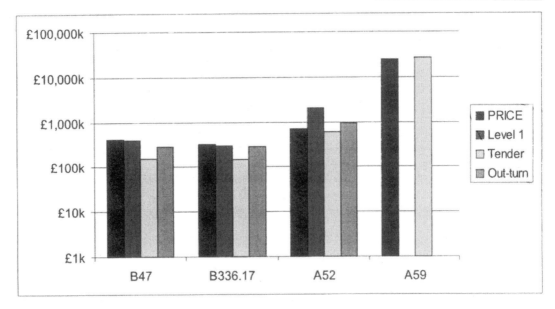

Figure 15-2. Comparison of Decommissioning Parametric Cost Estimates with Contractor Bid Prices and Final Contract Out-turn Costs.

through competitive tender for the works together with the out-turn costs for the completed decommissioning works. As more data is accumulated, the database will become an even more powerful tool. With such good estimating, there is always a dilemma for client organisations at the time of assessing bids from contractors for the works. Should the client inform the lowest and seemingly capable contractor during tender assessment that his bid is very low and perhaps he has forgotten to add in all necessary contingencies? Perhaps the contractor has an innovative solution for doing the work that the client has not considered, and this is what makes his tender so low in price?

If the contract is placed with a contractor who has a very low bid, it is highly likely to lead to problems later in the contract, as the contractor finds that he is not making sufficient profit or, indeed, is making a loss.

15-5. Reference

1. NEA/OECD. *A proposed standardised list of items for costing purposes in the decommissioning of nuclear installations*, NEA/OECD Publications, Z rue Andre Pascal, 75775 Paris Cedex 16, France (1999).

Chapter 16
Waste Management — Introduction and Overview

16-1. Requirements to Manage Radioactive Wastes

The systematic and progressive reduction of hazards as applied to the management of radioactive wastes involves treatment or conditioning of the wastes into passively safe forms, interim storage and, where waste routes exist, disposal. There are four fundamental requirements:

- Production of radioactive wastes should, where possible, be avoided. Where radioactive waste arisings are unavoidable, then the production should be minimised.
- Radioactive material and radioactive waste should be managed safely throughout its life cycle (from arising to recognised end-point) in a manner consistent with modern standards.
- Full use should be made of existing routes for the disposal of radioactive wastes, taking all factors (including social and economic factors) into account.

- Remaining radioactive material and radioactive waste should be put into a passively safe state for interim storage pending future disposal or other long-term solution.

Once the radioactive waste has been generated, its activity will not "go away" until the natural benefits of radioactive decay have taken their course. Packaging waste up in containers does not reduce the radioactivity. Any thoughts of radioactivity reduction by dilution of the activity per unit volume goes against the principles of waste minimisation. It would also be an expensive option to adopt, since nuclear waste storage and disposal costs are both activity and volume dependent. "Concentrate and contain" is preferred to "dilute and disperse" in accordance with IAEA RADWASS Fundamentals. Therefore, the driver for effective waste management, as seen in Chapter 14 for the decommissioning process in general, is more to do with hazard reduction. Figure 16-1

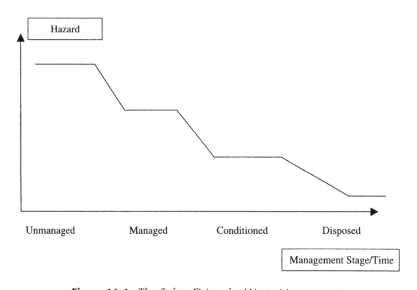

Figure 16-1. The Safety Driver for Waste Management.

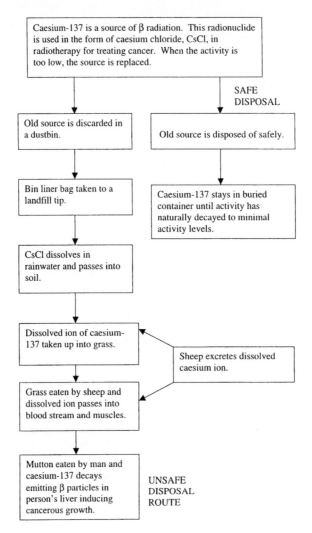

Figure 16-2. Examples of Safe and Unsafe Disposal of Radioactive Waste.

illustrates the theoretical safety benefits from hazard reduction brought about by effective waste treatment.

Most industries produce waste. The waste has to be disposed of safely to prevent the spread of contamination. The flow diagram in Figure 16-2 compares and contrasts the consequences of safe and unsafe disposal of caesium-137 contaminated waste.

16-2. Characterisation and Segregation

If they are to be managed safely, then all radioactive materials must first be characterised in terms of their physical and chemical form, radioactive content, origin, present state, current storage conditions, and other relevant information and properties. Segregation involves accumulating together those materials with similar characteristics, and avoiding mixing those with different characteristics. Segregation is most efficient if it is taken into account at the design stage, and it should be done as close to the point of waste generation as is reasonably possible. Experience has shown that this early characterisation and appropriate segregation can significantly contribute to the safe and economic management of radioactive materials. It is also technically difficult to remotely characterise (assay) bulk packaged wastes. Failure to take such steps can restrict disposal options and lead to significant technical difficulties and huge increases in waste management costs.

The legacies of the past include inappropriate bulk storage of uncharacterised and unsegregated wastes. Corrosion and general chemical degradation has turned many such wastes into difficult to handle, condition, and package sludges. An example is Magnox fuel hulls, which have corroded during storage under water leading to hydrogen emissions, corrosion to sludges, and expensive recovery operations. An example of the need for meticulous record keeping is illustrated at Dounreay in the UK, where a previously authorised ILW disposal shaft is now having to be emptied many years after its operational phase, so as to conform with more modern regulatory requirements. Uncertainty in the exact inventory necessitates caution, long program timescales, and increased expense. Sea disposal drums are an example of a case where changes in policy have led to the need for expensive repackaging.

Since wastes may be stored for a considerable period prior to eventual possible disposal, it is very important that the characterisation and storage records are capable of being maintained in secure facilities over similar lengthy timescales.

16-3. Passive Safety

Where medium-to-long-term interim storage of wastes is required pending an eventual disposal route, then there is a significant benefit from placing the radioactive materials and wastes into a passively safe form under passively safe conditions. Passive safety requires the radioactivity to be immobilised and packaged in a form that is physically and chemically stable and which minimises the need for control and safety systems, maintenance, monitoring, and human intrusion.

A set of engineering principles associated with the achievement of passive safety is given in Table 16-1 [1]. Passive safety may be achieved by a combination of provision of an adequate waste form, its packaging, and the storage conditions under which it is maintained. In general, the more hazardous the waste, then the more mobile will be its form and the greater will be the benefit from the early achievement of hazard reduction by placing the wastes in a state of passive safety.

Chapter 21 describes a formal process, whereby waste packaging proposals may be assessed for their suitability for interim storage and eventual disposal. In some cases, a conflict may arise from, on the one hand, the pressing need (for short-term safety reasons) to package wastes as soon as reasonably possible and, on the other hand, the need to avoid foreclosure of future options. An appropriate balance between current and future safety requirements has to be demonstrated and, in some cases, the need for short-term improvements may be over-riding.

16-4. Classification of Wastes

16-4-1. *Introduction*

Chapter 1 gives the definitions of various nuclear waste classifications, and these are illustrated in Figure 1-3.

Table 16-1. Practical Engineering Principles for the Achievement of the Passive Safe Storage of Radioactive Wastes

Principles
The radioactivity should be immobile
The waste form and its container should be physically and chemically stable
Energy should be removed from the waste form
A multibarrier approach should be adopted in ensuring containment
The waste form and its container should be resistant to degradation
The storage environment should optimise waste package life
The need for active safety systems to ensure safety should be minimised
The need for monitoring and maintenance to ensure safety should be minimised
The need for human intervention to ensure safety should be minimised
The storage building should be resistant to foreseeable hazards
Access should be provided for response to incidents
There should be no need for prompt remedial action
The waste packages should be such as to allow inspection
The waste packages should be retrievable for inspection or reworking
The lifetime of the storage building should be appropriate for the storage period prior to eventual disposal
The storage facility should be such as to enable retrieval of the wastes for final disposal (or restoring)
The waste packages should be acceptable for final disposal

For the purposes of implementing the European Basic Safety Standards Directive (Council Directive 96/29/Euratom), Waste Management Companies and Regulators must ensure that dose limits are complied with and that all radiation exposures to people resulting from radioactive waste disposals are as low as reasonable — having regard to *maximum* dose constraints of $0.3\,\mathrm{mSv}\ \mathrm{y}^{-1}$ from any new single source and $0.5\,\mathrm{mSv}\ \mathrm{y}^{-1}$ for the discharges from any complete site (which may include several sources (see Chapter 2)). This sets the upper limits on the radioactivity that may be discharged, and includes the additional separate requirement for Best Practical Means (BPM) to limit discharges to members of the public as ALARA. Furthermore, applications for disposal must make a case for the discharge limits proposed in terms of operational needs. In the UK, such applications for disposal are *not* set on the basis of environmental capacity or at levels of public exposure corresponding to a dose constraint.

In addition, there is an increasing awareness that the "decide and defend" decision-making process may not receive public confidence. Therefore, a close consultative decision-making process where the public are fully engaged is included in sensitive radioactive waste

Table 16-2. Some NORM Quantities (US-EPA-93)

Waste stream	Production rate (te/annum)	Total Uranium (Bq/g)	Total Thorium (Bq/g)	Total Radium (Bq/g)
Phosphates	5×10^7	Background to 3	Background to 1.8	0.4 to 3700
Petroleum production	2.6×10^5			Background to 3700
Water treatment	3×10^5			0.1 to 1500
Mineral processes	5×10^9	Background to 129	Background to 900	Background to 129

disposal decision-making. Prior to public consultation, public "surgeries" may be held where:

- the applicant can explain the application to dispose of radioactive waste;
- the public have an opportunity to express their concerns, support the application, and ask questions; and
- the Regulatory authorities may outline the consultation process and how the application will be determined, explain the role of the Regulators and describe what factors they will (and will not) be taking into account.

16-4-2. Exempt Materials

Internationally, there is a real difficulty with regard to the different standards being applied to the disposal of radioactive materials resulting from natural as opposed to man-made radioactivity. Industrial non-nuclear practices involving large volumes of material may generate wastes with natural activity levels that may be disposed of directly within legally allowable limits. Naturally Occurring Radioactive Material (NORM) may be considered exempt under the Radioactive Substances Act Schedule 1 or under conditional exemption orders. If the waste meets these criteria, then it may be disposed of to normal landfill sites. Table 16-2 illustrates some NORM quantities from industrial practices.

Consider, for example, the work of a German Company which melted 350 te of scrap metal from the natural gas industry. This process resulted in:

- 18 te slag: average specific activity, 93 Bq/g.
- 1 te of filter dust: average specific activity, 535 Bq/g.
- 3.6 te of floor sweepings: average specific activity, 255 Bq/g.

Practical and economic waste management practices were sought for this material and agreed with the authorities based upon an individual dose of $1 \, \text{mSv} \, \text{y}^{-1}$. At the same plant, slightly contaminated waste metal scrap arising from nuclear industry decommissioning waste is being regulated against a nuclear criterion of $10 \, \mu\text{Sv} \, \text{y}^{-1}$. It could also be used for road construction, but only if the dose to the critical group results in an uptake 100 times less than the NORM values. As far as the authors are aware, there is no evidence that the properties of NORM differ from the properties of any other radionuclides in

ways that would necessitate the development of different approaches to risk assessment. Estimates of absorbed dose in tissue are fundamental physical quantities that determine radiation risk for any exposure situation. There is no plausible rationale for any differences in risk, due to ionising radiation arising from naturally occurring and any other radionuclides. This is because absorbed dose in tissue depends only on the radiation type and its energy, not on the source of the radiation.

In the UK, exempt materials may contain man-made radionuclides, but with activity levels less than those stated in the Radioactive Substances of Low Activity (SoLA) Exemption Order 1986. This requires:

- Solids to have activity levels < 0.4 Bq/gm.
- Organic liquids, C-14 and H-3 to have activity levels < 4 Bq/ml. All other liquids containing man-made radionuclides are categorised as Low Level Wastes (or VLLW).
- Gases — to have half-lives < 100 s.

16-4-3. Clean Materials — Free Release

Free release materials are those which do not require an authorisation from the Regulators for their disposal. They are clean in the sense that they are neither contaminated nor activated above background levels. From a regulatory viewpoint, clean and exempt materials are, therefore, treated the same. Clean solids may be consigned for disposal to normal landfill sites and liquids or gases released to the environment. This is not to say, however, that no consideration may be given to the toxic nature of any such wastes which must be such as to comply with regulations before disposal at normal sites is possible.

16-4-4. Very Low Level Waste (VLLW)

This category is primarily intended for small volume nuclear waste arisings, and wastes in this category may be disposed of with ordinary refuse (dustbin disposal) for each $0.1 \, \text{m}^3$ containing less than $400 \, \text{kBq} \, \beta/\gamma$ activity or single items containing less than $40 \, \text{kBq} \, \beta/\gamma$ activity. It is not entirely clear in the UK if this classification is acceptable by the regulators for large volume decommissioning wastes.

16-4-5. *Low Level Waste (LLW)*

In most countries with nuclear programs, disposal facilities exist for this LLW category. Policy dictates that, where such routes exist, the onus is on the waste producer to employ them rather than keep wastes in temporary storage. The disposal facilities are usually specifically built structures with concrete bases and water run-off drainage systems. The water may be collected and monitored in catchment bunds and pits and, if clean, allowed to be discharged to natural drainage. If contaminated, it may be diverted to treatment plants prior to discharge.

16-4-6. *Intermediate Level Waste (ILW)*

A review of the current international status of ILW and other waste category disposal facilities is given in Appendix 1. No such disposal facility currently exists in the UK.

16-4-7. *High Level Waste (HLW)*

Also known as heat generating waste typically derived from fuel usage and treatments. Used fuel (fuel that has been irradiated in a nuclear reactor or test facility) may be reprocessed to recover useful fissile material if it makes economic sense so to do. Such reprocessing generates small quantities of HLW, which is generally encapsulated in glass of synthetic rock and stored for typically 50 years to allow for the heat to dissipate prior to eventual disposal.

Figure 16-3 is a flow diagram showing the possible routes or treatment of used fuels to recognised end points. In the UK, there is currently some debate about whether used fuel should be treated as a waste or, by extraction, a valuable fuel commodity. This is especially the case for plutonium bearing used fuels that may be mixed with uranium in an oxide form and used in power reactors. Recovered plutonium may also be stored ("energy in the bank" for use by a future generation) or treated as a waste if surplus to requirements and so contaminated that it does not make economic sense to treat it for future use.

See definitions for the terms used in Figure 16-3 in Chapter 1:

- No ILW disposal facility currently exists in the UK. UK Nirex Ltd. have a set of waste packaging specifications that are likely to meet final disposal requirements. Interim storage is the only currently available option for ILW in the UK.
- No HLW disposal facility or waste acceptance criteria for such a facility currently exists in the UK. Interim storage is, therefore, the only currently available HLW option.

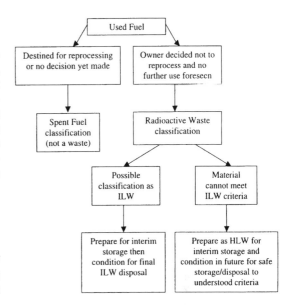

Figure 16-3. Routes to Recognised End Points for Used Fuel.

Table 16-3. Anomalies Between Current Regulatory Requirements for Disposal of Radioactive Wastes

Regulation for disposal	Regulatory requirement
SoLA (Substances of Low Specific Activity) Exemption Order	To be "exempt" from the Radioactive Substances Act (RSA 93) all man-made nuclides have to be <0.4 Bq/g activity.
Schedule 1 of the RSA 93	At levels of: <0.74 Bq/g Lead; <2.59 Bq/g Thorium; <0/37 Bq/g Actinium, Polonium, Protactinium and Radium; <11 Bq/g Uranium; then these radionuclides cannot be classed as radioactive.
BSS (Basic Safety Standards) Directive	Correctly does not distinguish between NORM (Normally Occurring Radioactive Substances) and man-made radionuclides with respect to "exempt" material. Further disposal is, again correctly to the authors' minds, based upon the effective dose to a member of the public.[a]
Phosphatic Substances and Rare Earths Exemption Order	Seven naturally occurring radionuclides are not classed as radioactive if they contain <14.8 Bq/g.

[a]This means that if the UK adopts the BSS Directive, the levels below which material will be classified as "exempt" would be nuclide-specific.

16-5. Summary

Most nuclear waste arises from the use of nuclear materials in power generation and weapons programs. A large proportion is already committed, and clear routes for disposal or longer term interim storage need to be in place. Waste needs to be managed so as to improve safety, and regulations (some conflicting) exist to ensure this. Wastes may be categorised in terms of their radionuclide content, activity levels, or anticipated dose to the critical group. The UK approach is as follows. "Clean" wastes are identified by pedigree and/or monitoring for off site disposal. "Exempt" material is classified in terms of SoLA/RSA93 Schedule 1 exemption criteria for interim storage and disposal (see Table 16-3). VLLW requires a specification and procedures for its disposal, but the category is not designed for bulk decommissioning wastes. VLRM is at the low end of the LLW category, but is, as yet, an officially unrecognised waste classification. LLW is an understood category which may undergo assay for existing disposal facility acceptance criteria and be packed in half height ISO containers for transport to disposal sites. Current LLW disposal site capacity is anticipated to be available to ~2050; however, a further expansion beyond this date will be necessary. There is currently no ILW (or HLW) disposal facility in the UK, so packaging and interim storage to well founded criteria is used. Encapsulated HLW is stored for at least 50 years to allow for cooling.

16-6. Reference

1. Williams, L., D. Mason, S. Blakeway and C. Snaith. "A regulatory view of the long term passively safe storage of radioactive waste in the UK," *Proceedings of an International Conference on Safety of Radioactive Waste Management,* IAEA, Cordoba (March 2000).

Chapter 17

Waste Management Strategy

17-1. Introduction

A waste management strategy defines a structured approach to the current and future management of radioactive waste from its production through to disposal. Under Government policy, producers and owners of radioactive waste are responsible for developing their own waste management strategies. This comes under the heading of strategic planning and is usually done in conjunction with the planning for decommissioning.

This chapter describes the requirements to be considered in developing a waste management strategy, the main components of a waste management strategy, and how to develop a new strategy taking account of all relevant factors. The importance of integration is considered together with a summary of the key principles of the waste management strategies for main licensees.

17-2. Waste Management Strategy Requirements

17-2-1. Regulations

Waste producers must develop their waste management strategy within an overall framework of Government policy, regulatory requirements, and international agreements (e.g., OSPAR/SINTRA). This includes the requirement to regularly review and update the strategy and to meet all safety requirements. In the UK (as in most other countries), there are significant differences in the activity levels at which Naturally Occurring Radioactive Material (NORM) and man-made artificial nuclides are exempt from nuclear regulations for the purposes of disposal. Chapters 16 and 17 give a description of general operational and environmental regulatory requirements in the UK.

17-2-2. Consultation

The strategy should be developed in consultation with all relevant stakeholders including the regulators, i.e., HM Nuclear Installations Inspectorate (NII) and the Environment Agencies (EA and SEPA), who have the duty to ensure that the Government policy is properly implemented. This will require discussion at various stages during the development of the strategy, as described in the following sections. Waste producers also need to consult Nirex (Nuclear Industries Radioactive waste Executive charged with the disposal of ILW and some long-lived LLW) if the waste is destined for deep waste disposal (see Nirex requirements below) or BNFL if the waste is destined for disposal as LLW to the UK disposal site at Drigg in North West England.

17-2-3. Completeness

The strategy should not just cover radioactive material which waste producers currently regard as *waste*. It should cover *all nuclear material* which has the potential to become radioactive waste in the future, e.g., spent fuel and other stocks of unwanted fissile material. The strategy should cover the complete life-cycle of the material and associated facilities, and should include routine discharges of liquid and gaseous radioactive wastes.

17-2-4. NII Requirements

NII has four fundamental expectations which should be met so far as is reasonably practicable and as described in Chapter 16, Section 16-1 [1].

Waste minimisation must be considered during planning and development, including facility design and during operations and decommissioning. Proper facility design with the correct use of materials can minimise contamination, allow easy decontamination, and, hence, the volume of waste produced. Management during operations can minimise the production of secondary waste, e.g., managing the quantity of materials taken into a facility which would become contaminated and have to be managed as LLW. Reuse and recycling must also be considered.

In many instances, the waste is already being produced or about to be produced, e.g., from decommissioning. The important issue in these circumstances is to get the waste into a passively safe form, using modern standard facilities, and providing a facility for interim or longer term storage if no waste route to a disposal site exists. Note that, as described in Chapter 16, Section 16-3, Government policy as expressed in White Paper Cm 2919 (1995) actually says " where it is practicable *and cost effective to do so,* . . . store it in accordance with the principles of passive safety." It should be noted that there is a potential conflict here with another regulatory requirement associated with avoidance of waste management actions that may foreclose future management options. Therefore, each situation has to be justified to the decommissioning company management and to the regulators on a case-by-case basis.

17-2-5. Environment Agencies' Requirements

SEPA and EA are interested in waste management strategy, as it affects discharges to the environment and disposals. In practice, this covers most aspects of the waste management planning. In particular, they expect waste production to be minimised, and for BPEO studies to be produced to cover the management strategies associated with waste arisings.

17-2-6. ILW Disposal Company (Nirex) Requirements

The currently proposed eventual destination for Intermediate Level Waste (ILW) and some low level waste (LLW) in the UK is to a Deep Waste Repository (DWR). UK Nirex Ltd. is the company currently charged with developing the deep waste repository concept. In the absence of detailed conditions for acceptance, Nirex provides guidance to waste producers on waste package acceptance criteria in the form of Waste Package Specifications and Guidance Documentation. This guidance documentation helps to ensure that waste packages produced now and in the near future will meet the requirements of long-term on-site storage, transport to, and disposal at a possible eventual DWR. Nirex issues "Letters of Advice and Comfort" to ILW producers, which essentially give some confirmation that the waste treatment, conditioning, and packaging proposals being adopted by the waste producers will probably be suitable at some distant time in the future for disposal. There is no equivalent organisation giving such advice for the packaging of High Level Waste or unwanted spent fuel. The management of ILW is described in Chapter 20 and HLW in Chapter 21.

17-2-7. LLW Disposal Company (BNFL, Drigg) Requirements

BNFL set down "Conditions for Acceptance" for waste destined for disposal at their Low Level Waste (LLW) disposal facility at Drigg in West Cumbria, North West England. This covers the activity limits to be applied to the waste, radionuclide content present, and the packaging required.

17-2-8. Integration of the Strategy

The strategy for managing radioactive wastes and materials should be developed alongside the strategy for decommissioning of nuclear facilities, including the *treatment* of *radioactive* contaminated *land* (in effect an accumulation of radioactive waste). This *integration* ensures that wastes from decommissioning and site restoration are fully accounted for in developing the strategy for other forms and arisings of radioactive waste. Consideration also has to be given to the other elements of site infrastructure and organisation.

A regulatory audit of the management of safety at UKAEA Dounreay in June 1998 covered a wide range of topics to assist clarity of the interpretation of Government policy. Two of the recommendations were:

- R45: "UKAEA should develop an integrated decommissioning strategy for Dounreay."
- R69: "UKAEA should develop a strategic plan for handling, treatment, storage, and disposal of all radioactive wastes on site, integrated with the plans for operation, POCO, C&M, and decommissioning."

A combined decommissioning and waste management strategy already existed before this date, but the requirement was to develop the lower site level programs in greater detail, fully considering the interactions between the detailed plans for individual projects. UKAEA produced the Dounreay Site Restoration Plan (DSRP) in September 2000 in response to the audit. This, coupled with other plans such as that developed for Hanford, US [2] in 2002, are generally accepted to be leading examples of their kind for others to follow in the future.

Figure 17-1 shows the inter-relationship of the constituent parts of the plan.

17-2-9. Costs

Complete development of the waste management strategy allows future lifetime management undiscounted and discounted costs, to be estimated such that sufficient

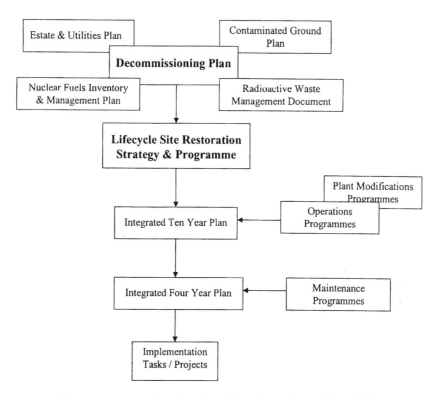

Figure 17-1. Inter-Relationship of Constituent Parts of the DSRP.

provisions may be put aside to fund the forward program.

17-3. Elements of a Waste Management Strategy

Figure 17-2 shows the basic components of a waste management strategy. Not all of the components might be present in any one strategy and, in some instances, there may be further components, e.g., pretreatment before interim storage to segregate wastes, or for transfer between facilities within a site or transport between sites.

17-3-1. *Waste Generation*

Waste can be generated during operations or from decommissioning. It can be in the form of solid, liquid, or gaseous waste. Wastes may also be generated as a by-product of further steps in the strategy.

17-3-2. *Interim Storage*

After generation, wastes can sometimes be stored in a temporary form, which is not the final form appropriate

Figure 17-2. The Main Elements of a Waste Management Strategy.

for disposal. This is the position for a number of historic wastes in store on UKAEA sites.

17-3-3. *Retrieval*

This involves the recovery of waste and waste packages from storage either for inspection purposes, repackaging, further storage, or disposal. It may require the installation of specialist retrieval equipment.

Nowadays, waste management considerations begin at the plant design stage, where radioactive waste generation can be minimised by appropriate design features. This continues through construction to operation, where practices can be optimised to reduce the quantities and impact of radioactive waste management.

In the early years of nuclear development, the implications of waste management were not fully thought through, and waste was often stored directly without prior treatment or even packaging. Wastes stored in this way from the 1950s and 1960s is now being retrieved and the waste fully conditioned to be acceptable for long-term storage to modern standards pending disposal.

Examples of this are the shaft and wet silo at UKAEA Dounreay (see Figure 17-3). Waste will be retrieved from these as soon as retrieval facilities and a new packaging plant and store are made available. The shaft has particular novel problems and will need considerable research to develop isolation and retrieval techniques.

Further examples are the silos at BNFL, Sellafield, used historically for ILW from reprocessing. Retrieval from water-filled silos has already been achieved, and equipment and facilities are now being designed and built to retrieve similar but older material, as well as further historic wastes, in the form of sludges, from other facilities such as early fuel storage ponds.

At UKAEA Harwell, waste is being retrieved from old tube stores and being repacked into Nirex acceptable 500 liter drums. The project to retrieve all the existing waste and to put it into a passively safe form is currently estimated to take to 2020. The difficulties associated with retrieval, even from a set geometry within an existing store, should not be underestimated.

17-3-4. *Treatment*

This involves changing the characteristics of the waste. Basic treatment concepts are volume reduction, radionuclide removal, and change of composition. Typical treatment operations include incineration or compaction (for volume reduction), evaporation, filtration or ion exchange of liquid waste (radionuclide removal), and precipitation or flocculation of chemical species (change of composition).

Treatment may include decontamination of the waste to reduce its waste categorisation, e.g., from ILW to LLW or even from LLW to free-release. One example is the clean-up of lead bricks which were used in the construction of shielded cells.

17-3-5. *Conditioning*

This involves transforming radioactive waste into a form suitable for handling, transportation, storage, and disposal. It may include immobilisation of waste, placing waste into containers, and providing additional packaging.

Figure 17-3. The Dounreay (Scotland, UK) ILW Wet Silo.

17-3-6. *Storage*

This may take place between and within the basic waste management steps. The intention is to isolate the radioactive waste, provide environmental protection, and to facilitate control. The usual requirement for storage is to act as a buffer between steps, e.g., for waste expected to go for eventual deep disposal.

17-3-7. *Disposal*

This consists of the authorised emplacement of packages of radioactive waste in a disposal facility. Disposal may also comprise of discharging radioactive waste.

17-4. Strategic Planning

Nuclear licensees/waste producers undertake strategic planning for the management of all their radioactive material and radioactive waste, i.e., deciding the best method for managing the waste through all stages described above. This is necessary to develop and build the required facilities on the appropriate timescales and to ensure that funding can be made available. The full strategy should contain the following elements:

- The licensees radioactive waste management objectives and policy;
- The current and future inventory of radioactive waste;
- The preferred option for managing each waste stream throughout its life cycle to disposal together with fall-back options;
- The justification of the preferred option showing consistency with Government policy and regulatory requirements;
- Programs showing the timescales for each element of the strategy;
- The arrangements for providing and maintaining the waste safely until its ultimate disposal;
- Identification of significant uncertainties and their impact;
- The approach to ensuring safety; and
- The costs of implementing the strategy.

Considering each element of the process in turn:

17-4-1. *Waste Inventory*

The first step in developing any strategy is characterisation of the waste. This involves determining the physical, chemical, and radiological properties. This will be required as basic input to the selection process for treatment or processing options.

17-4-2. *Evaluation of Treatment/Processing Options*

The second step is to develop the options for treatment or processing of the waste (and retrieval if necessary). There are two parts to this, defining the treatment process and determining the required product. Regulatory consultation is important early on in the planning to ensure that the final process chosen will be acceptable and to avoid nugatory expenditure.

The preferred treatment process is identified through strategic assessment and increasingly involves the use of BPEO studies. A range of options are identified and compared in terms of safety, environmental impact, practicality, cost, public acceptability, and must include the production and management of secondary wastes, e.g., gaseous and liquid wastes. Figure 17-4 shows the identification a range of options from the assessment some years ago into the possible replacement of the LLW incinerator at UKAEA Harwell.

Inputs to the strategic assessment may include a number of supplementary studies, e.g., design and development work or scale mock-ups to understand the feasibility and cost of the options.

The options considered technically feasible are subject to economic assessment. Guidance on option appraisal in central government, including the use of discounted cash flow analysis, is given in "The Green Book" [3]. Much of the guidance is relevant for any financial appraisal, as discussed in Chapter 13.

Product evaluation involves designing a wasteform which is suitable for storage, transport, and disposal. This covers the form of the waste itself, e.g., producing an immobile waste and the packaging.

Choice of the final option will involve risk assessment to fully evaluate the range on the cost and timing of the various options and to understand the potential impacts of significant risks.

17-4-3. *Reference Strategy*

The reference strategy for a particular waste is chosen by taking a balance between the relevant factors. This may involve a number of techniques including multiattribute decision analysis.

17-5. Integration and Costing

The reference strategy for one particular waste must be integrated with the strategies for other wastes and reviewed within the context of the whole companies strategy, including decommissioning, land restoration, and

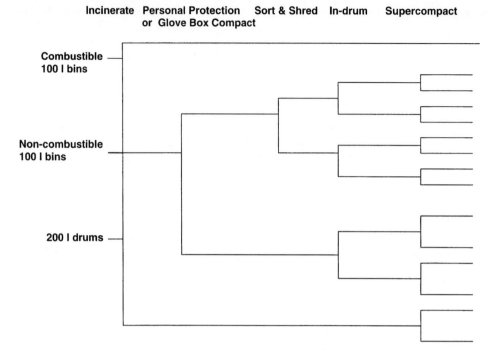

Figure 17-4. Options Identification/Decision Tree Example.

infrastructure. The timing chosen for processing a waste in isolation may not be possible when there are a number of wastes to be dealt with in a single plant with operational limits. The wastes will have to be prioritised and the full strategy reanalysed in terms of cost and plant optimisation.

Chapter 12, Figure 12-1 shows a UKAEA planning cycle, which is an iterative process picking up the latest strategy for particular wastes and decommissioning and incorporating it within the site and corporate plans. The strategy is reviewed on an annual basis.

Integration can result in an extremely difficult model of the number of interactions, and each licensee has developed tools to help with its strategic planning.

UKAEA uses its Strategic Planning System (SPS), an Oracle™ database computer system which allows multiple users and tracks changes centrally for auditing purposes. SPS allows UKAEA to ensure that all wastes produced can be managed within store capacities, processing limits of treatment plants, feasible transport, and disposal rates, i.e., that the strategy "works."

Chapter 12, Figure 12-2 shows a sample of the modeling which can be managed by SPS. Strategies for waste streams from a diverse range of facilities, which undergo a number of steps, including changes in volume as the waste is processed, may be analysed in this way. SPS also allows UKAEA to fully cost its decommissioning and waste management strategy.

Figure 17-5 shows a sample input screen for a waste management processing facility and Figure 17-6 a typical cost profile for the strategy envisaged arising from such modeling.

17-6. Review and Updating

NII requires that licensees regularly review and update their waste management strategies. Government policy requires that a review is undertaken of licensees decommissioning strategies every 5 years, by the HSE in consultation with the Environment Agencies. This is to ensure that they "remain soundly based as circumstances change." Decommissioning strategies necessarily covers the management of all related wastes and, hence, "a Quinquennial Review" (QQR) is effectively a review of the robustness of the waste management strategies.

As part of the Quinquennial Review, NII will assess the various elements of licensee's arrangements to determine the extent to which their four basic expectations are being met.

17-7. Fundamentals of Licensees' Strategies

UK nuclear licensees have all well developed waste management strategies, but these vary depending on the

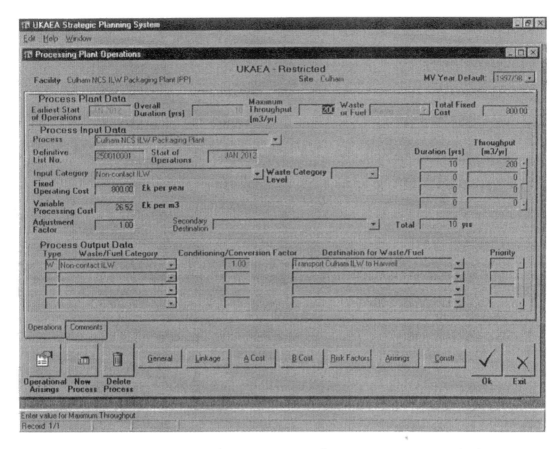

Figure 17-5. Sample Data Input Screen for a Processing Facility Using UKAEA's Strategic Planning System (SPS).

amount of wastes being produced and the stage reached in decommissioning.

17-7-1. *UKAEA*

UKAEA has completed its original mission to develop peaceful uses of atomic energy from fission. Fusion research continues. The bulk of UKAEA's facilities are now redundant and are in various stages of decommissioning. Waste is being retrieved from older waste management facilities, and newer facilities are being developed, especially at Dounreay. The strategy is extremely complex to decide the timing of decommissioning when waste management facilities are available.

UKAEA policy is to minimise new arisings of radioactive wastes. In each case, UKAEA looks to identify and assess all relevant options, so that the chosen option represents the best practicable balance of environmental, safety, economic, and stakeholder considerations. For existing wastes, and for future arisings of operational

and decommissioning wastes which cannot be avoided, UKAEA considers each waste stream on a case-by-case basis, sometimes considering further subdivisions of a waste stream where there are specific requirements or differences. The following general principles are applicable:

- *HLW.* Dounreay is the only UKAEA site with HLW. The strategy for this material is to vitrify, or otherwise immobilise the waste, and to store the vitrified product in a new, purpose built store at the site until a national disposal facility becomes available.
- *ILW.* As there is no National disposal facility for ILW in the UK, UKAEA policy is to condition solid and liquid ILW for long-term storage by conversion to a passively safe form, without unnecessarily foreclosing future options for disposal. Liquid ILW will be solidified as part of this strategy.
- *LLW.* Solid LLW at Southern sites is sent to the UK's LLW disposal facility at Drigg in North West England, save for those waste streams whose specific nuclide

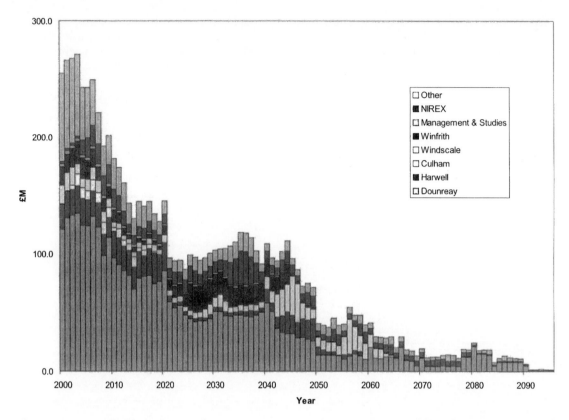

Figure 17-6. Cost Profile for a Particular Waste Management Strategy, as Derived from the Strategic Planning System (SPS) Modeling.

inventory is outside the acceptance criteria for Drigg. These waste streams are managed as ILW. The Dounreay LLW disposal facility is now almost full, and Dounreay are exploring disposing of LLW to Drigg pending the availability of new facilities providing a long term solution.

- *Low Level Liquid Effluent.* Liquid LLW is treated to remove as much of the radioactivity, as practicable, prior to discharging the treated liquid within authorised limits. Separated solids containing the radioactivity are immobilised and managed with other solid wastes on the sites.
- *Nuclear Materials.* UKAEA also has responsibility for the safe management of nuclear fuels which were associated with the development and operation of experimental prototype reactors. There are three principal strategies for dealing with surplus nuclear materials: (i) Condition and transfer from UKAEA sites to another licensed user, (ii) Store for future use by a licensed user (other than UKAEA), and (iii) Declare as waste and prepare for storage and future disposal.

17-7-2. BNFL

BNFL Magnox has various reactors going through decommissioning as well as a number of operational stations (see Chapter 1, Table 1-2). BNFL at Sellafield has a similar legacy to UKAEA of redundant facilities and older waste management facilities as well as a range of newer waste management plants.

BNFL disposes of its low level waste on arising to Drigg, in accordance with Government policy.

BNFL have developed a strategy for its ILW to provide a solid wasteform which is safe and convenient for storage. BNFL have constructed modern state of the art ILW storage facilities on site pending development of a possible future disposal facility for ILW in the UK. The conditioned, drummed, and stored wastes can either be transported directly to a future repository or undergo further treatment to convert the wasteform to a disposable product. Existing stocks of historic or legacy raw wastes together with currently arising wastes are being retrieved and packaged for interim storage in a massive and impressive campaign. Authorised disposal of liquid and gaseous effluents is undertaken after suitable treatment to minimise impacts on the public and the environment.

HLW from fuel reprocessing is converted to a glass wasteform to be followed by a period of at least 50 years storage prior to disposal.

17-7-3. *British Energy (BE)*

BE produce LLW and ILW from the operations of their AGR and PWR nuclear power stations, all of which are still operating. Hence, their strategy for producing and managing decommissioning wastes will depend on the operating life of the stations. As illustrated in Chapter 1, Figure 1-8, the more modern nuclear power stations produce relatively small volumes of waste in comparison with the early designs.

The operational LLW is predominantly trash, including paper and discarded protective clothing. Some process arisings such as HEPA filters, sludges, and ion exchange resins are also LLW. This is either incinerated or compacted before being sent to Drigg for disposal.

ILW is currently stored in vaults and tanks pending conditioning prior to disposal. Waste at Sizewell B is retrieved and packaged during operations. In general, at the AGRs, wet wastes are processed and packaged during the early stages of decommissioning, but some solid inert wastes in vaults will be left until the final stages of decommissioning.

Spent fuel from the AGR stations is sent to BNFL at Sellafield for reprocessing. Spent PWR fuel is currently stored on site.

17-7-4. *Liabilities Management Authority (LMA)*

The recent announcement (Cm 5552 [4]) proposing the formation of a Liabilities Management Authority (LMA) to take on the ownership and responsibility for the management of UK public sector civil nuclear liabilities

(BNFL and UKAEA) will have an impact on the waste management strategy of these two companies. The full impact will not be known for some years, but there may be scope for further integration/collaboration for certain waste streams.

17-8. Summary

Licensees develop radioactive waste management strategies to:

- Plan the future waste management infrastructure required to support their decommissioning and operational requirements;
- Provide a firm basis for costing the Company's liabilities; and
- Comply with regulatory requirements.

Significant work has been done by UK licensees in this area, but considerable work will be required as:

- Companies planned activities move from operational to decommissioning, where decommissioning programs get larger and where they begin to tackle the more difficult decommissioning challenges; and
- The need for a full understanding of the strategy will increase as the repercussions of the LMA become clearer.

17-9. References

1. Guidance for Inspectors on the Management of Radioactive Materials and Radioactive Waste on Nuclear Licensed Sites – contains guidance on the contents of waste management strategy http://www.hse.gov.uk/nsd/waste1.pdf
2. Performance Management Plan for the Accelerated Cleanup of the Hanford Site, US DOE Report, DOE/RL-2002-47.
3. "The Green Book" – Appraisal and Evaluation in Central Government". Treasury Guidance, HMSO, 1997.
4. Stationery Office. *UK Government White Paper Managing the Nuclear Legacy – A strategy for action*, Cm 5552, HM Stationery Office, London, UK (July 2002).

Chapter 18

Policy and Regulatory Aspects of Waste Management

18-1. Introduction

This chapter describes the policy and regulatory issues associated with the operational and environmental management of nuclear waste. On the operational side, there exists a relatively clear set of guidance documentation issued by the Regulators in the UK to their inspectors to assist with the interpretation of policy and legal requirements. The environmental regulation itself is as clear, but the interpretation is less specific, and allowance must be made for this in program development timescales and costs. This chapter also draws out differences between UK, European, and US approaches.

18-2. Nuclear Site Operations

18-2-1. Liability and Compensation for Nuclear Damage

The decommissioning activities, storage, and processing of nuclear waste on UK nuclear licensed sites is regulated by the Nuclear Installations Inspectorate of the Health & Safety Executive under the Nuclear Installations Act 1965 (as amended). Liability and compensation for nuclear damage is covered under this Act up to a current limit of £140M for 10 years after the incident. The limit is kept under review, but the damage is met by the Government for the next 10–30 years. OECD Paris and Brussels Conventions are implemented by the Act. For a major incident involving off-site releases, such compensation levels are potentially low and are under review. The International Nuclear Event Scale (INES) shown in Table 18-6 gives levels, descriptions, criteria, and examples to help define nuclear accidents and incidents. The Vienna Convention on Civil Liability for Nuclear Damage is a UN Convention revised in 1997. Additional compensation through this route is possible.

18-2-2. Operational Safety

The HSE NII are primarily concerned with the safe operations of nuclear plant, but their remit, because safety is embedded in and may be considered to extend to all aspects of nuclear site operations, is often very wide ranging. The regulator has powers to impose fines on noncompliant operators, and HSE NII's remit extends way beyond safe plant operations to analysis of the adequacy of staffing structures and forward decommissioning programs, as well as the funding mechanisms to support these.

The HSE NII have regulatory powers covering:

- Directions;
- Approvals;
- Specifications;
- Consents;
- Notifications; and
- Agreements.

These may be applied in connection with all aspects of nuclear site operations generally, as shown in Table 18-1.

A **direction** is issued by the NII when it requires the licensee to take a particular action. For example, Licence Condition 31 gives HSE NII the power to direct a licensee to shut down any plant, operation, or process. Such a direction would generally relate to a matter of major or immediate safety importance, the continued operation of which would pose unacceptable risks.

Approvals are used to freeze an arrangement, or part of such an arrangement, made by a licensee. Once approved, no change can be made without NII agreement, and the arrangement must be carried out as specified. Failure to do so would infringe the licence condition. For example, a licensee's emergency arrangements are approved to ensure they are not changed without the licensee first seeking NII's agreement to the change.

The standard licence conditions give the NII discretionary controls for a licensee's arrangements and these

Table 18-1. Topic Groupings Associated with UK Nuclear Site Licence Conditions

Topic	Licence Condition(s)
Interpretation	LC 1
Control of the Site and Nuclear Matter	LCs 2, 3, 4, 5, and 16
Quality Assurance and Control of Records	LCs 6 and 17
Investigation and Reporting	LC 7
Instruction, Training, and Authorisation of Persons on the Site	LCs 8, 9, 10, and 12
Emergency Preparedness	LC 11
Advice on Nuclear Safety	LC 13
Control of Safety Cases	LCs 14 and 15
Control of Plant Design and Status	LCs 19, 20, 21, 22, and 35
Control of Employee Doses	LC 18
Control of Operations	LCs 23, 24, 25, 26, 27, and 28
Plant Shutdown and Test Requirements	LCs 29, 30, and 31
Control of Waste	LCs 32, 33, and 34
Control of Organisational Change	LC 36

are implemented through Specifications. For example, if HSE so specifies, the licensee is required to refer operating rules to its nuclear safety committee for consideration.

A **consent** is required before the licensee can carry out various activities identified in the licence or which may be specified by the Inspectorate. For example, a consent is normally required before routine operations can start on a plant following commissioning. Before such consent is granted, the licensee must satisfy the NII that its proposed operation is safe and that all necessary procedures for control are in place.

The standard licence conditions give NII powers to require the submission of information from the licensee by **notification** of the requirement. For example, under Licence Condition 21(8), the licensee shall, if notified by HSE, submit a safety case and shall not begin operation of the relevant plant or process without the consent of HSE.

An **agreement** issued by the NII allows a licensee, in accordance with the licensee's own arrangements, to proceed with a planned course of action. For example, Licence Condition 22 requires arrangements to control modifications to safety-related plant. Such arrangements often require that, if any modifications could lead to serious safety implications if they were to be inadequately conceived or implemented, then they should not be carried out without the prior agreement of NII. Agreement in writing is only given after the submission of an acceptable safety case justifying the modification.

Similar arrangements apply internationally, as shown in Table 18-2 and Appendix 1. Acronyms are given in Table 18-2, so that more information may be found using

them when searching through the Internet. The coverage of the 36 UK nuclear site licence conditions is briefly summarised in Table 18-3.

During a decommissioning project, the HSE NII will, therefore, be looking for key safety-related documentation associated with the safe operation of plant or processes. The mechanism for their agreement to proceed to the next phase of the project lifecycle is through the production of safety reports and safety cases. Such key safety documentation is illustrated in Figure 18-1.

Note, a consent for plant operation to commence requires the formal completion of a Safety Case and the associated Operating Rules (ORs), Examination, Inspection, Maintenance, and Testing (EIMT) schedules, etc.

18-3. Environmental Policy and Regulation

18-3-1. *Introduction*

The role of the Environmental Agencies in the UK is the prevention of pollution and protection of the environment. The Environment Agency (EA) covers England and Wales, and the Scottish Environmental Protection Agency (SEPA) covers Scotland for redundant nuclear facility decommissioning, waste management, and site environmental remediation. The agencies have duties and powers to enforce environmental laws and regulations to protect the environment. Enforcement means taking action to ensure regulatory compliance ranging from site audits to formal enforcement or prohibition notices and prosecution. The different responsibilities between the HSE NII and the Environment Agencies regulatory system in the UK is illustrated in Figure 18-2.

18-3-2. *Specific Regulations*

The Environment Agencies are, therefore, responsible for granting air, sea, and land discharge authorisations, and these are designed to ensure minimal risk to the public and to ensure that any discharges are kept as low as reasonably practicable (ALARP). Treaties and agreements under which the framework of environmental legislation resides include:

- *Article 35 of the Euratom Treaty.* Member countries must establish facilities to monitor continuously the levels of radioactivity in the air, water, and soil, and to ensure compliance with standards.
- *Article 37 of the Euratom Treaty.* Article 37 applies to the planned disposal or accidental releases to the environment associated with certain activities. These activities include dismantling of reactors or reprocessing plant operations or storage of wastes arising from

Table 18-2. International Waste Management Regulation and Regulatory Bodies

Country	Framework and Regulatory bodies	Regulatory system
Belgium	Construction, operation and decommissioning radiological protection based upon BSS by Federal Agency for Nuclear Control (FANC). Clearance levels set at 0.4 Bq/cm^2 surface β emitters, 0.04 Bq/cm^2 surface γ emitters.	Legal framework classified into: Class 1 — fuel cycle and disposal facilities; Class 2 — nonfissile storage or other facilities Class 3 — relatively small quantities of radioactive materials involved.
Canada	Canadian Nuclear Safety Commission (CNSC) currently developing assessment proposals.	Nuclear Safety and Control Act (NSCA).
Finland	Radiation and Nuclear Safety Authority (STUK).	Nuclear Energy Act and Decree.
France	French Atomic Energy Commission (CEN) and newly established (2002) Nuclear Safety and Radiation Protection (DGSNR).	Article 6 ter (1963) as amended 1990.
Germany	Respective Federal State overseen by Federal Ministry of Environment, Nature Conservation, and Nuclear Safety (BMU), with technical support from Federal Office of Radiation Protection (BfS), etc.	Atomic Energy Act.
Hungary	Public Agency for Radioactive Waste Management (PURAM) and regulated through Ministries.	Act 116/1996 on Nuclear Energy, Govt. Decree 108/1997 and Nuclear Safety Regulations (NSR).
Italy	Generally follows BSS Directive 96/29/Euratom by National Agency for the Environment Protection (ANPA). Radionuclide clearance activity concentrations from the lesser of Euratom Directives 80/467 and 96/29 as well as consideration of dose.	Primarily Act 186 and Legislative Decree 230 (1995).
Japan	Nuclear Safety Commission Guidance. Clearance criteria under development.	Law for the Regulation of Nuclear Source Material, Nuclear Fuel Material, and Reactors (LRNR).
The Netherlands	Environmental Impact Assessments required. Three Ministries form the competent regulatory bodies with establishment of The Central Organisation for Radioactive Waste (COVRA).	Nuclear Energy Act, as revised (2000).
Spain	Operation under binding assessment reports by CSN.	Nuclear Installations Regulation (1993).
Sweden	Large number of regulatory bodies involved including Nuclear Power Inspectorate (SKI), Swedish Radiation Protection Institute (SSI), Swedish Environmental Protection Agency, etc.	Act on Nuclear Activities and Radiation Protection Act (SFS).
Switzerland	Federal Government is licensing authority with Swiss Federal Nuclear Safety Inspectorate (HSK).	Swiss Atomic Law (AtG) Clearance procedures based upon HSK guidance.
UK	HSE NII Regulate under Licence Conditions (see Tables 18-3 and 18-4).	Nuclear Installations Act (1965 as amended), Health & Safety at Work Act (1974 including Ionising Radiation Regulations (IRR) and associated enforcement powers for nuclear and conventional safety), Radioactive Substances Act (1993). Clearance under Substances of Low Activity (SoLA) fee release thresholds. Activity based clearance levels.
USA	Responsibility for Waste Management by US Department of Energy (DOE).	Nuclear Waste Policy Act (1982). Dose-based clearance levels.

these. Government has to supply an Article 37 submission to the European Commission to cover decommissioning activities involving discharges, so that they may be assessed for their impact on Member States. The submission includes descriptions of the processes, proposed waste routes, monitoring arrangements, and contingency plans, together with estimates of the wastes generated from the decommissioning activities, including estimates of the wastes generated.

• *OSPAR.* See Chapter 1, Section 1-6-2, essentially about working towards further reductions in radioactive discharges to the marine environment.

• *Government Radioactive Waste Management Policy.* UK Government 1995 White Paper, Command 2919,

Table 18-3. Summary of the UK HSE NII Nuclear Site Licence Conditions

Licence condition	Brief description
LC 1: Interpretation	Assigns defined meanings to commonly used terms used in nuclear operations such as: "commissioning," "excepted matter," the "Executive," "experiment," "installation," the "licensee" and the "site," "modifications," "nuclear matter" and "relevant site," "nuclear safety committee," "operations" including "operational" and "operating," "radioactive material" and "radioactive waste," "safety," and "safety cases."
LC 2: Markings of the Site Boundary	Associated with marking out the boundary of the nuclear site and the prevention of unauthorised persons entering the site together with the associated arrangements to achieve this and their maintenance.
LC 3: Restriction on Dealing with the Site	Dealing with prevention of transfer of possession or letting of the site or any part of it to third parties without prior consent.
LC 4: Restrictions on Nuclear Matter on the Site	To ensure that the licensee controls the introduction and storage of nuclear matter on licensed sites. This includes both new and used fuel as well as radioactive waste. Both carriage and storage of such materials are included and the arrangements must cover adequate safety cases, records of the nature of nuclear matter and its storage location on the site plan. Storage facilities must be suitable, separate, and secure with appropriate criticality controls.
LC 5: Consignment of Nuclear Matter	Associated with the off-site transfer of nuclear matter (other than excepted matter and nuclear waste) in the UK. Primarily covering adequacy of records of what, where, and how the consignment has been dispatched which must be retained for 30 years. Includes the need for a justification of the movement.
LC 6: Documents, Records, Authorities, and Certificates	To ensure that adequate records are made and retained in suitable storage conditions for a suitable period so as to demonstrate continuous historical compliance. This is a general requirement covering every Licence Condition, with typically 30 year record retention requirements.
LC 7: Incidents on the Site	To ensure that incidents are adequately reported and recorded. In this context, "incident" means any matter which may affect the site operations or safe condition of a plant and, consequently, applies to not only incidents and occurrences, but also events of safety interest or concern. These include human errors or failures of plant or in procedures which cause near misses or abnormal occurrences. Lessons learnt from these and other incidents on other sites are also to be considered. The licensee must have a system in place for the classification of incidents according to their severity and type and recording this information. Timely notification of such occurrences and the arrangements for suitably qualified personnel to carry out investigations and report their findings through the appropriate nuclear safety committee for consideration and advice. Includes reviews on a regular basis.
LC 8: Warning Notices	To assist with safety of personnel on the site by the provision and maintenance of adequate warning notices and signs. Thereby, help ensure that staff may respond without delay to an incident or emergency situation.
LC 9: Instructions to Persons on the Site	To ensure that every person authorised to be on the nuclear site receives adequate instructions as regards the risks and hazards associated with the plant and its operations, so as to enable them to take appropriate precautions and to respond adequately and without delay to an incident or emergency situation.
LC 10: Training	The purpose is to ensure that all those people on the site who have responsibility for an action which may affect safety are appropriately trained for that purpose. The condition covers not only those who control and supervise operations but also extends to persons carrying out the operations. Compliance requires a comprehensive program for each person or group of persons on the site. Topics include induction, site familiarisation, general health and safety, radiation and hazardous substances, incident and emergency responses, together with job and postspecific training.
LC 11: Emergency Arrangements	Detailed plans so as to respond effectively to any incident ranging from minor on-site incident to a major release off-site of radioactive material. The plans should make provision for an off-site facility where measures to protect the public can be co-ordinated and match into area planning arrangements. Rehearsals, with regulator observer status, form part of the emergency planning and preparedness.

Continued

Table 18-3. Continued

Licence condition	Brief description
LC 12: Duly Authorised and Other Suitably Qualified and Experienced Persons	To ensure that only suitably qualified and experienced personnel (SQEPs) perform duties which may affect safety of operations on the sites. Job competency requirements have to be linked with training records and personnel selection procedures. A register of all such persons is to be maintained which covers names, details of authorised duties, qualifications, training, and experience.
LC 13: Nuclear Safety Committee	A senior level Committee which can consider and advise upon all matters which may affect safety on or off the licensed site. The Committee must have members who are adequately qualified and experienced to perform the task. Independent members are required and it may be appropriate for them to be employees of other licensees. The role of the Nuclear Safety Committee (NSC) is embedded within the site licensee's management arrangements, but is advisory in nature without direct responsibility for peer review and independent safety assessment. The Terms of Reference (ToR) for the NSC and the arrangements for dealing with important safety proposals are so important that they require the approval of the HSE NII.
LC 14: Safety Documentation	Covering the preparation and assessment of safety-related documentation comprising safety cases, so as to ensure that the licensee justifies safety during design, construction, manufacture, commissioning, operation, and decommissioning. The safety case will comprise of a predefined suite of documentation covering the different stages of the project or plant lifecycle. The arrangements will cover the peer review and independent nuclear safety assessment (INSA) and whether or not it will be submitted to the Nuclear Safety Committee (NSC). When considering safety, it is not the considered risk, but the potential hazard arising directly or indirectly during or after the activities under consideration that will matter and which must be addressed. This includes any hazard arising from inadequacy in conception or execution.
LC 15: Periodic Review	To ensure that, throughout the declared plant lifetime, it remains adequately safe and safety cases are being kept up-to-date. Such safety cases should be periodically reviewed in a systematic manner to meet the following objectives: • to review the current safety case for the plant and confirm that it is still adequate, • to compare the case against current standards for new plant, evaluate any deficiencies, and implement any reasonably practicable improvements to enhance safety, • to identify aging processes which may limit the life of the plant, and • to revalidate the safety case and the next periodic safety review subject to the outcome of routine regulation.
LC 16: Site Plans, Design, and Specifications	To provide a detailed site plan and schedule of all buildings, plant areas, and associated operations which may affect safety. The buildings and plant so identified to be on the basis of safety significance. This is not confined to nuclear safety issues, but also to conventional safety associated with storage of inflammable or explosive material, etc.
LC 17: Quality Assurance	To set out the managerial and procedural arrangements that will be used to initiate control and monitoring of those actions that may affect safety. International modern management systems are process-based following ISO 9001:2000 and/or IAEA NUSS 50-C-QA Code requirements.
LC 18: Radiological Protection	To ensure that the licensee assesses the average dose equivalent to specified groups of employees and notifies the HSE NII if these doses exceed the level specified. This is in accordance with Ionising Radiation Regulations (IRR) 1999. The arrangements should set out the classes of person the licensee is distinguishing in the calculations of average effective dose equivalent and the means for checking them.
LC 19: Construction or Installation of New Plant	To ensure that adequate arrangements are in place for the control of construction or installation of new plant which may affect safety. Close cooperation with those responsible for conventional safety aspects of regulation is important during construction phases.
LC 20: Modifications to Design of Plant under Construction	No modifications to plant may be made without due consideration to the effect of such modifications on the safety case. The actual process of construction is covered under LC 19.

Continued

Table 18-3. Continued

Licence condition	Brief description
LC 21: Commissioning	To ensure that adequate arrangements exist for the commissioning of new or modified plant or process. The arrangements should provide for a system of categorisation and control of commissioning on the basis of safety significance, and then for the production of a document which identifies the testing to be carried out by SQEPs in support of the safety case. The planning, implementation, control, and recording of the commissioning tests will all require to be checked for adequacy. Such testing should be in a structured systematic manner with appropriate controls at all stages. The purpose of the commissioning is to ensure that the plant performs in the way expected by the designer, and which was assumed in the plant's safety case. The work includes identification of the operating rules, safety mechanisms, devices and circuits, and maintenance schedules. It is normally carried out in two parts — namely nonactive commissioning (prior to the introduction of radioactive materials) and active commissioning (with radioactive materials present).
LC 22: Modifications or Experiment on Existing Plant	To ensure that adequate arrangements are in place to categorise and control all modifications and experiments. This should cover all stages of the proposed work. The modification may require personnel to undergo elements of additional training in accordance with LC 10. A series of minor modifications could have a significant cumulative effect on safety, that such work should be seen as an overall plan rather than as small works in isolation.
LC 23: Operating Rules	All operations that may affect safety must be supported by an adequate safety case. This safety case must identify the conditions and limits that will ensure that the plant or process is kept in a safe condition. The safety cases should distinguish between limits and conditions, which are necessary because they define the safety envelope, and those which are desirable. For example, those which may prevent unnecessary reductions in the life expectancy of plant components, but which have no immediate effect upon safety. Operations must be within Operating Rules (ORs), and that these reflect the requirements of the current safety case. OR parameters should be physically and preferably directly measurable, but derived information may exceptionally be used.
LC 24: Operating Instructions	To ensure that all operations which may affect safety are undertaken in accordance with written operating instructions. Such instructions should be clear and unambiguous and should be consistent with the safety case and its assumptions. They should highlight Operating Rules (ORs) and require operations to be undertaken in accordance with them.
LC 25: Operational Records	To ensure adequate records are kept regarding operation, inspection, and maintenance of any safety-related plant. Normally, such records include plant operational logs kept by plant managers and supervisors, together with records of maintenance schedules and activities.
LC 26: Control and Supervision of Operations	No operations may be carried out which may affect safety, except under the control and supervision of suitably qualified and experienced personnel (SQEPs).
LC 27: Safety Mechanisms, Devices, and Circuits (SMDCs)	No plant should be operated, inspected, maintained, or tested unless suitable and sufficient safety mechanisms, devices, and circuits are properly connected and in good working order. The suitability and sufficiency of SMDCs should be identified in the safety case and established in the appropriate Operating Rules (ORs). Such requirements apply to the totality of the operating system, so as to ensure a system-wide approach and operation within the plant's safe operating envelope.
LC 28: Examination, Inspection, Maintenance, and Testing	To ensure that all plant that may affect safety, as identified in the safety case, receives regular and systematic examination, inspection, maintenance, and testing (EIMT) by and under the control of SQEPs in accordance with the plant maintenance schedule. The licensee should have a general program covering all aspects of EIMT for all plant on the site. The LC covers the arrangements for updating or amending the maintenance schedules.
LC 29: Duty to Carry Out Tests, Inspections, and Examinations	The LC enables the HSE NII, following consultation with the licensee, to require the licensee to perform any tests, inspections, and examinations that they may feel required and to be provided with the results.

Continued

Table 18-3. Continued

Licence condition	Brief description
LC 30: Periodic Shutdown	To ensure that, where necessary, any licensee periodically shuts down plant in order to carry out those requirements of LC 28. Such shutdowns require a shutdown plan and a statement of completion of the works. If an extension of the operating period between shutdowns is required, the licensee must provide the justification for this in accordance with LCs 14 and 28.
LC 31: Shutdown of Specified Operations	This LC gives discretionary powers to the HSE NII so as to require plant or process to be shutdown within a given period and to require a consent for start-up of any process shutdown under this condition. Necessary actions during the shutdown may involve plant modifications, improvements, and the preparation of or revision to safety cases in respect of the plant, operations, or processes.
LC 32: Accumulation of Radioactive Waste	This allows enforcement of adequate arrangements for waste minimisation and the total quantities of radioactive wastes accumulated on the site at any time and for recording the wastes so accumulated. Wastes should be disposed of via authorised routes where they exist or to recognised (perhaps interim) "end points" such as interim storage in a safe passive form.
LC 33: Disposal of Radioactive Waste	The LC gives discretionary power to the HSE NII to direct that radioactive wastes be disposed of by the licensee in a specified manner. This is also related to the powers of the UK Environment Agencies, where disposal is covered under the Radio Substances Act (RSA 1993). Once such disposal routes are established, there is the presumption that they should be utilised at the earliest opportunity commensurate with the safe handling of the radioactive waste.
LC 34: Leakage and Escape of Radioactive Material and Radioactive Waste	To ensure that radioactive material is adequately controlled or contained so that it cannot leak or otherwise escape from such control into the environment. This condition relates to the potential hazard associated with radioactive material and not to the risk. Consequently, it may apply even if there is no immediate effect on safety.
LC 35: Decommissioning	This requires the licensee to make adequate provisions for decommissioning and to give discretionary powers to the HSE NII to direct that decommissioning of any plant or process may be initiated or halted. All other conditions also apply to decommissioning, albeit that some may have reduced impact depending upon the hazard remaining until the site is delicensed. In general, the most hazardous materials, for example spent fuel, should be removed from the reactor at the earliest stage in the decommissioning process. Where the timescales are lengthy, wastes should be stored in a safe passive form and this should be identified in the decommissioning program. The decommissioning program should be based upon the systematic and progressive reduction of hazards and that decommissioning should take place as soon as reasonably practicable, taking all factors (including safety and economics) into consideration. Such a decommissioning program requires close liaison with the Environment Agencies since they control and authorise radioactive discharges from the site and the disposal of radioactive wastes.
LC 36: Control of Organisational Change	This LC allows the HSE NII to give a judgment upon the adequacy of the licensee's human resource strength and organisational structures. It provides guidance for HSE NII inspectors on judging the adequacy of the licensee's arrangements to control and change its organisational structure or resources which may affect safety. Particularly important is the transparency of the process. It applies to all changes to organisation, structure, and resources, without prejudging if the change will affect safety. It also requires a "baseline" submission on resources from the licensee, from which future changes will be evaluated.

entitled "Review of Radioactive Waste Management Policy — Final Conclusions" sets out the following policy requirements for waste producers:

(a) can deal with the wastes that they create using current technologies,

(b) characterise and segregate the waste and store it "in accordance with the principles of passive safety," and

(c) plan and develop programs to dispose of accumulated waste and for the decommissioning of redundant plant.

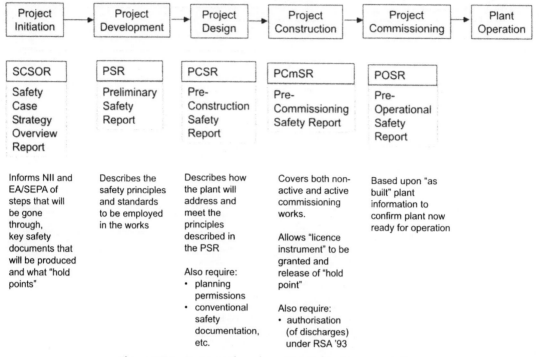

Figure 18-1. Project Lifecycle — Key Safety Documentation.

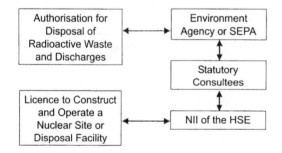

Figure 18-2. Regulatory System Showing the Different Responsibilities of the Safety Regulator (HSE NII) and the Environmental Regulators (EA and SEPA).

Cm 2919 also strongly emphasises the principles of sustainable development in relation to radioactive waste management policy.

• *International Treaties and Conditions on Transboundary Shipments of Radioactive Waste.*

    (a) *The Fourth ACP-EEC Convention (Lomé Convention).* Approved by the EC in 1991 such that the Community shall prohibit all direct or indirect export of hazardous or radioactive waste to the African, Caribbean, and Pacific (ACP) regions — mostly former colonies of European

countries. These provisions do not prevent a Member State to which an ACP State has chosen to export waste for processing from returning the processed waste to the ACP State of origin.

    (b) *The Bamako Convention.* The Organisation of African Unity (OAU) adopted the Bamako Convention on the ban of the import into Africa and the control of transboundary movement and management of hazardous wastes within Africa.

    (c) *Directive 92/3/Euratom on the Supervision and Control of Shipments of Radioactive Wastes between Member States and into and out of the Community.* This applies to shipments whenever the quantities and concentrations exceed the levels laid down.

• *EC Directive 85/337/EEC (as amended by 97/11/EC) on Environmental Assessment.* Adopted in 1997 concerning the assessment on the effects of certain public and private projects on the environment (known as the "EIA Directive"). It requires that, before development, consent is given to projects likely to have a significant effect on the environment and they should be subject to an assessment and that this assessment is integrated into the consent procedure. Figure 18-3 illustrates a process for carrying out an Environmental Impact Assessment which follows European guidance

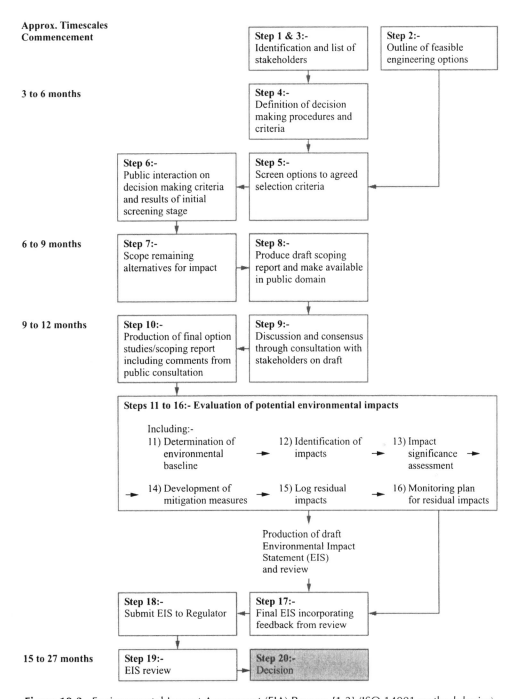

Approx. Timescales
Commencement

Step 1 & 3:-
Identification and list of stakeholders

Step 2:-
Outline of feasible engineering options

3 to 6 months

Step 4:-
Definition of decision making procedures and criteria

Step 6:-
Public interaction on decision making criteria and results of initial screening stage

Step 5:-
Screen options to agreed selection criteria

6 to 9 months

Step 7:-
Scope remaining alternatives for impact

Step 8:-
Produce draft scoping report and make available in public domain

9 to 12 months

Step 10:-
Production of final option studies/scoping report including comments from public consultation

Step 9:-
Discussion and consensus through consultation with stakeholders on draft

Steps 11 to 16:- Evaluation of potential environmental impacts

Including:-
11) Determination of environmental baseline

12) Identification of impacts

13) Impact significance assessment

14) Development of mitigation measures

15) Log residual impacts

16) Monitoring plan for residual impacts

Production of draft Environmental Impact Statement (EIS) and review

Step 18:-
Submit EIS to Regulator

Step 17:-
Final EIS incorporating feedback from review

15 to 27 months

Step 19:-
EIS review

Step 20:-
Decision

Figure 18-3. Environmental Impact Assessment (EIA) Process [1,2] (ISO 14001 methodologies).

documentation and ISO 14001 Standards incorporating a series of steps with indicative timescales. Although these timescales may seem to introduce a delay into the "nuts and bolts" of the decommissioning and waste management program, this consultation is necessary so as to gain public confidence and, thereby, avoid the "decide and defend" criticism that has plagued the nuclear industry in the past.

- *International Rules on Sea Disposal.*

  (a) *London Dumping Convention 1972.* Originally adopted a global ban on the dumping at sea of high level radioactive wastes. In 1983, this was extended to a moratorium on the dumping of all radioactive wastes at sea.

(b) *UN Conference on Environment and Development Agenda 21.* The document has no legal effect, but represents a significant advance in international cooperation in the implementation of global environmental policies. It sets out an environmental action plan for sustainable development and seeks support for the safe and environmentally sound management and disposal of radioactive wastes.

(c) *United Nations Law of the Sea Convention 1982.* "States shall take all measures necessary to ensure that activities under their jurisdiction or control are so conducted as not to cause damage by pollution to other States and their environment . . . " [3].

• *International Convention on the Safety of Spent Fuel and Radioactive Waste management.* The convention was negotiated under the aegis of the IAEA, but contains general "motherhood" statements of good intent. These include:

(a) achievement and maintenance of high levels of safety world-wide including safety related technical cooperation,

(b) ensuring that during all stages of spent fuel and radioactive waste management, effective defences against potential hazards to individuals, society, and the environment are protected from the harmful effects of ionising radiation, and

(c) prevention of accidents with radiological consequences.

• *The Environmental Protection Act (EPA)1990.* This is a wide-ranging piece of general environmental legislation consisting of several parts. It is not specifically applicable to radioactive substances, but three of these sections are particularly important to nuclear decommissioning and waste management operations, namely:

(a) Industrial control, in particular Integrated Pollution Control (IPC);

(b) Waste Management Regime; and

(c) Statutory Nuisance (for example, noise, odor, etc.).

They are all enforced through regulations.

Part 1 of the EPA states that no person shall carry out a prescribed process . . . except under the authorisation granted by the enforcing authority (in the UK, this would be SEPA or the EA, but may also involve the Planning Authorities and HSE NII). Generally, the more polluting processes are covered by Integrated Pollution

Control (IPC). Discharges to air, water, or land are regulated and the principle of Best Practical Environmental Option (BPEO) applies. Less polluting processes are regulated for air emissions, and the principles of BPEO do not apply to these. Therefore, a nuclear fuel fabrication plant on a nuclear licensed site will require an authorisation under IPC because of the nonferrous metal processing operations involved. The scope of such an authorisation may also cover the conventional heating boilers and the aqueous effluent handling facilities on the site.

Part II is concerned with industrial, commercial, and domestic solid and liquid waste production. Since 1996, all wastes being disposed of at landfill sites have been subject to a "landfill tax" in order to encourage waste minimisation. Part IIA covers contaminated land regimes.

Part III covers nuisance pollution associated with noise, odor, dust, or any other such disruption to lives or the operation of a business.

18-3-3. *Assessment Terminology*

The approaches adopted to demonstrate compliance include:

• *Best Available Techniques not Entailing Excessive Cost (BATNEEC).* BATNEEC was introduced under the Environmental Protection Act 1990. It must be used to prevent or, if this is not practicable, to minimise release of prescribed substances into the environment, and to render harmless any substances that are released. The "best available technique" is the most effective technique for preventing, minimising, or making harmless polluting releases that can be achieved by the site operator. Techniques can include plant, processes, staff training, working methods, etc. There is a duty on the nuclear site operator and the Environment Agency Chief Inspector to keep up-to-date with developing technology and techniques as the "best available technique" may change over time. The cost of applying this technique should not be disproportional to the environmental benefits gained (cost benefit argument). BATNEEC generally applies to the nonradioactive discharges from a nuclear site.

• *Best Practical Environmental Option (BPEO)* (see also Chapter 1, Section 1.8). BPEO is the outcome of a systematic consultative and decision-making procedure, with emphasis on the protection and conservation of the environment across land, air, and water. The BPEO covers all aspects of the option for the delivery of raw materials to the final disposal. In other words, this seeks to minimise the overall environmental impact through consideration of the way in which a process should be managed in terms of issues such as the type of resources to be used and the final disposal route for any wastes.

- *Best Available Techniques (BAT).* BAT was introduced in the European Directive (96/61) for Integrated Pollution, Prevention, and Control (IPPC). This will eventually replace Integrated Pollution Control (IPC). The concept of BAT will replace BATNEEC and BPEO. Therefore, protecting the environment as a whole must be considered when determining BAT (this is how BPEO is taken into account). BAT also correctly includes economic considerations to be taken into account (as does the BATNEEC technique).
- *Best Practical Means (BPM)* (see also Chapter 1, Section 1.8). BPM relates to the means used to minimise the production and the release of radioactive wastes to the environment. The means include the engineering and management options in the same way as included for in the BATNEEC technique as applied to nonradioactive releases. BPM is a condition of waste disposal authorisations under the Radioactive Substances Act 1993, and is used to exert downward pressure on the discharge of radioactive waste to the environment. In essence the issue of BPM is the requirement to optimise a particular process specifically to ensure that radioactive wastes are not created unnecessarily. Both BPM and BPEO need to be considered in undertaking any activity involving radioactive materials — primarily to ensure that the overall environmental burden is minimised through the consideration of BPEO, and then that the preferred waste management option is optimised to minimise radioactive releases through the consideration of BPM.
- *Waste.* Waste and controlled waste are defined in Section 75 of the Environmental Protection Act (EPA) 1990. Waste is "any substance which constitutes a scrap material or an effluent or other unwanted surplus substance arising from the application of any process, and any article which requires to be disposed of as being broken, worn out, contaminated or otherwise spoiled." Controlled waste means "household, industrial, and commercial waste." Radioactive waste is classed as "Special Waste" controlled under the Radioactive Substances Act.

18-3-4. *Assessment Criteria*

The Environment Agency is charged with ensuring that all radiation exposures to people resulting from radioactive waste disposals as low as reasonable, having due regard for the *maximum* dose constraints of 0.3 mSv y^{-1} from any new source and 0.5 mSv per annum for the discharges from any single site. Authorisations for discharges, as granted under RSA 93 by the Environment Agency, must, therefore, be within these upper bounds, but they must also be accompanied by a Best Practical Means (BPM) study to demonstrate that discharges and the resultant

exposure to the public is kept as low as reasonably achievable (ALARA). The limits and the BPM requirements are separate and both must be complied with.

In the past, authorisations for discharges were granted separately for discharges of liquid and gaseous effluent, for solid waste incineration, and for offshore transfer of radioactive waste. Such authorisations are now being integrated into a single multimedia authorisation under which all radioactive waste disposals will be controlled.

This should also be seen in the context of the Government's commitment to the OSPAR Strategy, whereby the UK is working towards achieving further substantial reductions or elimination of radioactive discharges. By the year 2020, the OSPAR Commission will attempt to ensure that discharges, emissions, and losses of radioactive substances are reduced to levels where the additional concentrations in the marine environment above historic levels, resulting from such discharges, emissions, and losses, are close to zero. Of course, the decommissioning process by its very nature generates wastes. Just as has been seen in Chapter 14 for hazard, there will also be inevitable increases in discharges as a function of the decommissioning process. With modern techniques, it is anticipated that it will be possible to reduce discharges within current overall authorisations. However, there could well be certain increases in radionuclide specific activity levels until that particular aspect of the decommissioning work is completed. The presumption that increases in discharge levels will be exceptional during intensive periods of decommissioning is probably incorrect. Discharges should, however, be within current overall discharge envelopes, and time-limited increases should be allowed so as to reduce risks associated with historic waste legacies/redundant plant. Discharge limits should reflect operational and decommissioning business needs and not be set at the dose constraint level.

Chapter 2 covers issues associated with the protection of man from adverse effects of radiation. In the past, it was felt that if protecting humans from the harmful effects of radiation then other species should also be protected. Now there is a growing view that environmental radiation protection should include species other than, and as well as, humans. Further research (see 5th European Framework R&D Programme project FASSET [4]) is required, and is being carried out, in this area.

18-4. Environmental Management System (EMS)

Compliance with such a plethora of regulation requires a well-developed process which would normally cover the following key areas:

- commitment from senior management;

- defined environmental objectives and targets that are annually reviewed;
- establishment of a management program which demonstrates how environmental objectives and targets are to be achieved. This may be done under the framework of ISO 14001 [5,6];
- improved environmental awareness and competence through appropriate training;
- a structure for internal and external communication and reporting;
- assessment of the environmental impacts and risks from processes on site;
- identification of all environmental legal requirements directly attributable to the nuclear site activities;
- maintenance of an effective Environmental Management System (EMS) documented control procedure;
- a system of environmental operational control (including procurements and contractors);
- maintenance of a system of emergency preparedness and response; and
- an audit program and a system of management review.

Such an EMS, therefore, requires involvement of the Site Director, the site Environmental Manager, and Group managers. The appointment of a single point of responsibility to manage the EMS together (depending on the size of the site), a site EMS co-ordinator, the senior Suitably Qualified and Experienced Person (SQEP) plant operator (sometimes known as the Authority to Operate or ATO Holder) and the plant operators themselves so that the culture runs through the whole organisation. Independent assessment against a site ISO 14001 Environmental Management System (EMS) is carried out in a similar way to assessments against Quality Assurance ISO 9001 requirements.

In particular, a decommissioning site must:

- carry out an analysis of plant discharges to the environment;
- ensure that the results of the analysis are properly recorded and retained for at least 5 years and that copies are available to the environmental regulators;
- have procedures in place for both plant operation and managerial processes and that all staff are given the appropriate training;
- have maintenance procedures including Unusual Occurrence Reporting (UNOR) to allow monitoring of plant performance. These systems should allow trends to be spotted before authorised discharge levels are likely to be breached;
- safety critical plant and equipment to be maintained through Examination, Inspection, Maintenance, and Testing (EIMT) schedules including Critical Environmental Protection (CEP) systems and Environmental Protection (EP) systems;

- for new plant and processes, including decommissioning, IPC implications to be considered by using appropriate project management checklists; and
- provision of environmental monitoring work in accordance with regulatory requirements.

See also various websites [5–9].

18-5. Organisational Framework

Figures 1-6 and 18-4 describe the nuclear decommissioning and radioactive waste policy, advice, regulation, and operation arrangements in the UK. The complexity and numbers of bodies involved makes it understandable why progress on approvals to move forward with decommissioning in the UK has been so slow. Acronyms, which are not spelled out in Figure 18-4, are also explained in Chapter 1, Section 1-9.

The currently accepted dose levels in the UK are set out in Table 18-4.

18-6. Tolerability of Risk

Chapter 14 describes the difference between risk and hazard. The concept of Tolerability of Risk (ToR) is associated with both conventional and nuclear practices. It defines risks which are so high they are intolerable, and risks that are so low that they may be considered as broadly acceptable, such that no further regulatory pressure to reduce the risks further would be applied. Between these levels, the risks must be reduced in accordance with the principles of As Low As Reasonably Practicable (ALARP), as illustrated in Chapter 14, Figure 14-1. Before carrying out a decommissioning or waste management activity, the licensee has to demonstrate that the risks involved in the methods being adopted or arising from plant or decommissioning activities are as low as reasonably practicable. Although UK regulation is non-prescriptive, publicly available guidance (intended for the regulator's inspectors) on how to apply and demonstrate ALARP principles [10] has been published. In certain cases, the ALARP or tolerability region has been translated into numerical values in the form of Basic Safety Limits (BSLs) and Basic Safety Objectives (BSOs). The following nine principles are likely to need to be addressed in any analysis:

- The application of ALARP can only be to those risks that the nuclear site licence holder controls.
- Affordability. Whether the Decommissioning Management Company is in a position to afford the costs is not a legitimate factor in the ALARP argument, though the costs themselves are.

Nuclear Decommissioning Authority (NDA)	The NDA is a NonDepartmental Public Body (NDPB) with a formal Board responsible to Government Ministers. (Anticipated to be formally constituted by 2005.)

NDA Customers	Customer Interests / Responsibilities
DTI	Funding, sponsorship of BNFL and UKAEA, security regulation, decommissioning policy, and accounting to Parliament for nuclear safety.
MoD	Funding.
DEFRA	Radioactive waste policy, environmental, and planning regulation in England.
Department for Transport	Transport of radioactive waste.
Treasury	Public Expenditure.
Scottish Executive	Radioactive waste policy, environmental, and planning regulation for Scotland.
Regulators	HSE (safety), EA & SEPA (environmental), and OCNS (security).
National Assembly for Wales	Clean up of Wylfa and Trawsfynydd sites and management of associated wastes in Wales.

BNFL   - British Nuclear Fuels Limited
DEFRA - Department for Environment, Food, and Rural Affairs
DTI   - Department for Trade and Industry
EA   - Environment Agency
HSE   - Health and Safety Executive
MoD   - Ministry of Defence
OCNS   - Office for Civil Nuclear Security
SEPA   - Scottish Environment Protection Agency
UKAEA - United Kingdom Atomic Energy Authority

Figure 18-4. Relationships between the proposed future Nuclear Decommissioning Authority (NDA) and other Public Sector Bodies in the UK.

Table 18-4. Recognised Annual Dose Levels Associated with UK Regulation

Dose	Applicability/Comments
2.2 mSv y^{-1}	Average radiation dose to members of the UK population from natural background radiation (see Chapter 14, Table 14-2).
1 mSv y^{-1}	Recommended maximum dose to members of the public from man-made radiation.
0.5 mSv y^{-1}	Target maximum dose to members of the public from any single nuclear site (irrespective of the size of the site and number or type of nuclear installations on it).
0.3 mSv y^{-1}	Target maximum dose to members of the public from a new nuclear installation.
10 μSv y^{-1}	De minimus level for regulation (equivalent to an annual risk of death of 10^{-6} or 1 in a million).

- Simplistic applications of the ALARP argument must not be used to argue against meeting legislative or regulatory requirements and declared Government policy.
- ALARP demonstrations must consider the various options open which could improve the level of safety. The timescales for implementation may be a factor in the choice of options. On the one hand, the environmental regulator may wish to delay any actions so as not to foreclose options. The safety regulator may wish to see more immediate action, so as to improve safety and reduce risks (already likely to be very low) or hazards.
- Options may include partial and full implementation of one or more particular measure to arrive at an overall solution.

- For existing plants, it is necessary to compare the plant with modern standards, examine shortfalls, and what options exist for improvement. Older plants may meet the ALARP requirement at higher risks than new ones. This is especially relevant to redundant facilities where it would be pointless, within an acceptable safety envelope, to modernise a plant prior to decommissioning and demolishing it so site environmental remediation could continue.
- A consideration of the costs in relation to the effects of a possible resulting accident, although not a full Cost Benefit Analysis.
- The ALARP case should be fit for purpose. If the risks are high, then a demonstration of ALARP would need to be more rigorous than if the risk is low. The degree of rigor should also depend upon both the probability and consequence levels involved. The sensitivity of an analysis over a range of uncertainties should, where appropriate, be considered.
- The demonstration of ALARP employs a comparison of costs and risk reduction benefit prior to ruling out an improvement. Legal interpretation on this subject in the UK is based upon the concept of "gross disproportion." Unfortunately, decommissioning does not bring a benefit in terms of a useful commodity or revenue stream from the end product, and it is, therefore, extremely difficult to justify decommissioning activities in these terms. Rather, the costs for decommissioning are all outgoing leaving, ultimately, a green field site for free and unrestricted use.

An ALARP checklist for those reviewing the case is summarised in Table 18-5.

Table 18-5. ALARP Checklist — For Use by Regulatory Inspectors When Assessing ALARP Arguments

No.	Levels of Risk and ALARP (as low as reasonably practicable) Basic Points for Consideration
1	Has the full range of health and safety detriments been considered adequately?
2	Does the ALARP argument refer only to those risks which the licensee controls?
3	Affordability is not a legitimate factor in the assessment of costs.
4	ALARP cannot be used to argue against statutory duties or government policy.
5	Have all relevant options been considered by the licensee?
6	Does the licensee's study of options begin with the safest (as opposed to the cheapest) option?
7	If measures are deemed not reasonably practicable, has partial implementation been considered? Need also to be wary of "deluxe" measures unduly inflating the cost.
8	If implemented measures do not make the risks broadly acceptable, has implementation of additional measures been considered?
9	For measures deemed not reasonably practicable, have the licensees demonstrated gross disproportion, taking due account of aversion, and that the higher the consequences the more weight they should have in the decision?
10	The ALARP arguments should explicitly consider qualitative features related to engineering and other types of relevant good practice.
11	For cases relying solely upon good practice, are the requirements acceptable (up-to-date, most stringent of good practice, not of a minimum requirement, which good practice option has been employed, etc.).
12	Are all relevant engineering Safety Assessment Principles (SAPs) met? If not, have the licensees identified and considered any deficiencies from an ALARP perspective?
13	Application of Safety Assessment Principles (SAPs) and moving up the SAP hierarchy (avoid hazard, maintain safe conditions by passive rather than active means if possible, initiate automatic protection in preference to manual systems).
14	Quantitative ALARP requires the reduction in risk to be estimated.
15	All health and safety effects of the modification must be considered in considering the change in risk terms.
16	A Cost Benefit Analysis (CBA) on its own is not acceptable as an ALARP case.
17	The value of a life should not be below £2M (2001 money values) for cancer or radiation-induced deaths.
18	Have adequate (all inputs to the CBA) sensitivity studies demonstrating the robustness been carried out? Are there uncertainties such that a precautionary approach is appropriate?
19	Costs of implementation cover all aspects (fabrication, training, loss of revenue, etc.) and should be offset by gains in production, etc., other than safety.
20	Temporary shutdown costs are legitimate, but if inclusion of these costs indicates an improvement is not called for, then consideration ought to have been given by the licensee to delayed or phased implementation.
21	The discounting of costs and benefits is acceptable, but it is important to make sure such claims are reasonable and to use Government guidelines on discount rates (currently 6%).
22	Discounting over long periods (in excess of 50 years) is problematical and needs careful consideration.
23	Have the guidelines on Cost Benefit Analysis been followed?
24	ALARP applies to all times, and arguments employing Time at Risk may need special consideration.
25	Reverse ALARP arguments for increased risk are only allowable in special circumstances.
26	Dose sharing: has the licensee given adequate consideration to changing working methods, engineering controls, or other means of restriction?
27	Sharing the risk, from accidental exposure, between groups of workers is not allowable.
28	Have occupancy factors in assessments of worker risk been properly considered?
29	For long-term risks, good practice and the Safety Assessment Principles (SAPs) hierarchy with emphasis on "control of the hazard" are important, as is the need to consider the full-life cycle of the installation.

Table 18-6. International Nuclear Event Scale (INES) (UNO-IAEA & OECD-NEA)

Level	Description	Criteria	Examples
Accidents			
7	Major accident	External release of a large fraction of the reactor core inventory typically involving a mixture of long- and short-lived fission products (typically radiologically equivalent to more than tens of thousands of TBq ^{131}I). Possibility of acute health effects. Delayed health effects over a wide area, possibly involving more than one country. Long-term environmental consequences.	Chernobyl, Ukraine, 1986
6	Serious accident	External release of fission products (typically radiologically equivalent to thousands to tens of thousands of TBq ^{131}I). Full implementation of local emergency plans likely to be required so as to limit serious health effects.	
5	Accident with off-site risks	External release of fission products (typically radiologically equivalent to hundreds to thousands of TBq ^{131}I). Partial implementation of local emergency plans required in some cases to lessen likelihood of health effects. If reactor initiated, then severe damage to a large fraction of core due to mechanical and heating effects.	Three Mile Island, USA, 1979; and Windscale, Pile 1, UK, 1957
4	Accident mainly in installation	External release of radioactivity resulting in a dose to the most exposed individual off-site of the order of a few mSv. Unlikely to be necessary to provide off-site protective actions except for possible local food controls. If reactor initiated, then some damage to core due to mechanical and heating effects.	St. Laurent, France, 1980
Incidents 3	Serious incident	External release of radioactivity above authorised limits resulting in a dose to the most exposed individual off-site of the order of tenth mSv. Off-site protective measures not required. High levels of on-site radiation and/or contamination due to equipment failures or operational incidents. Over exposure of workers with individual doses exceeding 50 mSv. Incidents in which a further failure of safety systems could lead to accident conditions, or a situation in which safety systems would be unable to prevent an accident if certain initiating events were to occur.	Vandellos, Spain, 1989
2	Incident	Technical incidents or anomalies which, although not directly or immediately affecting plant safety, are liable to lead to subsequent re-evaluation of safety provisions.	
1	Anomaly	Functional or operational anomalies which do not pose a risk but which indicate a lack of safety provisions. This may be due to equipment failure, human error or procedural inadequacies. (Such anomalies should be distinguished from situations where operational limits and conditions are not exceeded and which are properly managed in accordance with adequate procedures.) Such anomalies are typically classified as "below scale."	
Below scale 0	No safety significance		

18-7. References

1. www.iso14000.net/isolink.html [AQ 18-6].
2. www.abev.ac.uk/-eiawww/ecomate/links.htm [AQ 18-7].
3. *United Nations Law of the Sea Convention*, Article 194(2) (1982) [AQ 18-8].
4. Euratom R&D Project, "Framework for Assessment of Environmental Impact (FASSET)," Information available on the FASSET website, www.fasset.org [AQ 18-9].
5. www.iso14000.net/isolink.html [AQ 18-10].
6. www.abev.ac.uk/-eiawww/ecomate/links.htm [AQ 18-11].
7. www.defra.gov.uk [AQ 18-12].
8. www.sepa.org.uk [AQ 18-13].
9. www.environment-agency.gov.uk [AQ 18-14].
10. HSE NII Technical Assessment Guide, Demonstration of ALARP, T/AST/005, Issue 1 (2002).

Chapter 19

Management of Low Level Wastes (LLW)

19-1. Introduction

Almost all materials are, strictly speaking, radioactive, because they contain traces of naturally occurring radionuclides. The term radioactive waste is reserved for particular classes of waste, defined in guides to National regulations, which contain concentrations of radioactivity above the levels specified in those guides. Low Level Waste (LLW) is defined in Chapter 1, together with the typical relative volumetric arisings between the different classes of wastes. Chapter 16 describes some of the conflicts within the guides at the low end of the Low Level Waste activity spectrum. These are in particular need of clarification for large volume low activity wastes and for incorrect technical distinctions between disposal requirements for naturally occurring and man-made radionuclides.

This chapter describes the sources of LLW, its treatment, packaging, and disposal.

19-2. Sources of LLW

19-2-1. Introduction

Low Level Waste may contain natural radionuclides, generally uranium, thorium, and the products into which they decay, and man-made radionuclides. Most of the man-made radionuclides result from the fission of uranium in nuclear reactors: they are either the fission products themselves and their radioactive decay products, or activation products, which are produced when neutrons released during the fission process are absorbed by atomic nuclei, for example in materials that make up the structure of the reactor. LLW may be protective clothing and filters that have come into contact with such radionuclides and so become contaminated to the extent that the activity falls within the LLW classification. There is a third class of radionuclides that is particularly important from the point of view of the management of radioactive wastes from the nuclear industry. These are the actinides, a group of heavy elements including thorium and uranium (which

occur naturally), and man-made elements such as plutonium, americium, and neptunium, which result from the absorption of neutrons by uranium or thorium, and a succession of subsequent reactions. These are generally long-lived alpha species and, as such, are very limited in their activity levels within the LLW classification.

Some of the radionuclides used in medicine, industry, and research, which in turn appear in radioactive wastes, are produced in particle accelerators rather than in nuclear fission reactors. Medical and industrial radionuclides are of the same general type as those associated with the nuclear industry. For example, ^{60}Co is an activation product, ^{137}Cs a fission product, and ^{241}Am an actinide. Half-lives vary from seconds to many thousands or even millions of years. Activation and fission products emit mainly beta and gamma (β and γ) radiation, whereas actinides are mainly alpha (α) emitters and are much more radiotoxic if they enter the human body. LLW classifications generally more severely limit the activity levels associated with long-lived alpha emitters than the shorter lived beta/gamma (β/γ) species.

Sources of LLW are described in Sections 19-2-2–19-2-5.

19-2-2. Fuel Manufacture

Starting from uranium ore concentrates, uranium is processed into metal or oxide form and fabricated into fuel elements. LLW arises from small quantities of unirradiated fuels and scraps, contaminated handling equipment, and protective clothing involved, together with filters and effluent treatment products used in the manufacturing process. A simplified flow diagram for LLW (and some ILW) production from fuel manufacture and enrichment is shown in Figure 19-1.

19-2-3. Nuclear Power Generation and Decommissioning

The fuel elements so produced are utilised in nuclear power stations. The different types of reactor, as described

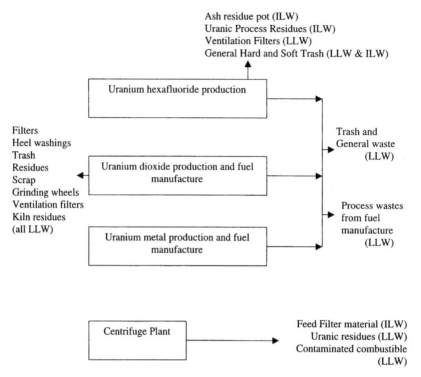

Figure 19-1. Simplified Flow Diagram for Radioactive Waste Production From Fuel Manufacture and Uranium Enrichment.

in Chapter 1, produce different waste streams. In general, though, LLW is produced from contaminated reactor building items, spent fuel storage pond and water treatment plant filters (the sludges and ion exchange material more likely to be ILW—see the Case Study in Chapter 20), general operational and maintenance activities producing combustible and noncombustible products such as incinerator ash, laundry, effluent treatment plant wastes and building contamination, evaporator concentrate accumulation facility waste, and some secondary side ion exchange resins, which may include some ILW.

Power station decommissioning wastes (assuming a safestore strategy as described in Chapter 6) are illustrated in Figure 19-2.

19-2-4. *Fuel Reprocessing*

After utilisation in the power station, the spent fuel may be reprocessed to recover unused uranium and plutonium, and to separate the highly radioactive fission products and actinide wastes. Alternatively, as discussed in Chapter 16, the used fuel may be declared as High Level Waste. Low Level Waste represents much the largest volume of waste arising from reprocessing. Operational waste from commercial reprocessing consists of a wide range of soft

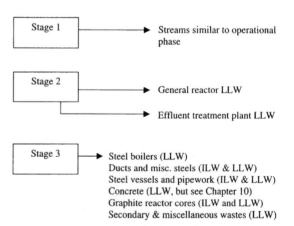

Figure 19-2. Power Station Decommissioning Wastes (assuming a safestore decommissioning concept).

and hard trash from routine operations and maintenance. Waste items include discarded protective clothing, paper towels, general tools, filters, plastic bags and sheeting, pipework, cabling, glassware, redundant equipment, concrete, rubble, and soil. Redundant fuel transport flasks and fuel storage pond furniture also contribute to the total LLW arisings. Decommissioning LLW would include ductwork, pipework, ventilation systems, cells, glove boxes, radiation shielding, and building structures.

19-2-5. *Other Sources*

Apart from the nuclear fuel cycle, radioactive wastes are also produced from nuclear industry research and development (R&D) activities, but in much smaller volumes. On sites having been involved in fast reactor research, postirradiation examination (PIE) and PCM LLW is generated from the operation and maintenance of plants. The normal operations at other research sites will produce contaminated equipment and materials. As noted in Chapter 16, minor waste producers may be able to dispose of their LLW at higher disposal activity thresholds than nuclear site power station operators.

Defence sites involved with weapon assembly, disassembly, and refurbishment, but which do not manufacture plutonium or uranium components, only produce LLW or below threshold wastes, but may have quantities of depleted uranium (DU) for shell casings. Plants handling DU will have DU contaminated filters, target materials, target washings, and redundant equipment waste streams. The major radioactive contaminants at weapons sites are plutonium and uranium and smaller quantities of tritium and other $\beta\gamma$ emitting radionuclides. The solid wastes take the form of soft waste, such as coveralls, tissues, gloves, and sludges. The hard wastes are redundant contaminated machinery and decommissioning building rubble. Tritium filtration equipment may also provide a significant waste stream. Nuclear propulsion plant generates similar waste streams to PWR nuclear power generation plant.

Instrument dial manufacture involving radium luminising operations is no longer normal practice. However, there continues to be refurbishment work on such equipment and a requirement to dispose of older unwanted instruments. In the UK, the MoD has some 250,000 gaseous tritium light devices, with a total activity of some 5,000 TBq listed in their inventory [1].

Other minor waste producers include educational and research establishments, health authorities, Government departments, and Industrial companies. Examples of soft and hard LLW are shown in Figures 19-3 and 19-4, respectively.

19-3. LLW Disposal

19-3-1. *Regulatory Controls*

The regulatory aspects of authorisation for the disposal of LLW involve the requirement for immobilised and passively safe waste forms. Therefore, only solid wastes are normally acceptable for disposal and the construction of a long-term safety case (as described in detail in Chapter 20 for ILW disposal) necessitates that consignments must include details of activity levels.

Best Practical Means (BPM, see Chapter 18, Section 18-3-3) is used to:

- minimise waste volumes (usually by compaction or incineration);

Figure 19-3. Typical Soft LLW.

Figure 19-4. Typical Hard LLW.

Table 19-1. UK LLW Disposal Site Annual Activity Limits (GBq)

Radionuclide(s)	Annual Radiological Limits (GBq)
U	300
^{226}Ra and ^{232}Th	30
Other Alpha Emitters	300
^{14}C	50
^{129}I	50
^{3}H	10, 000
Others (inc. ^{60}Co)	15, 000
^{60}Co	2, 000

- limit activity migration (usually by grouting of the wastes into concrete blocks or within ISO type containers or drums and the provision of near surface engineered bunkering facilities);
- collect and monitor leachate (maintain necessary discharge and sampling equipment so as to monitor and keep appropriate records and then be able to take any preventative action to avoid unacceptable radioactive seepage into the biosphere);
- limit marine discharges and meet stream concentration limits (marine discharges may be limited by the volume and rate of discharge, the α, β, and ^{3}H content, the chemical oxygen demand (COD), the suspended solids content, the pH range, the total iron content, and assurance that it is free from oils and greases);
- monitor the wastes and the surrounding environment; and
- the maintenance of records.

Authorisations for disposal are reviewed on a regular basis and, in some cases, annual disposal limits are placed on the disposal company running the disposal site. Annual radiological limits for the LLW disposal site in Drigg, West Cumbria, UK are given in Table 19-1. Post-closure long-term safety cases are based upon a risk to any individual of $<10^{-6}$ and assessed against the radiological capacity of the disposal facility, taking into account future waste arisings.

LLW repository operational safety requires compliance with licence conditions typically as those described in Chapter 18 and includes attention to:

- management arrangements;
- training;
- operating conditions;
- inspection and maintenance requirements;
- dose assessments;
- record keeping; and
- emergency procedures.

The operational dose uptake to the critical group may be assessed through an environmental monitoring program agreed with the Regulators and might typically be equal to or less than 300 μSv y^{-1} to the critical group.

19-3-2. *Waste Control Systems*

All LLW accepted for disposal must normally comply with the disposal company wasteform and procedural specifications; which in turn must comply with operational and long-term safety-case regulatory requirements. The wastes must, therefore, be generated and consigned under an approved quality control regime. As such, waste generators may be subject to a program of audits by, or on behalf of, the repository company, and accept a degree of waste monitoring upon receipt at the disposal site.

The repository company will normally enter into a formal and legally binding contract with the waste producer involving technical waste acceptance criteria that may include:

- the definition of solid LLW (to place activity bands upon the wastes);
- materials to be specifically excluded (to avoid compromising the long-term safety case for the disposal facility);
- radioactivity limits;
- fissile content limits;

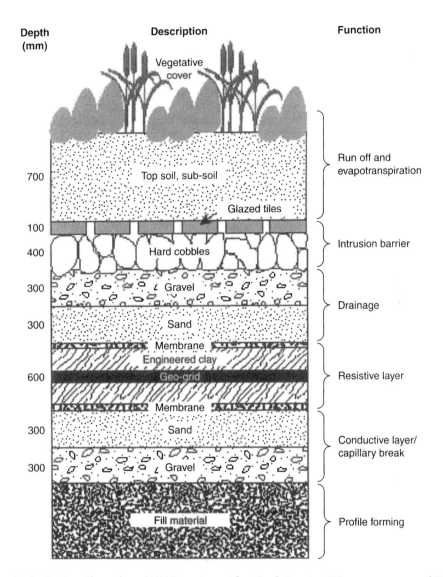

Depth (mm)	Description	Function

Vegetative cover

700 — Top soil, sub-soil — Run off and evapotranspiration

Glazed tiles

100
400 — Hard cobbles — Intrusion barrier

300 — Gravel — Drainage
300 — Sand

Membrane
Engineered clay
600 — Geo-grid — Resistive layer
Membrane

300 — Sand — Conductive layer/ capillary break
300 — Gravel

Fill material — Profile forming

Figure 19-5. Section Through an LLW Repository After Final Capping (Diagram courtesy of BNFL).

- waste conditioning requirements;
- quality assurance aspects;
- procedural and documentation requirements;
- a system of prior notification (so as to be ready for the delivery and to meet annual regulatory activity limits if applicable); and
- allocation of radiological disposal capacity.

Further guidance to the waste consigner may include the essential design features for the disposal containers (so as to ensure efficient space emplacement in the repository with common mechanical handling equipment) and other specifics, such as details of ion exchange resin acceptance.

Wastes may be characterised into wastestreams on a physical basis (e.g., combustibility, compactability, etc.)

or radionuclide composition associated with a waste "fingerprint." The physical, chemical, radiological, and toxic content of the waste will be required to be included in the general waste description and management records. The information that may be required includes:

- wastestream number and name;
- description of the process giving rise to the LLW;
- physical and chemical composition including how either prohibited materials are made safe or excluded;
- details of the conditioning and packaging of the LLW; and
- method and basis of radioactivity assessment. (For example, a dose rate conversion, fully referenced derivation, limitations, and how nonconforming wastes

are assessed, including consideration of potential uncertainties. Alternatively, a radionuclide fingerprint and whether determined by sampling or analysis. Short-lived radionuclides may be given exemption unless they are not in equilibrium, and some radionuclides may be excluded if, for example, they are below de-minimis levels.)

A typical waste receipt monitoring campaign might include a set of levels whereby, at Level 1, 100% of consignments are given radiation, contamination, and weight checks. At Level 2, perhaps, some 5% of consignments may be subjected to nondestructive testing using real time X-ray analysis, high resolution gamma spectrometry, and both passive and active neutron monitoring of the container. At Level 3, perhaps, 1% of consignments may be subjected to intrusive examination and destructive analysis.

19-4. LLW Disposal Practices

A description of world-wide nuclear disposal is given in Appendix 1. Practices differ, but general tumble tipping into unlined and unmonitored trenches is normally not acceptable practice. Trenches cut into clay layers which may act as an impermeable ground layer may be adequate, depending on the disposal site's projected waste inventory and how this matches with long-term safety case requirements. More normal modern practice is to provide purpose-built concrete bunker or vault type arrangements with a system of monitored drains and break tanks such

that sampling may occur before any effluent is discharged directly or routed for treatment. Uncontrolled lateral effluent discharges are prevented by the concrete bunker walls. The vaults may be covered with a soil surface capping and any run-off water collected in the monitored drainage system. Vault capping may be grassed over and trees planted so as to improve the environmental impact of the LLW disposal site. However, it is recognised that a soil cap has a limited life and, for the longer-term (prior to site closure), such a temporary soil cap will need to be replaced with a far more durable and impermeable barrier. Figure 19-5 shows a cross-section through a typical LLW disposal facility after final capping.

Self-contained concrete cubes, full size and half height ISO (International Standards Organisation) containers are often used for LLW waste disposal because they may be handled and transported using conventional equipment and lorries. Leachate generation is eliminated during the period of container integrity. Figure 19-6 shows full height ISO containers of LLW in engineered vaults prior to capping. Figure 19-7 shows the grouting of compacted wastes within an ISO container and Figure 19-8 is a drawing of such a grouting facility.

19-5. LLW Conditioning Facilities

The typical LLW strategy is, therefore:

• waste minimisation;
• waste characterisation;
• volume reduction;

Figure 19-6. Stacked Full Height LLW ISO Containers in Concrete Vaulted Disposal Facility (Photograph courtesy of BNFL).

Figure 19-7. Grouting of LLW in ISO Containers (Courtesy BNFL).

Figure 19-8. Grouting Facility for Containerised LLW (Courtesy BNFL).

- vault design and operation; and
- long-term radiological performance.

High force compaction and grouting is an efficient way to achieve waste volume reduction and immobilisation. Hard wastes may be loaded directly into the containers or size reduced using shredders or hydraulically operated cutting tools. Soft wastes are normally precompacted within their initial containers (often standard mild steel 200 l drums) and then loaded into larger ISO containers. The containers are then transported to a grouting facility before disposal. A low viscosity grout made up from pulverised fuel ash (PFA), cement, and super plasticiser is used. The grouting fills internal voidage and provides a cap across the external upper face of the container. In this way, a uniform load distribution is generally achieved and the grouting allows only limited settlement of the repository final site closure cap.

Technical support work helps to underpin the engineering and waste management developments and provide information for numerical modeling involved in radiological impact assessments and comparative risk assessments. The overall repository postclosure, or long-term, safety case includes consideration of groundwater, gaseous other potential pathways associated with human intrusion, and the return of radionuclides to the biosphere.

This, in turn, requires an understanding of the hydrogeology in the vicinity of the site.

19-6. Reference

1. UK Nirex/DoE. *UK National Radioactive Waste Inventory*, UK Nirex/DoE Report DOE/RAS/96.005, Electrowatt Engineering Services (UK) Ltd., Horsham, W. Sussex, May 1996.

Chapter 20

Management of Intermediate Level Wastes (ILW)

20-1. Introduction

This chapter describes some of the sources of Intermediate Level Wastes (ILW), together with their processing, storage, and handling criteria. Bearing in mind that ILW disposal facilities are still being developed around the world, this chapter also covers a phased disposal approach for such wastes, together with the development of appropriate waste package specifications and the suitability of proposals to meet such criteria. The phased disposal concept covers the management of the waste from generation to eventual possible disposal in a deep waste repository and includes:

- treatment and conditioning,
- waste packages,
- transport systems,
- interim storage,
- receipt and transfer underground,
- package transfer to underground vaults,
- extended monitored storage and issues of retrievability,
- backfilling of the vaults, and
- repository closure and geological isolation.

A waste packaging case study is included to illustrate, in a simplified manner, some of the factors that need to be addressed for production of a waste package suitable to meet the phased disposal concept.

20-2. Regulatory Requirements for ILW

The Intermediate Level Waste category is defined in Chapter 1, together with projected volumes of waste arisings. Regulatory requirements include the conversion of the ILW into a product which is safe and convenient for engineered storage. In essence, this means that the wastes must be immobilised and contained in passively safe forms.

Liquid wastes may be volume reduced by treatment through ion exchange columns, chemical separation, or filtration, such that the resultant less or nonactive liquors may be discharged. The remaining concentrated active constituents are immobilised in a matrix using cement grout, or other materials which do not enhance their solubility.

Solid intermediate level wastes may consist of contact handleable Plutonium Contaminated Materials (PCM), decommissioning contaminated or activated structural steels, and building rubble, together with highly active and γ emitting operational wastes and fuel housings. Again, treatments involve reducing the mobility of the wastes coupled with volume reduction.

Gaseous wastes outside direct discharge authorisations are normally contained within controlled envelopes and captured by the ventilation system absorbers or filters and then treated as for solid intermediate level waste.

Data on the product must be recorded and maintained so as to ensure that it may be safely transported from interim storage to a final repository. Such data must, therefore, also capture the parameters required to meet National radioactive waste management requirements including anticipated future disposal requirements. Characterisation instrumentation is described in Chapter 2.

Within a particular waste management option for packaging and eventual disposal, the Best Practical Means (BPM, see Chapter 18) is that level of management and engineering control that minimises, as far as practicable, the radiological impact of the option whilst taking account of a wider range of factors, including cost effectiveness, technological status, operational safety, social, and environmental factors. In determining whether a particular aspect of the waste management proposal represents BPM then, on the basis of the "polluter pays" principle, the owner of the waste will have to incur expenditure. It is recognised that such expenditure, whether in terms of money, time, or trouble, should not be disproportionate to the benefits likely to be derived. A balance has, therefore, to be struck between radiological and other factors when considering phased disposal. Where it is demonstrated that BPM has been applied, and radiological doses and risks from the waste as a source term of exposure are consistent with the relevant dose or risk standards, the level of protection may then be said to be optimised and may be regarded as low as reasonably achievable (ALARA, see Chapter 1).

20-3. Sources and Processing Requirements

ILW arises from fuel fabrication as uranium scraps, enrichment processes, and as a by-product of commercial reprocessing. Military operational waste arises from weapons manufacture as plutonium, plutonium contaminated material, tritium (^{3}H), filters, and liquid wastes, in addition to decommissioned nuclear propulsion plant reactor cores and operational wastes. Medical, industrial, and minor user wastes include a wide variety of sealed sources, ^{60}Co sources, ^{3}H, and ^{14}C, which are returned to manufacturers or sent to National disposal services. Reactor operational wastes consist of miscellaneous activated components, fuel element debris, filters, and ion exchange resins, sludges, evaporator concentrates, and graphite core materials, etc. Decommissioning wastes include contaminated and activated structural materials such as concrete, reinforcing steel bar, and rubble, reactor core components including graphite, control rods, and flux flattening bars. Some of these materials are illustrated in Figures 20-1–20-4.

Once immobilised, the internationally accepted approach to phased waste disposal is based upon the provision of "multiple barriers" to delay or prevent the migration of the waste material from its interim storage facility or eventual disposal site back to the environment. The physical and chemical form in which the waste will be disposed of, including any conditioning media but excluding the container, is known as the "waste form."

The processing requirements must take into consideration:

- Simple and proven technologies;
- Versatile plant which is able to deal with a range of wastes;

Figure 20-1. Fuel Element Debris (Photograph Courtesy of BNFL Magnox).

Figure 20-2. ILW from Magnox Decanning Operations (Photograph Courtesy of BNFL Magnox).

Figure 20-3. Light Water Reactor (LWR) Hulls (Photograph Courtesy of BNFL Magnox).

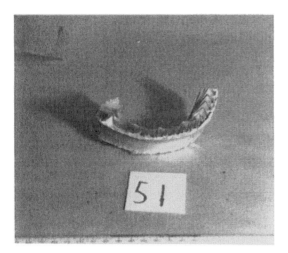

Figure 20-4. Magnox Swarf (Photograph Courtesy of BNFL Magnox).

- Safe operations;
- Ambient or near ambient temperature working;
- The minimum production of secondary waste; and
- Economic plant with low capital and operating (material and personnel) costs.

The grout mixes that are used to produce solid blocks of waste may be formulated with a high pH (alkaline) so as to provide a medium in which long-lived radionuclides are less soluble and, therefore, have less probability of returning to the surface from a deep repository by water transport. Grout formulation is a specialist technology which involves mixes designed to allow good flow around the waste and adequate strength without voids, but having, at the same time, a sufficiently low water content to avoid slump and uncontrolled cracking. One example of

Figure 20-5. Resin Solidification Plant Showing Drums on the Inlet Conveyor (Photograph Courtesy of BNFL Magnox).

the detail involved for grouts around PCM ILW includes rejection of grout plasticisers made from long chain polymer materials that could enhance plutonium solubility many hundreds and thousands of years into the future in a wet deep-waste repository environment. Another technology is to use polymers for solidifying power station operational ion exchange resin waste in self-shielded drums. The in-drum mixing uses "lost paddles" (paddles which are discarded after use within the solidified waste form) and the drums are self-shielded (external volume 760 l, internal volume 450–200 l, depending upon degree of shielding required) (Figure 20-8).

The further confinement of the radioactive waste material so as to prevent or limit its dispersal is known as its "containment." The wastes and its waste container, as prepared for interim storage and eventual disposal, are collectively called the "waste package."

Figure 20-6 illustrates a supercompactor for the compression and volume reduction of standard 200 liter mild steel (oil) drums containing PCM wastes. The compacted pucks so produced are then placed in a cage within a standard ILW 500 liter stainless steel container with annular grouting, as illustrated in Figure 20-7.

Figure 20-8 is an illustration of a paddle incorporated within the waste drum to achieve homogeneity of the waste form within the waste package. Figure 20-9 is a cross-section through a standard 500 l ILW container of Magnox swarf simulant.

20-4. Standard Waste Packages and Specifications

20-4-1. *Waste Package Specification*

Following treatment and immobilisation, the second of the multistage barriers is the container itself. Each country

Figure 20-6. Supercompactor for Volume Reduction of 200 l Solid Drummed Wastes (Photograph Courtesy of BNFL Magnox).

Figure 20-7. Cross-section of Compacted PCM Pucks in Standard 500 l Stainless Steel Container (Photograph Courtesy of BNFL Magnox).

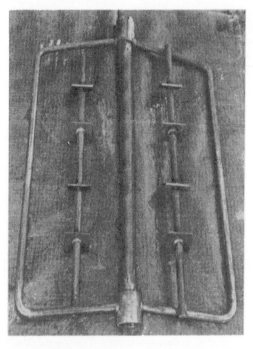

Figure 20-8. In-drum Mixing Process Using a Paddle to Assist Homogeneity of the Waste Form (Photograph Courtesy of BNFL Magnox).

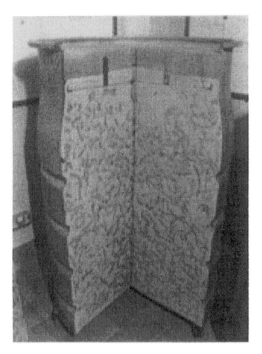

Figure 20-9. Cross-section of a 500 l Container of Immobilised Magnox Swarf (Photograph Courtesy of BNFL Magnox).

has developed its own particular standardised waste container designs. In the UK, four different stainless steel designs, aimed at being suitable for different types of waste arisings, have been specified. They are:

500 l drum	for most operational ILW, for either in-drum mixed sludge type wastes or for encapsulated solid items,
3 m³ box	a larger container for operational and decommissioning solid wastes,
3 m³ drum	a larger container intended for in-drum mixing and solidification of sludge type wastes, and
4 m box	a standard dimensioned self-shielded container intended typically for the less active large item decommissioning wastes.

Each waste package is governed by detailed specifications [1] covering:

- Dimensions (to allow common performance and materials handling);
- Manufacturing materials (for quality control);
- Manufacturing methods (for quality control);
- Lid sealing and fixings (for longevity performance and quality control);

- Lifting features (for common handling);
- Stackability (for cost effective storage);
- Gas venting (during curing and for the low ILW heat generation effects);
- Identifiers (for QA records and monitoring over time);
- Package mass (for materials handling);
- External dose rate (to avoid contamination — surface effects — and safety); and
- Heat output (storage and repository design parameter).

Figure 20-10 shows variants of the standard 500 l drum being used for homogeneous, heterogeneous, and super-compacted wastes. In France, one such standard ILW package consists of a self-shielded concrete block with cast-in handling catchments as used at the surface disposal facility at Centre de l'Aube. Such standardised specifications allow for common materials handling equipment to be adopted on a National basis.

20-4-2. Storage

Storage criteria for the waste packages include consideration of:

- Strength (of the containers for handling and stacking purposes);
- Stable products (in terms of chemical, radiation, thermal, and mechanical effects);
- Retention of activity (under normal and accident conditions involving attention to impact and fire resistance);
- Package volume minimisation (for cost effective storage, transport, and disposal);
- Minimum corrosion (longevity of package); and
- Chemical stability (for safety and to avoid possibility of future expensive repackaging).

Buffer stores (and interim storage facilities prior to an eventual resolution of the issue with regard to deep disposal) continue to be built to match waste production volumes. Figure 20-11 shows four standard 500 l drums which are contained in a stillage, so that the drums may be stacked in the store (in this case to nine high). Such facilities require remote handling when dealing with unshielded waste packages.

20-4-3. Transport

Transport containers or packages are designed to meet the standards laid down in the IAEA (International Atomic Energy Agency) regulations for the safe transport of radioactive material [2,3]. This subject is covered in detail in Chapter 22.

A transport safety assessment is based upon a deterministic approach which ensures an appropriate response

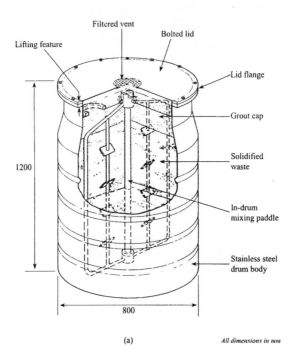

(a)   *All dimensions in mm*

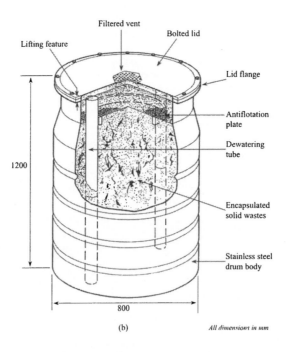

(b)   *All dimensions in mm*

Figure 20-10. Variants on the UK Standard 500 l ILW Container for (a) Homogeneous, (b) Heterogeneous, and (c) Supercompacted Wastes (Drawing Courtesy of UK Nirex Ltd).

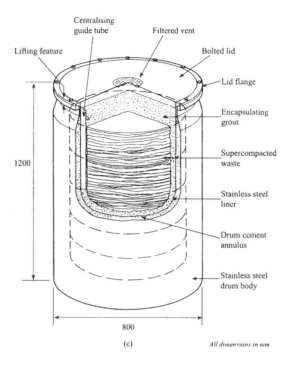

Centralising guide tube

Filtered vent

Lifting feature

Bolted lid

Lid flange

Encapsulating grout

1200

Supercompacted waste

Stainless steel liner

Drum cement annulus

Stainless steel drum body

800

(c)

All dimensions in mm

Figure 20-10. Continued

Figure 20-11. Storage of Waste Packages (Photograph Courtesy of BNFL Magnox).

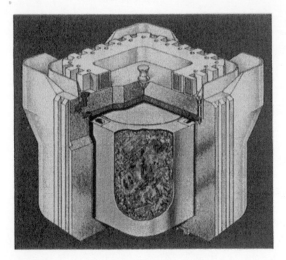

Figure 20-12. Transport Container with 3 M^3 ILW Box (Photograph Courtesy UK Nirex Ltd).

to normal and accident conditions during transport. Since transport to a possible future deep ILW repository of some 200,000 cu m capacity would involve a sizeable transport campaign, a risk assessment must be carried out to show that the risks are as low as reasonably practicable (ALARP, see Chapter 1). A reusable shielded transport container (RSTC) developed by UK Nirex for carrying four 500 l unshielded waste drums or one 3 cu m box is illustrated in Figure 20-12.

20-4-4. *Disposal*

Continuing with the multibarrier concept, current International policy, based upon technical studies, is for the eventual disposal of ILW should this receive public acceptance. Disposal criteria for consideration include:

- Precautions to minimise the solubility of long-lived radionuclides (actinides) by using high pH buffers, redox potential, and absorption considerations;
- Low permeability sites (so as to reduce pathways for return of radionuclides to the surface);
- Long-term stability of geological formations;
- Containment of short-lived radionuclides;
- Checks on the effects of possible criticality events;
- Waste package heat output (will not lead to thermal runaway, etc.); and
- Chemical compatibility of waste package treatment and conditioning with possible future repository environment.

Both operational and postclosure repository safety assessments are necessary in order for a repository to receive Regulatory approval. In addition, the approval of such a waste disposal concept is not just a technical issue. In a democracy, it requires political will, a sound economic case, together with public understanding and support if it is to become a reality.

The operational safety case may involve a variety of techniques to analyse the situation including:

(1) Hazardous Operational (HAZOP) reviews to identify potential faults and hazards, the frequencies of occurrences and their consequences, identification of options to eliminate, protect, and mitigate the effects;
(2) A design basis accident (DBA) analysis as a deterministic investigation into the level of robustness of the design against impacts such as fires, criticality events, etc.;
(3) A probabilistic safety assessment (PSA) to show compliance with the risk criteria and to illustrate that no particular class of accidents dominates the risk;
(4) An operational dose assessment (ODA) considering the dose update by workers during normal routine operations;
(5) A routine off-site dose assessment including aerial discharges, etc.;
(6) A criticality safety assessment; and
(7) A conventional safety assessment.

The repository postclosure performance assessment [4] looks at the robustness of the multibarriers built into the system against the return of radionuclides as a dose to the critical group. Three pathways for exposure include groundwater, gas, and human intrusion, with a systematic identification of the features, events, and processes which could affect multibarrier performance. A base case is generated with scenario variants, together with computer modeling of probabilistic safety assessments.

Groundwater pathway modeling looks at information on solubility and sorption of radionuclides in the "near field" engineered system (the waste package itself — although this has a limited life in comparison with the half lives of the radionuclides involved — the repository vault backfill grout, sealing of man-made repository entrances, etc.), and the "far field" rock, salt dome, or clay bed parameters. The three basic parameters are:

- The groundwater flux through the repository (which can carry the radionuclides back into the biosphere);
- The travel time between the repository and the biosphere (which if short negates the advantageous effects arising from radioactive decay); and
- The mixing flux, or dilution, of water carrying radionuclides by water in overlying rocks.

For radionuclides reaching the biosphere, the environment itself and the activities leading to dose uptake by the most exposed group have to be considered for time periods many thousands of years into the future (with ice ages in between).

Gas pathway and human intrusion modeling take into account gas migration (which could be faster than groundwater movement) of such species as gases containing ^{14}C. A possible scenario for the human intrusion case needs to consider a future geotechnical worker drilling a bore hole into the repository, and a site occupier making a living directly above the facility.

20-5. ILW Conditions for Acceptance for Interim Storage and/or Eventual Disposal

In order to reduce the risk of incompatibility of waste packaging proposals with long-term waste management (interim storage and eventual disposal) requirements, it is necessary to provide a basis on which packaging may be carried out (now) and, thereby, avoid expensive nugatory work (in the future). In addition, the availability of "conditions for acceptance" (or some equivalent assurance of the suitability of waste packaging) for an existing or future waste disposal facility allows regulators to have confidence in the designs and fund providers to make due allowances accordingly.

In the UK, a staged process of "Letter of Comfort" (LoC) submissions by the waste producer to an independent industry funded group responsible for deep waste disposal (UK Nirex Ltd.) allows such a degree of assurance that the waste packaging criteria will be acceptable. The processes, the principles of which are appropriate both in the UK and overseas, are described in Table 20-1.

Waste packaging proposals are subjected to 16 technical areas of assessment, as described in Table 20-2. Since ILW waste disposal costs are very high, it is important that the overall assessment process is:

- simple to operate;
- systematic and meets the safety and regulatory standards required;
- transparent and consistent;
- independent;
- takes into account stakeholder perceptions;
- has a defensible technical foundation; and
- demonstrates evidence of appropriate QA and data recording arrangements.

All this rigor would lead one to believe that one is dealing with other than "waste." However, the approach gives a clear audit trail leading to the possibility of making a postclosure repository safety case. Despite this, there are still some difficult ILW packaging areas. These include:

- The addition of organic materials in the waste package that may increase radionuclide mobility after disposal;
- Dealing with complex hetrogeneous wastes, segregation, and treatment;
- Predictability of performance;
- Criticality safety cases and limits arising (to detect 50 gm Pu in a 500 l grouted drum is on the limits of available measurement technology);
- Adequacy of radionuclide inventory; and
- Stakeholder perception.

It is essential that the postclosure performance of a deep waste repository is robust. Should it be susceptible to such small quantities of plasticisers in the grout or such small fissile loadings, then it may be necessary to

Table 20-1. The Staged Letter of Comfort Process for the Acceptability of ILW Packaging to Suit Eventual Deep Waste Disposal

	Stage	Key information	Purpose
1	Conceptual stage	(i) Detailed description of the waste (ii) Outline of packaging concept (iii) Assurance that necessary research and development work will be carried out, that QA will be applied to all activities, and that a realistic and justifiable inventory will be recorded for each waste package	Waste producer describes what they plan to do. Disposal Company gives assurance that packaging concept is feasible.
2	Precommitment stage	(i) Results from R&D work (ii) Detailed information on package properties and performance, and data recording processes	Provision of key information to show consistency with waste package specifications.
3	Preoperational stage	(i) Evidence (e.g., inactive commissioning plant results) that product meets Repository requirements (ii) Evidence that QA and data recording systems fully in place	Assurance that waste packaging plant is capable of making the specified product and is consistent with plans for a future (or existing) repository.

Table 20-2. ILW Packaging — Areas of Technical Assessment

The 16 waste package assessment areas	Data involved
(A) *Technical Assessment*	
(A1) Nature and Quantity of Waste	Where is the waste, how much is there, and how many packages will arise? Consistent information will be extracted from submission information and elsewhere and used by all assessors. Also assess:
	• Whether the information is consistent;
	• The expected variation in activity and other materials between packages.
(A2) Wasteform	Are the waste and proposed encapsulant compatible? And what are the properties? Consider:
	• Wasteform behavior during transport; prolonged storage and under disposal conditions;
	• Definition of the "product envelope" (the bounds on the quantity of waste and encapsulant; and
	• Acceptable features relating to: Immobilisation of particulates; Immobilisation of liquids; Active and nonactive gases; Treatment of hazardous materials; Stability and aging; Thermal properties and heat; and Exclusion of prohibited materials.
(A3) Criticality	Do the packages raise criticality issues during transport, prolonged storage, and following emplacement in a deep waste repository after allowing for degradation processes? Approach requires:
	• short-term and long-term special criticality safety cases, unless covered by a generic case for packages with < 50 gm total fissile materials; and
	• criticality compliance assurance documentation for all packages.
(A4) Container Design	Is the container consistent with standard designs and performance requirements? Consistency with Waste Packaging specifications as per items listed in Section 20-4-1.
(A5) Corrosion	Does the container have adequate corrosion performance to permit future use following prolonged storage? Consideration given to:
	• performance of the container in the store and under repository conditions;
	• materials that may increase both internal and external corrosion, i.e., galvanic coupling, chlorides, etc.; and
	• corrosion mechanisms and rates — general, pitting, and crevice corrosion.
(A6) Impact Performance	Standard drop tests or finite element analysis. Does the package have low and predictable releases under impact conditions? Consideration given to:
	• Expected radionuclide releases as particulates in impacts, for use in the operational safety case;
	• Flat surface and aggressive feature impacts; and
	• Container behavior, e.g., lid retention.
(A7) Fire Accident Performance	Does the package have low and predictable releases under fire accident? Consideration given to:
	• Expected radionuclide releases in fires, for use in the operational safety assessment;
	• Expected releases of toxic gases from pyrolysis of wastes, for use in the operational safety assessment; and
	• 1000°C, 1 hour fire.
(A8) Quality Assurance	Are the packages being produced and supporting activities affecting product quality being performed under appropriate Quality Management System (QMS)? Requirement to apply recognised QMS to all activities affecting product quality:
	• compliance with QA standard (ISO 9000);
	• QA program and plans;
	• Waste product specification(s);
	• Independent verification of quality system permitted; and
	• Arrangements for auditing.

Continued

Table 20-2. Continued

The 16 waste package assessment areas	Data involved
(A9) Data Recording	Are appropriate data/information on packages being recorded to inform future waste management decisions? Identify important data for packages and establish systems for their acquisition and retention:
	• Radionuclides; • Physical/chemical properties; and • Process variables.
	>100 radionuclides may be relevant. For example, for operational and transport safety consider ^{60}Co for gamma dose and heat, ^{239}Pu for criticality, ^{3}H and ^{222}Rn for gaseous discharges. For disposal postclosure safety ^{129}I and ^{36}Cl for groundwater pathway considerations and ^{14}C for gaseous discharges.
(A10) Physical Protection	Are any special security measures required to protect the packages during future transport operations? Includes consideration of:
	• Consistency with security plan(s); and • Based on contents (e.g., fissile materials) and accessibility of package contents.
(A11) Safeguards	Are any special safeguard measures required for packages during storage or following emplacement? Includes consideration of:
	• Safeguards status of the wastes; and • Commitment to meet safeguards requirements (IAEA/Euratom).
(A12) Nonnuclear Environmental	Do the packages represent an appropriate use of resources for package manufacture and transport? At the conceptual stage only focusing on:
	• Optimising the use of available facilities; • Minimising number of packages and transport movements by choice of waste container; and • Comparative environmental impact of any options for waste treatment.
(B) *Packaging Principles*	
(B1) Policy	Are the packages within the remit of the disposal company? Issues relating to the waste that could affect suitability for disposal due to inconsistency with waste disposal company, National or International waste management policy, for example:
	• Resource potential; • Classification outside disposal company remit; and • Overseas origin.
(C) *Phased Disposal Concept*	
(C1) Transport Safety	See Chapter 22.
(C2) Operational Safety	See Section 20-4-4.
(C3) Postclosure Safety	See Section 20-4-4.

reconsider the whole concept for certain types of waste in certain geological environments and concentrate, in the meantime until a more acceptable site or solution is found, on safe and secure surface or near surface interim storage.

20-6. Case Study — Waste Packaging Exercise

20-6-1. *Introduction*

The purpose of this exercise is to introduce the reader to the application of the concepts for ILW waste packaging introduced in this chapter. The exercise introduces two hypothetical wastes stored on a waste producer's site. One is solid ILW, the other an intermediate level sludge. The case study examines a number of waste packaging issues including:

• Why certain radionuclides may be more important for consideration than others;
• The importance of information about the wastes;
• How the transport impact accident performance of a package is assessed;

- The hazards presented by the wastes and how some of these may be mitigated or reduced by the waste packaging process; and
- Long-term storage issues.

Questions are given in Section 20-6-5, with specimen answers to the questions in Section 20-6-7, and general Case Study data is included in Section 20-6-6.

20-6-2. *Waste Descriptions*

Solid Waste

Solid operational power station waste is stored on site in raw form. The waste is activated fuel element debris removed at the site before the spent fuel itself was transported for reprocessing. In addition, there is a smaller quantity of laboratory waste from Post Irradiation Examination (PIE) cell operations. Relevant information about these wastes includes:

(i) The wastes are stored in a vault below ground level in nominally dry conditions, but dampness has permeated into the storage area in the past;

(ii) The fuel element debris waste is largely graphite, with some stainless steel and some Magnox (Magnesium Alloy) items. Radionuclides in the fuel debris are present as activation products created by neutron irradiation in the site reactor;

(iii) Laboratory waste is small scale fuel element microscope samples. Records from the laboratory suggest that there will be small quantities of powdered fuel element materials, laboratory chemicals — probably in the form of solids or liquids on swabs — and various other materials such as corrosive chloride compounds in bags used as a metal fire extinguisher, rubber gloves, etc. The fuel in the laboratory samples is the same as that used in the power station reactor;

(iv) The laboratory waste is all contained in thin walled, painted, mild steel cans, as originally used for transfer to, and deposition in, the vault. The laboratory records are reasonably good, although it is not known which cans contain which specific waste items;

(v) All waste items have been dropped into the vault via gamma gate access ports in the roof. The laboratory cans are mixed with the power station fuel debris; and

(vi) The volume of waste is some $100 \, m^3$ in total of which $2 \, m^3$ is laboratory waste.

The power station management needs to improve its waste storage conditions and convert the waste into a passively safe form. Option studies have indicated the most appropriate process for this work and consideration

Table 20-3. Average Solid ILW Radionuclide Concentrations (Assume Valid at Time of Reading this Case; 1 TBq = 10^{12} Bq)

Radionuclide	Activity (TBq/m^3)	Radionuclide	Activity (TBq/m^3)
^{3}H	3	^{235}U	2×10^{-7}
^{14}C	1×10^{-1}	^{236}U	4×10^{-6}
^{36}Cl	2×10^{-3}	^{238}U	6×10^{-6}
^{60}Co	1×10^1	^{237}Np	4×10^{-6}
^{59}Ni	1	^{238}Pu	2×10^{-2}
^{63}Ni	1×10^2	^{239}Pu	6×10^{-3}
^{90}Sr	1	^{240}Pu	1×10^{-2}
^{99}Tc	2×10^{-4}	^{241}Pu	1
^{137}Cs	1.33	^{242}Pu	2×10^{-5}
^{234}U	2×10^{-5}	^{241}Am	2×10^{-2}

has been given to:

- short- and long-term environmental impacts;
- doses to workers and safety implications;
- costs;
- project risks; and
- waste storage arrangements, etc.

for each option. A project team has been set up to implement the option. The estimated average radionuclide concentrations in the combined wastes in the vault are as shown in Table 20-3, and they are not expected to vary by more than a factor of three smaller or greater than listed. Assume that the ^{3}H, ^{14}C, ^{36}Cl, ^{60}Co, ^{59}Ni, and ^{63}Ni only arise in the fuel element debris. All other radionuclides, including the ^{137}Cs, arise in the laboratory wastes.

Liquid Effluent Sludge Waste

The nuclear power plant uses a liquid effluent treatment plant to remove certain radionuclides before discharge of the effluent to the sea. The treatment process has resulted in the formation of an iron hydroxide rich sludge containing the removed radionuclides from the effluent. Sludge and storage data includes:

(i) The sludge volume is approximately $100 \, m^3$ and is stored in a large tank;

(ii) The effluent and the effluent treatment process have changed over the years and, because no mixing equipment has been fitted to the sludge storage tank, the sludge is present in the tank in layers of differing composition;

(iii) Limited sampling campaigns have been carried out over the years and the radionuclide composition of the sludges is only approximately known; and

(iv) However, the data suggests that the sludge is certainly categorised as ILW waste and is unsuitable

Table 20-4. Average ILW Sludge Radionuclide Concentrations (Assume Valid at Time of Reading this Case; 1 TBq = 10^{12} Bq)

Radionuclide	Activity (TBq/m^3)	Radionuclide	Activity (TBq/m^3)
^{60}Co	1.2×10^{-4}	^{236}U	1.2×10^{-6}
^{65}Zn	6×10^{-6}	^{238}U	3×10^{-5}
^{90}Sr	6×10^{-3}	^{237}Np	9×10^{-4}
^{99}Tc	9×10^{-4}	^{238}Pu	1.2×10^{-2}
^{129}I	3×10^{-4}	^{239}Pu	6×10^{-2}
^{137}Cs	6×10^{-4}	^{240}Pu	3×10^{-2}
^{144}Ce	9×10^{-5}	^{241}Pu	6×10^{-1}
^{234}U	1.2×10^{-4}	^{242}Pu	9×10^{-6}
^{235}U	1.2×10^{-6}	^{241}Am	1.5

for currently available near surface disposal facilities intended for LLW and VLLW on the power station site.

The power station team has carried out good option studies to assess the most appropriate way to treat the sludges and turn into a passively safe form for interim storage and eventual deep disposal as ILW. The same project team as for the solid wastes has been set up to take this work forward. The estimated radionuclide concentration in the sludges is not known with great accuracy, but there is justified confidence to believe that the inventory is unlikely to be more than a factor of five smaller or greater than the values given in Table 20-4.

20-6-3. *Solid Waste Packaging Concept*

The project team plans to retrieve the solid wastes from the vault using manipulators working from the top downwards to the base of the vault. Batches of the waste will

then be placed in skips. The skips will be emptied at the packaging plant and the waste encapsulated with an encapsulating cement grout in 3 m^3 stainless steel boxes designed to waste disposal company standards. From previous research and development, an approximate specification is available for the waste package, and it is anticipated that there will be approximately 2700 kg of grout per box. The packages will then be placed in a purpose-built store for an as yet undetermined period. A summary of the anticipated waste packages and contents is described in Figure 20-13 and in Table 20-5.

20-6-4. *Sludge Waste Packaging Concept*

The project team plans to retrieve the sludges from the tank and to take samples so as to obtain better data on the radionuclide inventory, physical/chemical content, and characteristics. They plan to package the wastes into 500 liter drums which meet the disposal company specifications. The sludge will be mixed with cement powders to form an homogeneous solid product. From previous research and development, an approximate specification is available for the waste package and it is anticipated that there will be approximately 333 liters of sludge mixed with 500 kg of cement powders per package. A summary of the waste packages and contents after further sampling is described in Figure 20-14 and in Table 20-6.

20-6-5. *Questions and Hints to Answers*

The reader will need to refer to Section 20-6-6 for the additional background information necessary to derive answers to questions 5, 7, 13, 14, and 15.

(1) If the solid wastes were packaged without any form of waste treatment or conditioning/encapsulation,

Title:	Power Station Site X — Vault Solid Wastes.
Nature:	Fuel element debris from the power station, and mixed laboratory wastes stored in cans. Vault stored raw solid ILW.
Waste Volume:	98 m^3 of Fuel Element Debris (FED) and 2 m^3 of laboratory wastes.
Proposed Encapsulant:	Cement Y.
Package Type:	3 m^3 box.
Waste Package Mass:	6.3 te (600 kg container, 3000 kg waste, and 2700 kg cement grout).
Number of Packages:	34
Radionuclide Inventory	
Reference Date:	2003 (or at time of reading this Case Study).
Total Package Activity:	0.168 TBq α and 352 TBq $\beta\gamma$

Figure 20-13. Summary Sheet.

Table 20-5. Average Solid Waste Package Radionuclide Inventory at Current Date

Radionuclide	Activity Conc. (TBq/m^3)	Activity per package (TBq)	A_2 Multiples per package	Heat (W) per package	Fissile Mass (g) per package
^3H	3	9	2.25×10^{-1}	8.18×10^{-3}	
^{14}C	1×10^{-1}	3×10^{-1}	1.5×10^{-1}	2.38×10^{-3}	
^{36}Cl	2×10^{-3}	6×10^{-3}	1.2×10^{-2}	2.63×10^{-4}	
^{60}Co	1×10^{1}	3×10^{1}	7.5×10^{1}	1.25×10^{1}	
^{59}Ni	1	3	7.5×10^{-2}	3.33×10^{-3}	
^{63}Ni	1×10^{2}	3×10^{2}	1×10^{1}	8.22×10^{-1}	
^{90}Sr	1	3	3×10^{1}	5.43×10^{-1}	
^{99}Tc	2×10^{-4}	6×10^{-4}	6.67×10^{-4}	9.72×10^{-6}	
^{137}Cs	1.33	4	8	5.44×10^{-1}	
^{234}U	2×10^{-5}	6×10^{-5}	6×10^{-2}	4.67×10^{-5}	
^{235}U	2×10^{-7}	6×10^{-7}	0	4.48×10^{-7}	7.50
^{236}U	4×10^{-6}	1.2×10^{-5}	1.2×10^{-2}	8.83×10^{-6}	
^{238}U	6×10^{-6}	1.8×10^{-5}	0	1.23×10^{-5}	
^{237}Np	4×10^{-6}	1.2×10^{-5}	6×10^{-2}	9.48×10^{-6}	
^{238}Pu	2×10^{-2}	6×10^{-2}	3×10^{2}	5.36×10^{-2}	
^{239}Pu	6×10^{-3}	1.8×10^{-2}	9×10^{1}	1.51×10^{-2}	7.83
^{240}Pu	1×10^{-2}	3×10^{-2}	1.5×10^{2}	2.52×10^{-2}	
^{241}Pu	1	3	3×10^{2}	2.57×10^{-3}	7.86×10^{-1}
^{242}Pu	2×10^{-5}	6×10^{-5}	3×10^{-1}	4.78×10^{-5}	
^{241}Am	2×10^{-2}	6×10^{-2}	3×10^{2}	5.42×10^{-2}	
Totals	1.18×10^{2}	3.52×10^{2}	1.26×10^{3}	1.46×10^{1}	1.61×10^{1}

Title:	Power Station Site X — Effluent Treatment Sludges.
Nature:	Iron Hydroxide based sludges used for removal of radionuclides from site liquid effluents.
Waste Volume:	$100 \ m^3$.
Proposed Encapsulant:	Cement Z.
Package Type:	500 liter drums.
Waste Package Mass:	1030 kg (130 kg container, 400 kg sludge, and 500 kg cement powders).
Number of Packages:	301.
Radionuclide Inventory Reference Date:	2003 (or at time of reading this Case Study).
Total Package Activity:	0.534 TBq α and 0.203 TBq $\beta\gamma$

Figure 20-14. Summary Sheet.

what hazards could the packages present for future waste management, from storage to final disposal, and why? (*Hint*: Note that, in this case, standard containers are typically thin-walled stainless steel. Consider possible consequences of accidents during handling, the effects of long-term storage, and the barriers provided in the disposal concept.)

(2) If the sludge wastes were packaged without any form of waste treatment or conditioning/encapsulation, what hazards could the packages present for future

waste management, from storage to final disposal, and why? (*Hint*: Note that, in this case, standard containers are typically thin-walled stainless steel. Consider possible consequences of accidents during handling, the effects of long-term storage, and the barriers provided in the disposal concept.)

(3) For the packages of solid wastes, will the maximum possible waste package radionuclide inventory be limited to a factor of three times that given in

Table 20-6. Average Sludge Waste Package Radionuclide Inventory at Current Date

Radionuclide	Activity Conc. (TBq/m^3)	Activity per package (TBq)	A$_2$ Multiples per package	Heat (W) per package	Fissile Mass (g) per package
^{60}Co	8×10^{-5}	4×10^{-5}	1×10^{-4}	1.7×10^{-5}	
^{65}Zn	4×10^{-6}	2×10^{-6}	1×10^{-6}	1.9×10^{-7}	
^{90}Sr	4×10^{-3}	2×10^{-3}	2×10^{-2}	3.6×10^{-4}	
^{99}Tc	6×10^{-4}	3×10^{-4}	3.3×10^{-4}	4.9×10^{-6}	
^{129}I	2×10^{-4}	1×10^{-4}	0	1.4×10^{-6}	
^{137}Cs	4×10^{-4}	2×10^{-4}	4×10^{-4}	2.7×10^{-5}	
^{144}Ce	6×10^{-5}	3×10^{-5}	1.5×10^{-4}	1.2×10^{-5}	
^{234}U	8×10^{-5}	4×10^{-3}	4×10^{-2}	3.1×10^{-5}	
^{235}U	8×10^{-7}	4×10^{-7}	0	3.0×10^{-7}	5.0
^{236}U	8×10^{-7}	4×10^{-7}	4×10^{-4}	2.9×10^{-7}	
^{238}U	2×10^{-5}	1×10^{-5}	0	6.9×10^{-6}	
^{237}Np	6×10^{-4}	3×10^{-4}	1.5	2.4×10^{-4}	
^{238}Pu	8×10^{-3}	4×10^{-3}	2×10^{1}	3.6×10^{-3}	
^{239}Pu	4×10^{-2}	2×10^{-2}	1×10^{2}	1.7×10^{-2}	8.7
^{240}Pu	2×10^{-2}	1×10^{-2}	5×10^{1}	8.4×10^{-3}	
^{241}Pu	4×10^{-1}	2×10^{-1}	2×10^{1}	1.7×10^{-4}	4×10^{-2}
^{242}Pu	6×10^{-6}	3×10^{-6}	1.5×10^{-2}	2.4×10^{-6}	
^{241}Am	1	5×10^{-1}	2.5×10^{3}	4.5×10^{-1}	
Totals	1.5	7.4×10^{-1}	2.7×10^{3}	4.8×10^{-1}	1.37×10^{1}

Table 20-5? Explain your reasoning. (*Hint*: See the description of the solid wastes, but note that the question is a little more complex than at first sight, since a factor of 3× is referring to the average concentrations of radionuclides.)

(4) What benefits might arise if the project team decided to resuspend the sludge in its existing tank, by fitting the appropriate equipment, to fluidise and homogenise the sludge waste? (*Hint*: Consider how you might get this old and settled sludge out of the tank, and how many samples would you need to take to get realistic and justifiable waste package inventories?)

(5) Will the packages of sludge waste meet the Waste Package Impact Accident criterion in the Waste Packaging Specifications? Explain your reasoning. (*Hint*: See Section 20-6-6.)

(6) Will the packages of solid waste meet the Waste Package Specification criterion for "Heat Output"? Explain your reasoning. (*Hint*: Consider what the waste package specification requires and the package inventory in Table 20-5. Even if you have thought about the inter-package variability, and taken a maximum package inventory view, you should also consider what the possible sources of heat are.)

(7) For the sludge waste package radionuclide data recording, would you record the levels of ^{60}Co, ^{129}I,

^{235}U, and ^{238}U, respectively, and if so which should be recorded on a package specific basis and which only on a package collection basis? (*Hint*: Use the data in Section 20-6-6 and Table 20-6.)

(8) If the inventory of radionuclides in the solid wastes can all be related to the inventory of the strong gamma emitters ^{60}Co (activation product in irradiated materials) and ^{137}Cs (fission product in fuel), how might the radionuclide inventory of packages of solid waste be determined? (*Hint*: Consider how the levels of ^{60}Co and ^{137}Cs might be determined.)

(9) The Waste Package Specifications assume a long period of storage for the packaged wastes on the power station site, followed by transport and then a further period of storage at a phased disposal facility. How might you prove the adequacy of the packages for transport from the site to the repository and its suitability for acceptance there? (*Hint*: Think about what information might need to be recorded on waste package contents and properties.)

(10) Will the sludge waste packages require a criticality safety assessment and a criticality compliance assurance document? (*Hint*: Does each package need to be "safe" with regard to criticality?)

(11) For convenience, the waste disposal company has produced a general criticality safety assessment for 500l drum packages, making a number of

pessimistic assumptions concerning waste materials, packaging arrangements, and long-term rates of waste package degradation after disposal. Assume that this assessment has derived a safe fissile mass (SFM) of 50 g per package for the total mass of the radionuclides ^{233}U, ^{235}U, ^{239}Pu, and ^{241}Pu. Is the general assessment applicable to the sludge waste package or will it be necessary to produce a package specific assessment? (*Hint*: See Table 20-6, but also consider inter-package variability and how you might take into account the uncertainties in package inventory prior to packaging. Assume the inventory of ^{233}U is negligible.)

(12) In the generic phased disposal concept, what are the barriers to radionuclide migration after repository closure? In simple terms, explain how the barriers are intended to work. (*Hint*: See Section 20-4-4.)

(13) In the generic phased disposal concept, why should radionuclides such as ^{137}Cs and ^{90}Sr be the main contributors to the near-field flux at early times after disposal? Why don't these radionuclides appear in the far-field as main contributors to the flux? (*Hint*: See Section 20-6-6, giving general radionuclide characteristics. Think about the mobility and the rate of radionuclide decay.)

(14) In the generic phased disposal concept, why are ^{129}I and ^{36}Cl major contributors to the far-field flux, but only more minor contributors to the near-field flux? (*Hint*: See the description of radionuclide characteristics in Section 20-6-6. Think about the mobility and rate of radionuclide decay.)

(15) In the generic phased disposal concept, why do relatively short-lived radionuclides such as ^{210}Pb or even ^{226}Ra appear as major contributors to far-field flux? (*Hint*: Look up ^{210}Pb and ^{226}Ra in a science text book — (e.g., [5–7]) and consider how these radionuclides arise. Are they necessarily in the waste at the time of disposal and, if not, why not? Where do they come from and how can they appear in the far-field?)

20-6-6. *General Case Study Data*

Introduction

This section provides data on the generation and release of particulates from cemented wasteforms as a result of impact accidents. It also gives details of the levels of significance of radionuclides in waste packages (for use when considering data recording requirements), together with the characteristics of radionuclides under disposal conditions.

Impact Accident Release Fractions

Assessments of the impact accident performance of waste packages involves the estimation of the quantity of radioactive materials that may be released under such conditions. Of particular importance is the quantities released as particulates in respirable sizes. For the purposes of this Case Study, it may be assumed that tests have been carried out to demonstrate that, when a 500 l drum of cemented sludges undergoes an impact associated with a 10 m free fall drop, the release fraction is 0.025 (i.e., 2.5% of the particulates are released from the package upon impact). The significance of this on health to the public and what is meant by A_2 releases is discussed in Chapter 22 dealing with nuclear transport.

Radionuclide Recording Levels

Data needs to be recorded on waste packages associated with radionuclide inventory. There may be a large number of radionuclides, but many may be at levels that are not significant to safety. Guidance on the concentrations of radionuclides that are of significance for future waste management planning is, therefore, important. With a knowledge of the expected inventory in a waste package, the waste producer can use the recording levels and guidance as an early indication for the waste packaging project to estimate those that are likely to need to be recorded in each case.

"Package specific" and "package collection" data sets have been prepared. When a radionuclide concentration falls below its "package specific" quantity, its safety impact is deemed to be sufficiently small in relation to the safety limits which arise for individual packages (e.g., often transport safety related). If the radionuclide concentration is also small in relation to its associated "package collection" quantity, then it may be treated as being present in insignificant quantities in relation to all safety limits (i.e., often in relation to disposal conditions, where large numbers of packages are accumulated) and, hence, are unlikely to be subject to detailed determination.

The two sets of data also help to indicate whether a radionuclide inventory needs to be determined for each package, or can be averaged across a collection of packages, thereby potentially easing the data recording requirements. Generally, the "package collection" quantities are lower in value than the "package specific" quantities. The two sets of quantities are shown in Table 20-7 for the four radionuclides used in the Case Study.

Table 20-7. Case Study Package Collection and Package Specific Radionuclide Inventory Data

Radionuclide	"Package Collection" Recording Quantity (TBq/m^3 of packaged waste)	"Package Specific" Recording Quantity (TBq/m^3 of packaged waste)
^{60}Co	3×10^{-3}	2×10^{-1}
^{129}I	4×10^{-8}	4×10^{-4}
^{235}U	1×10^{-6}	3×10^{-8}
^{238}U	2×10^{-7}	4×10^{-4}

Table 20-8. Radionuclide Characteristics under Disposal Conditions

Radionuclide	Half-life (years)	Expected behavior under disposal conditions
^{36}Cl	3.0×10^5	High mobility
^{90}Sr	29	High mobility
^{99}Tc	2.1×10^5	Low mobility
^{129}I	1.6×10^7	High mobility
^{137}Cs	30	High mobility
^{210}Pb	22	Moderate-to-low mobility
^{226}Ra	1.6×10^3	Moderate-to-low mobility
^{233}U	1.6×10^5	Low mobility
^{234}U	2.5×10^5	Low mobility
^{238}U	4.5×10^9	Low mobility
^{239}Pu	2.4×10^4	Low mobility
^{240}Pu	6.6×10^3	Low mobility
^{241}Am	4.3×10^2	Low mobility

Radionuclide Characteristics Under Disposal Conditions

The safety of the deep waste disposal concept relies upon the multiple physical and chemical barriers (the engineered system and geological characteristics and solubility and sorption in nature) to the migration of radionuclides to the biosphere, as explained in Section 20-4-4. Table 20-8 summarises the expected behavior of some radionuclides in terms of their half life, and their mobility (determined by their solubility and sorption characteristics).

20-6-7. *Suggested Answers to the Case Study Questions*

(1) There would be potential for:

- Loss of a fraction of the radioactive contents during long-term storage, especially that present in particulate form, should container degradation occur;

- Loss of a fraction of the radioactive contents during accidents such as impacts, especially that present in particulate form;
- Risk of combustion and dispersion of radionuclides if the package is exposed to a fire; and
- Chemical compounds in the laboratory wastes may accelerate corrosion of the waste container and wastes, increasing the risk of radionuclide dispersion.

This could lead to the contamination of transport and storage systems, risks of exposure of workers, and off-site contamination. It may also cause operational difficulties in the vault, storage, or disposal facility; including problems for retrievability. In the longer-term, there would be a reduced "barrier" to radionuclide migration. It should also be noted that the formation of uranium hydride (a pyrophoric material) on uranium fuel samples in punctured cans may occur, and they would pose a fire hazard if rapidly exposed to air. This is one reason why waste package specifications refer to immobilisation of radionuclides and loose particulates so as to make hazardous materials safe and avoid wasteforms that could burn.

(2) There would be a potential for:

- Loss of radioactive contents during long-term storage due to leaks of liquid waste; and
- Loss of radioactive contents during accidents such as impacts.

This could lead to contamination of storage and transport systems and be a potential risk of exposure to workers and could also involve off-site contamination. The relatively thin walled waste containers were designed to anticipate a solid monolithic wasteform. A heavier gauge container would still have the potential for leakage over an extended storage period or under accident conditions.

(3) The maximum package inventory could be more than three times that given in Table 20-5, due to the uncertainty in the waste inventory, and also due to the heterogeneity within the waste. For example, the cans in the solid waste may not be evenly distributed within the fuel element debris. Treatment of the cans in the packaging plant may also lead to segregation. The maximum waste inventories are likely to be important for some safety considerations. For example, compliance with criticality safety and for peak external gamma dose rates. It should be noted that, in some cases, early estimates of radionuclide inventories may not be sufficiently realistic for project planning, and further waste sampling may, therefore, be required.

(4) In the short-term, it may be necessary to re-suspend the waste to be able to retrieve it and transfer it to a packaging plant. This would depend on the waste itself and the proposed transfer system. Homogenisation may reduce the amount of sampling required to provide waste inventories, but there would still be those who would argue that the samples were not representative. Homogenisation may also simplify wasteform product quality control in the packaging plant by reducing the variability in water/solids content and chemical contents between different batches of sludges. This should make it easier to produce a consistent process for the grout mix (neither too high in water content resulting in residual liquid or too solid with a danger of voids forming).

(5) Yes. The mass of respirable material released would be $0.025 \times 0.5 \, kg = 0.0125 \, kg$. The mass of the wasteform is $900 \, kg$, so the fraction released would be $0.0125/900 \, kg = 1.389 \times 10^{-5}$. The quantity of A_2 multiples released from an average waste package would be $1.389 \times 10^{-5} \times 2700 \, A_2$ multiples. The specification required five A_2 multiples. The maximum waste package inventory would lead to a slightly greater release, but this should still be well within bounds.

It should be noted that, even if all of the $500 \, g$ of respirable materials generated in the impact within the package was released, there would still only be a release of $1.5 \, A_2$ multiples from the average package. Homogeneous cemented wasteforms are also considered predictable in their behavior, since, for more severe impacts (for example drops from greater heights onto "aggressive features"), their break-up and release fractions should increase progressively with impact energy. This is mainly due to the uniform distribution of activity in the wasteform and its solid nature.

(6) The radiogenic heat output for an average solid waste package (Table 20-5) is expected to be about $15 \, W$. This is well within the $200 \, W$ limit for the $3 \, m^3$ box. Radioactive decay during storage will reduce the heat output further. The packages that are likely to meet the requirements as long as the main radiogenic heat producing items (probably the pieces of irradiated stainless steel which will contain the majority of the short-lived radionuclides) are not distributed very heterogeneously between different waste packages. Radiogenic heat may not be the only heat source in wastes. Chemical reactions (for example, from metal corrosion or during grout curing) and physical processes (for example, Wigner energy release from graphite) can also generate heat in certain circumstances. Appropriate R&D will aid

an understanding of the processes involved and help to determine heat generation effects. It is interesting to note that, by definition, in the UK, ILW is not considered to be sufficiently heat generating to affect its storage and disposal.

(7) The recording quantities (Table 20-7) indicate that it should not be necessary to make special provision to record ^{60}Co in this waste. However, the recording quantities indicate that it should be necessary to record ^{235}U at the "package specific" level and ^{129}I and ^{238}U at the "package collection" level. These recording quantities are derived from a series of safety scenarios, which are to some extent dependent upon the actual waste form and nature of the waste. They are, therefore, guidance rather than a set of hard and fast rules.

(8) The project team would probably consider measuring the levels of these two gamma emitters using some form of gamma spectroscopy on batches of the waste. These measurements could be combined with "fingerprints" for the waste where concentrations of radionuclides are related to the levels of ^{60}Co in the fuel element debris and ^{137}Cs in the fuel-related wastes. The fingerprint information could be derived from validated computer codes that estimate the levels of radionuclides in irradiated materials. However, the use of such tools requires details of the original reactor irradiation and the elemental composition of the materials, including their impurities. Such information is not often available. Knowledge of impurities is important to inventory determination. ^{36}Cl is an important driver when determining a wet repository long-term safety case. It is produced through neutron activation of stable ^{35}Cl, which is an impurity in materials such as graphite and reactor fuels.

(9) This is a difficult question. It is likely that a combination of evidence would be required including historical information, information from monitoring during storage, and measurements made just prior to transportation. Information may come from R&D reports on waste package performance, container manufacturing data, information on the wastes at the time of packaging and also information from quality checks made during an interim storage period.

(10) Yes. All packages will require derivation of their Safe Fissile Mass (SFM). A criticality compliance assurance document would normally introduce a Safe Working Limit (SWL—lower in value than the SFM) taking into account measurement errors/tolerances and other uncertainties in the inventory. Such safety assessment and compliance assurance documentation is normally developed on a waste stream basis. You will note, however, that

Question 11 refers to a general criticality case which may cover several waste streams.

(11) Even though the anticipated inventory is only some 13.7 g, it is possible that packages could exceed the SFM of the general safety assessment involving 50 g fissile material per drum. This is due to current uncertainties in the waste inventory (factor of 5), potential inter-package variability and because the SWL could be less than 50 g and depends upon potential measurement errors in the packaging plant assay system.

(12) Barriers can include:

- the waste itself (e.g., for radionuclides within solid materials);
- any encapsulant used in the wasteform (e.g., cement grouts);
- the container;
- the repository backfilling material used to surround the waste packages; and
- the materials of the host geology.

The waste packages provide a combination of both physical and chemical barriers. However, over the timescales envisaged, the major barrier must be considered to be the host geology itself.

(13) A combination of high mobility (high solubility, low sorption), and high disposal inventory/short half-lives, makes ^{137}Cs and ^{90}Sr significant to the near-field repository flux shortly after disposal. These radionuclides do not appear in the far-field at significant concentrations due to their relatively short half-lives compared to the ground water return times in a wet repository environment (i.e., they decay in transit). Current repository modeling in the UK is pessimistic, in that it does not take into consideration the effects of the package on the long-term safety case.

(14) ^{129}I and ^{36}Cl have high mobility (high solubility, low sorption) and long half-lives. This allows them to migrate to the far-field at higher levels of activity than most other radionuclides; even those with long half-lives. The inventory of such radionuclides as ^{129}I and ^{36}Cl are, therefore, of particular interest to the designers of wet deep disposal facilities. They are typically found in fuel reprocessing wastes. They have only a minor contribution to the repository near-field flux, because there will be much larger inventories of short-lived mobile radionuclides such as ^{137}Cs and ^{90}Sr.

(15) Such short-lived radionuclides arise in the far-field due to migration of long-lived ^{238}U and its decay daughters (see Appendix 5). ^{238}U daughters themselves have relatively short half-lives and would not appear in the repository far-field except by this process and from natural ^{238}U in the geosphere. This process, whereby the inventory of daughter radionuclides arise from their parents, is known as "in-growth." A number of radionuclides arise in the ^{238}U decay series including ^{226}Ra and ^{210}Pb; again as described in Appendix 5. Uranium fuels are highly purified before use and contain very low levels of in-grown daughters. However, over the very long periods of time (many thousands to millions of years) associated with repository modeling, daughters will in-grow. Any process that enhances the mobility of ^{238}U is, therefore, of great interest to those involved with the long-term repository safety cases.

20-7. References

1. Nirex. *Derivation of Waste Package Specification for Intermediate Level Waste*, Nirex report N/008 (October 2000).
2. International Atomic Energy Agency. *Regulations for the Safe Transport of Radioactive Material, 1996 Edition (Revised)*, IAEA Safety Series No. TS-R-1, IAEA, Vienna (1996).
3. International Atomic Energy Agency. *Advisory Material for the IAEA Regulations for the Safe Transport of Radioactive Material*, IAEA Safety Standards Series No. TS-G-1.1 (ST-2), IAEA, Vienna (2002).
4. Nirex. *Deep Waste Repository "Generic Post-Closure Performance Assessment,"* Nirex Report N/030 (2001).
5. International Commission on Radiological Protection. "1990 Recommendations of the Commission. ICRP Publication 60," *Annals of the ICRP*, 21(1–3) (1991).
6. International Commission on Radiological Protection. *Recommendations of the International Commission on Radiological Protection, ICRP 60*, Pergamon (1991). ISBN 0-08-041144-4.
7. Higson, D. "Resolving the Controversy of Risks from Low Levels of Radiation," *The Nuclear Engineer*, 43(5): 132–137 (September–October 2002).

Chapter 21
Management of High Level Wastes (HLW)

21-1. Introduction

High Level Wastes are those wastes in which the temperature may rise significantly as a result of their radioactivity so that this factor has to be taken into account in the design of wasteforms and storage or disposal facilities. Such wastes arise as the result of burning nuclear fuel in reactors, which typically involves the production of heat generating minor actinides and fission products such as ^{137}Cs and ^{90}Sr radionuclides. The category may, as explained in Chapter 16, Section 16-4-7, include spent or used fuel if such material is declared as a waste, and high level waste (HLW) separated from spent fuel during reprocessing operations. The aqueous raffinate from the solvent extraction cycle of such reprocessing generally contains some 97–99% of the fission product activity of the irradiated used fuel. This chapter describes the sources, treatment, conditioning, packaging, and routing of such wastes to recognised end points.

21-2. Origins and Disposition of HLW

Spent fuel continues to contain a low percentage of fissionable radionuclides which, if economically sensible, may be extracted and reused in newly fabricated fuel elements. A reprocessing operation separates the useful fissionable material (^{235}U, ^{239}Pu, and ^{241}Pu) from the relatively short-lived fission products. The recovered $^{238}U/^{235}U$ may be blended with enriched uranium and used for further fuel manufacture. The plutonium may be recycled as Mixed Oxide (MOX) fuels or stored for possible future use. The unwanted fission products are disposed of as either HLW or ILW, depending upon their characteristics. In addition, the long lived (>100,000 years) activation products, produced as actinides in the reactor by neutron absorption in heavy metals, and generated as a waste during the reprocessing operations, need to be treated as HLW. Figure 21-1 is a simplified diagram of the typical waste streams arising from reprocessing oxide fuels and the associated common services facilities.

Current World spent nuclear fuel holdings are in excess of 230,000 tonnes (HM) and their distribution is shown in Table 21-1 [1].

At present there is much debate about the desirability, or otherwise, of continuing to reprocess spent fuel.

Reprocessing extracts reusable uranium and plutonium from the irradiated fuel and is a technology operated by a number of countries, with a consistent approach to both process and management of the resulting wastes.

An example of the "useful" product and waste arisings from a representative reprocessing operation is shown in Figure 21-2, and the typical overall process in Figure 21-3.

It should be noted that reprocessing capacity worldwide is limited to around 4000 te (HM) per year and it would take some tens of years to reprocess current holdings of spent fuel with the plants available, even if this was judged desirable.

21-3. Spent Fuel

21-3-1. Introduction

By 2020, the total quantity of spent fuel generated is projected to be of the order of 445,000 te (HM).

An alternative option to early reprocessing of spent nuclear fuel is its long-term storage for direct disposal as waste, or reprocessing at some time in the future.

The UK Government believes that the question of whether to reprocess (and if so, when), or to seek alternative spent fuel management options should be a matter for the commercial judgment of the owners of the spent fuel, subject to meeting the necessary regulatory requirements.

In many cases, countries have not yet made up their minds whether to operate an open (irradiate and dispose) or closed (irradiate, reprocess, and recycle) fuel cycle. Interim safe storage of spent fuel allows them to keep their options open (see Table 21-2).

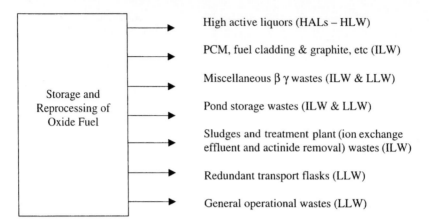

Figure 21-1. Simplified Flow Diagram for Radioactive Waste Production from Reprocessing.

Table 21-1. Some Estimates of World Spent Fuel Holdings (tonnes, heavy metal)

Region	NPP pool storage capacity, te HM	AFR wet storage capacity, te HM	AFR dry storage capacity, te HM	Total storage capacity, te HM
Western Europe	28,265	32,270	10,416	70,951
Eastern Europe	11,913	20,788	1,471	34,172
America	94,662	1,712	6,342	102,716
Asia and Africa	27,924	1,725	1,737	31,386
Total	162,764	56,495	19,966	239,225

NPP: Nuclear Power Plant; AFR: Away From Reactor; te HM: tonne Heavy Metal.

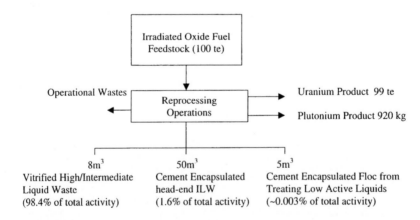

Figure 21-2. Representative Reprocessing Product and Waste Arisings (For Typical AGR Fuel Through THORP) [2].

21-3-2. *Storage*

The role of spent fuel storage, as a buffer for any management option for the back end of the fuel cycle, has continued to expand globally due to the growing inventory of spent fuel in almost all those IAEA Member States with nuclear power production. At the end of 2000, roughly two-thirds of the total amount of 230,000 Mte (HM) of spent fuel discharged from nuclear power reactors in the world was in storage either "at-reactor" or

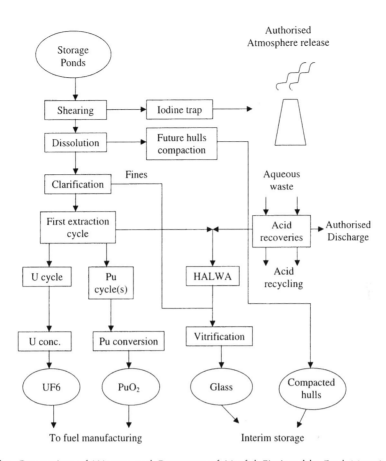

Figure 21-3. The Generation of Wastes and Recovery of Useful Fissionable Fuel Material From Typical Reprocessing Operations.

Table 21-2. World Decisions Concerning Open or Closed Fuel Cycle Operations

Countries which *reprocess* spent fuel	Countries *committed* to disposal of used fuel as HLW	Countries *undecided* on whether to reprocess spent fuel or disposal as HLW
Russian Federation	Sweden	Bulgaria
France	Finland	France
UK	USA	Korea
Japan	Italy	Lithuania
India	Germany	Romania
	Canada	Ukraine
	Czech Republic	
	Slovak Republic	

There is currently no clear favorite method for long-term storage of spent fuel. Most spent fuel (∼90%) is currently stored in water filled pools/ponds, as this is generally a feature of the fuel discharge routes of most reactor types. This provides cooling, shielding, and is cost effective. Continued long-term storage in pools is an option, but "dry" storage utilising monolithic concrete structures, vaults, and storage casks is also utilised.

Storage casks currently have a high profile as they are now a proprietary item, available from several international vendors, and marketed as a "fit and forget" solution for long-term fuel storage. Such flasks may be designed as single purpose (storage only), dual purpose (storage and transport), or triple purpose (where the same flask can be utilised for storage, transport, and ultimate disposal of the spent fuel).

Massive concrete and metal casks are used in Europe for the transport, interim storage, and, in some cases, eventual disposal of spent fuel. Figure 21-4 illustrates the German "CASTOR" triple purpose cask which weighs some 131 te. Its 4 × 2.7 m body is made from 0.37 m

"away-from-reactor" storage systems, with >90% in ponds or "wet" storage. The trend is likely to continue for the foreseeable future as the current situation of limited reprocessing, deferred decisions, and pending disposal continues.

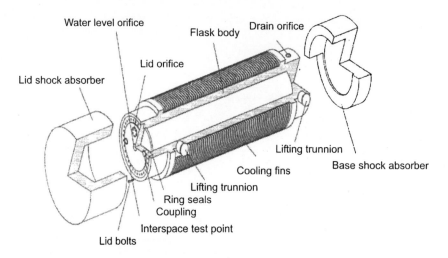

Figure 21-4. German "CASTOR" Spent Fuel Flask.

thick cast iron, with a stainless steel bolted lid containing a pressure monitoring device.

21-3-3. *Security and Safeguards*

The stored fuel is "special nuclear material." The organisation storing the fuel is responsible for meeting safety, security, and independently verified international safeguard requirements throughout the storage period. Because the spent fuel is irradiated, it is intensely radioactive. It may be considered "self-protecting" from a safeguards point of view. However, it must be recognised that, with *very* long-term storage, the spent fuel will become progressively less self-protecting due to decay of fission and activation products. Verification and inspection are part of the safeguards requirement, and this has to be considered at the design stage.

From a security point of view, access control, intruder detection, and counter measures need to be provided. This may involve provision for physical inspection, seals, and fixed cameras.

21-3-4. *Conditioning for Disposal*

Several designs exist for the canisters which will be used for spent fuel disposal. Each canister is typically designed to accommodate several reactor elements (21 commercial PWR assemblies in the case of a US Yucca Mountain design) and weigh some tens of tonnes when fully loaded. The canister may be manufactured from copper, cast iron, cast steel, or alloy, with a wall thickness of greater than 50 mm. Final closure of the canister can be by either welded or bolted end cap. Some repositories have been designed to accept spent fuel in triple-purpose casks for direct disposal.

Canisters, similar to those designed for spent-fuel disposal, can be utilised for the disposal of vitrified HLW waste forms.

21-4. HLW Characteristics and Inventory Data

A proportion of High Level Waste (HLW) is the heat generating waste that remains from the reprocessing of spent fuel.

The industry standard for the treatment of such HLW liquors is to further concentrate by reduced pressure evaporation for interim storage in double contained stainless steel tanks. Such tanks are heavily shielded with multiple cooling and agitation systems (see Figures 21-5–21-7).

The use of such High Active Liquor (HAL) tank storage facilities is proven in the short-to-medium term (>50 years). There has to be a degree of uncertainty regarding longer term (>100 years) tank integrity, due to corrosion and settlement effects of heat generating particles on nonagitated tanks. In addition, such facilities require a high degree of security and management. Therefore, both France and the UK have adopted a vitrification process for immobilisation of the activity in glass pucks. On the basis of hazard reduction (see Chapter 14), there is Regulatory pressure to reduce HAL storage to buffer stock quantities (approximately 200 cu m per annum at BNFL in the UK) between waste arisings from the reprocessing operations and its vitrification in modern purpose-built plants. The capital cost of a recently built vitrification facility in the UK is some £320M.

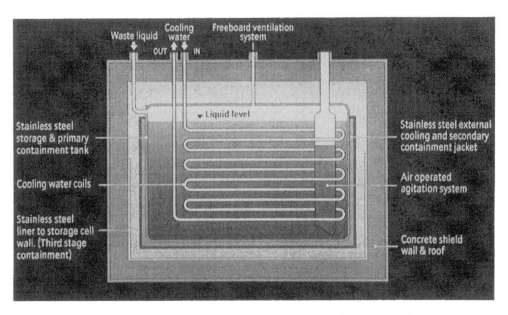

Figure 21-5. Principle of High Active Waste Storage Tank (Courtesy of BNFL).

Figure 21-6. Internal View of HLW Storage Tanks (Photograph Courtesy of BNFL).

Prior to vitrification, the HALs must be homogeneous and undergo a rigorous assay regime to determine the characteristics and properties of the material. Typically, this will involve Inductively Coupled Plasma (ICP) spectroscopy, Atomic Emission Spectroscopy (AES), and Mass Spectroscopy (MS), as well as feedstock temperature, density, lithium content (which interferes with glass formation and leach resistance), and pH as flowsheet parameters plus:

• Physical and chemical properties of the waste including an assessment of any hazards associated with the waste, e.g., corrosive, flammable, toxic, and presence of organic compounds;

Figure 21-7. Internal View of HLW Storage Tanks Showing Cooling Tubes and Giving an Indication of their Physical Size (Photograph Courtesy of BNFL).

- Radionuclide inventory;
- Volume of waste; and
- Compatibility of the waste stream with the proposed encapsulation matrix.

Once a form has been chosen, the performance of the encapsulated or immobilised waste must be assured, including:

- Mechanical properties;
- Physical properties;
- Physical/chemical stability;
- Impact performance;
- Thermal stability; and
- Radiation stability.

The vitrification process involves drying and chemical conversion (calcination) of the highly active liquor concentrate to a fine dry powder known as "calcine." The calcine is mixed with crushed glass in a ratio of about 25% waste to 75% glass, and heated to 1150 °C in an induction furnace. The glass melts and the calcine dissolves, creating a molten mixture of glass and fission products. The vitrified waste glass product is poured into 150 liter stainless steel containers (Figure 21-8), solidified by cooling, and enclosed by the fitting of a suitably welded lid.

The vitrified product waste containers are held in naturally ventilated air cooled stores (Figure 21-9) prior to disposal (or if part of a commercial operation in readiness for return to customers having due responsibility for the management of their own wastes).

It is important to emphasise the state of the art technology involved in such a process. The system requires remote operations, with extremely high reliability equipment operating in harsh conditions. The UK has three

French technology-based process lines at BNFL, Sellafield in West Cumbria. France uses a similar process at Cogema La Hague, with six lines in operation. Other world vitrification operations are indicated in Table 21-3.

Alternative processes for HALs immobilisation include "locking" the wastes into ceramic forms, and Synroc (synthetic rock) has been considered for US weapons plutonium wastes.

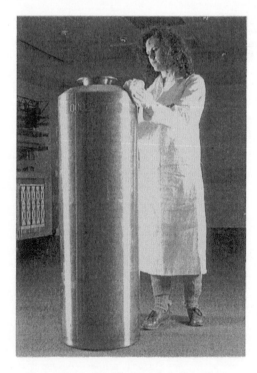

Figure 21-8. Vitrified Product Container.

Figure 21-9. Vitrified Product Store.

Table 21-3. HALs Vitrification Plants and Processes

Country	Process	Status
Belgium	PAMELA	Now decommissioning
China	German technology	
France	AVH	6 lines at Cogema, La Hague
Germany	VEK	Due operational 2003–2005
India	por	
Japan	JCM	Tokai
Russia	JCM and CCM	
USA	JCMs	West Valley, complete Savannah River, ongoing Hanford, not yet started
UK	AVM (French process)	3 lines at BNFL, West Cumbria

JCM: Joule heated, Ceramic lined, Melter (German design); and CCM: Cold Crucible Melter (French design).

21-5. HLW Current World Disposal Status

Direct disposal of spent fuel offers an alternative to reprocessing. A storage period of some 50 years is considered necessary to allow for heat dissipation and to take advantage from natural radioactive decay, thus simplifying future disposal. If direct disposal of spent nuclear fuel is the chosen management option, there is currently no HLW disposal facility in the world available to accept this material. The US DoE's facility at Yucca Mountain could become available for nuclear fuel disposal as a waste in 2007, and both Sweden and Finland plan to have deep underground fuel repositories available by perhaps 2010. World-wide approaches to HLW storage and intentions for eventual disposal are described in Appendix 1.

21-6. References

1. International Atomic Energy Agency. *Long term storage of spent nuclear fuel – Survey and recommendations*, IAEA Technical Report Series 1293, IAEA, Vienna (May 2002).
2. Philips/Miliken – WM 2000. Reprocessing as a waste management and fuel recycling option, Tucson, Arizona (February 2000).

Chapter 22

Transport

22-1. Introduction

A key aspect in the development of waste management and decommissioning strategies is the consideration of transport of the nuclear wastes. This may involve movement to an alternative location such as a long-term or interim store on the waste arising site. Alternatively, it may involve transport through the public domain to a separate nuclear licensed site for storage or disposal. This chapter describes the Regulatory aspects of nuclear materials transport that must be complied with and gives examples of the application of the appropriate National and International Standards.

22-2. Regulatory Requirements for Transport

22-2-1. Regulations

The IAEA Transport Regulations [1] form the basis for the regulations governing the transport of radioactive materials in the UK, as well as for the regulatory requirements set by international organisations, regional bodies, international agreements, or conventions governing the international transport of radioactive materials by sea, air, road, rail, and inland waterways.

The regulations recognise that a wide range of radioactive material requires to be transported, of varying characteristics, and aim to provide a uniform level of safety that is commensurate with the inherent hazard presented by the radioactive material being transported. Safety features are built into the design of the package, as far as is feasible, thereby placing primary reliance on the package design and preparation, rather than on the need for any special actions during carriage.

Within the UK, the current legislation is based on the 1996 edition of the IAEA Transport Regulations. The relevant UK legislation is shown in Table 22-1. For international transport by sea and air, the following regulations apply:

- International Maritime Organisation (IMO): International Maritime Dangerous Goods (IMDG) Code, Amendment 30-00; and

Table 22-1. Applicable UK Land, Sea, and Air Transport Regulations

Mode of transport	Legislation
Road	The Radioactive Material (Road Transport) Regulations 2002 SI 2002 No. 1093.
	The Radioactive Material (Road Transport) (Definition of Radioactive Material) Order 2002 SI 2002 No. 1092.
Rail	The Packaging, Labeling and Carriage of Radioactive Material by Rail Regulations 2002 SI 2002 No. 2099.
	The Carriage of Dangerous Goods (Classification, Packaging and Labeling) and Use of Transportable Pressure Receptacles Regulations 1996, SI 1996 No. 2092.
	The Carriage of Dangerous Goods (Amendment) Regulations 1999, SI 1999 No. 303.
Sea	The Merchant Shipping (Dangerous Goods and Marine Pollutants) Regulations 1997, SI 1997 No. 2367.
	Merchant Shipping Notice No. M 1755(M), The Carriage of Dangerous Goods and Marine Pollutants in Package Form — Amendment 30-00 to IMDG Code.
Air	The Air Navigation Order 2000, SI 2000 No. 1562.
	The Air Navigation (Dangerous Goods) Regulations 1994, SI 1994 No. 3187 and Amendment 2001, SI 2001 No. 918.

- International Civil Aviation Organisation (ICAO): Technical Instructions for the Safe Transport of Dangerous Goods by Air, 2001–2002 Edition.

For road and rail transport within Europe, the following regulations are applicable:

- European Agreement Concerning the International Carriage of Dangerous Goods by Road (ADR), 2001 Edition; and
- Convention Concerning International Carriage by Rail (COTIF) Appendix B, Uniform Rules Concerning the Contract for International Carriage of Goods by Rail (CIM), Appendix 1, Regulations Concerning the

International Carriage of Dangerous Goods by Rail (RID), 2001 Edition.

These international regulations cover all types of hazardous goods, including radioactive material. These are divided into the following material types:

- Class 1: Explosives;
- Class 2: Gases (compressed, liquified, dissolved under pressure, or deeply refrigerated);
- Class 3: Flammable liquids;
- Class 4: Flammable solids; solids liable to spontaneous combustion; substances which, on contact with water, emit flammable gases;
- Class 5: Oxidising substances; organic peroxides;
- Class 6: Toxic and infectious substances;
- Class 7: Radioactive material;
- Class 8: Corrosive substances; and
- Class 9: Miscellaneous dangerous substances and articles.

The IAEA Transport Regulations were first published in 1961, and since then there have been further editions published in 1964, 1973, 1985, 1985 (amended in 1990), 1996 and 2001 (1996 Revised) [1].

The international regulations governing the transport of hazardous goods are published every 2 years, and the IAEA has also decided to move to a 2-yearly publication cycle.

The IAEA also produce advisory material to assist in the application and understanding of the IAEA Transport Regulations. The most recent advisory material was published in 2002 [2] for the 1996 Revised edition of the IAEA Transport Regulations. The following IAEA publications are also helpful in respect of specific aspects of the transport of radioactive material:

- TS for emergency arrangements [3],
- QA [4], and
- Compliance assurance [5].

The remainder of the discussion of the IAEA Transport Regulations is based on the 1996 Revised edition and the associated Advisory Material.

22-2-2. *General Requirements*

The IAEA Transport Regulations are based around the following primary safety requirements for the packaging of radioactive material:

- prevention of release of the radioactive contents;
- limit radiation levels on the outside of the package;
- prevent criticality events occurring; and
- heat.

Table 22-2. IAEA Package Types

Package type	Permitted contents
Excepted package	Very small quantities of radioactive material
Industrial package	Low specific activity material or surface contaminated objects
Type A package	Less than one A_2 of radioactive material
Type B package	Generally limited by the package design, unless being transported by air, in which case certain limits apply
Type C package	Generally limited by the package design, and only required for air transport
Package containing UF$_6$	UF$_6$ within the limits dictated by the package design

The regulations recognise that a wide range of radioactive material requires to be transported, of varying characteristics, and aim to provide a uniform level of safety that is commensurate with the inherent hazard presented by the radioactive material being transported. Safety features are built into the design of the package, as far as is feasible, thereby placing primary reliance on the package design and preparation, rather than on the need for any special actions during carriage.

The result is a series of package types, with progressively more stringent design requirements as the inherent hazard of the radioactive contents increases. The package types and the types of material that can be transported within them is presented in Table 22-2.

The quantity A_2 referred to in Table 22-2 is specified in the regulations for all radionuclides, and is determined such that in a severe accident the release of material from a damaged package would not lead to excessive dose uptake by a person in the vicinity of the accident. The determination of the A_2 quantities takes account of external photon and beta doses, inhalation dose, skin and ingestion dose due to contamination transfer, and submersion dose.

Additional requirements apply for any packages carrying fissile material, although it should be noted that the term "fissile material" is specifically defined in the regulations. Most of the identified package types may carry fissile material.

Special arrangement shipments may also be carried out, in which the package design does not meet all the applicable requirements of the regulations, but the overall approach to the shipment should mean that an acceptable overall level of safety is achieved. However, the use of the special arrangement provision should be the exception. It may be applicable for the disposal of old equipment containing radioactive material where there is no reasonable way to transport the radioactive material in an approved package and where the hazard associated with repackaging and handling the radioactive material could outweigh the advantage of using an approved package.

For certain package types, approval of the package design by the relevant competent authority is required. The term "competent authority" means a national or international regulatory body or authority designated to carry out the functions of a competent authority as laid down in the regulations. In the case of the UK, functions of the competent authority are carried out by the Radioactive Materials Transport Department (RMTD) of the Department for Transport (DfT). Competent authority approval is required for the following package designs:

- packages containing 0.1 kg or more of UF_6;
- all package containing fissile material; and
- Type B and Type C packages.

Competent authority approval is also required for special arrangements and for certain shipments. The DfT expect organisations preparing applications for package approval to comply with their Applicants Guide [6].

The following section provides a summary of the regulatory requirements for the various package types, with the exception of the Type C and UF_6, as it is unlikely that these package types would be used for radioactive waste transport.

22-2-3. Package-Specific Requirements

Excepted Packages

Solid, liquid, and gaseous A_2 activity limits are specified for excepted packages containing radioactive material other than natural or depleted uranium. For transport by post, a total activity limit of one tenth of the specified levels for other transport medium for each package is specified.

Industrial Packages

The permitted contents of industrial packages are low specific activity (LSA) material or surface contaminated objects (SCO). LSA and SCO categories are themselves broken down into LSA-I, LSA-II, LSA-III, SCO-I, and SCO-II.

LSA-I primarily covers ores and material that has an activity concentration only 30 times that for transport in excepted packages: as the activity concentration in ILW is significantly greater than this, LSA-I is not applicable to ILW.

For LSA-II material, the following criteria must be met:

- the activity must be distributed throughout the material;

Table 22-3. Contamination Limits for Surface Contaminated Objects (SCOs)

	Contamination Limits: Bq/cm^2	
Criterion	SCO-I	SCO-II
Nonfixed contamination on accessible surfaces:		
• $\beta\gamma$ and low toxicity α emitters	4	400
• all other α emitters	0.4	40
Fixed contamination on accessible surfaces:		
• $\beta\gamma$ and low toxicity α emitters	4×10^4	8×10^5
• all other α emitters	4×10^3	8×10^4
Fixed plus nonfixed contamination on inaccessible surfaces		
• $\beta\gamma$ and low toxicity α emitters	4×10^4	8×10^5
• all other α emitters	4×10^3	8×10^4

- the average specific activity must not exceed 10^{-4} A_2/g for solids and gases, and 10^{-5} A_2/g for liquids; and
- water may have a tritium concentration up to 0.8 TBq/l.

The LSA-III material criteria are:

- the activity must be distributed throughout a solid or a collection of solid objects, or is essentially uniformly distributed in a solid binding agent such as concrete;
- the average specific activity must not exceed 2×10^{-3} A_2/g; and
- the loss of radioactive material by leaching, if placed in water for 7 days, must not exceed 2×10^{-3} A_2/g.

The requirements for SCO-I and SCO-II are summarised in Table 22-3.

It is possible for these criteria for LSA material or SCO to be met and result in an excessively high dose rate from the unshielded material. This is of significance because there is no performance requirement under accident conditions for industrial packages, and it would be possible for an individual to be exposed to a high dose rate. An additional requirement is, therefore, imposed such that the total quantity of radioactive material in an industrial package, whether LSA material or SCO, is restricted such that the external radiation level at 3 m from the unshielded radioactive material does not exceed 10 mSv/h.

There are three industrial package groups, each with different design requirements, and these are designated IP-1, IP-2, and IP-3. The allocation of the LSA material and SCO to these package types is specified in Table 22-4. The term "exclusive use" that is used in Table 22-4 means the sole use, by a single consigner, of a conveyance or of a large freight container, where all initial, intermediate,

Table 22-4. Industrial Package Requirements for Low Specific Activity (LSA) Material and Surface Contaminated Objects (SCO)

	Industrial package type	
Radioactive contents	Exclusive use	Not under exclusive use
LSA-I:		
• solid	IP-1	IP-1
• liquid	IP-1	IP-2
LSA-II:		
• solid	IP-2	IP-2
• liquid and gas	IP-2	IP-3
LSA-III	IP-2	IP-3
SCO-I	IP-1	IP-1
SCO-II	IP-2	IP-2

and final loading and unloading is carried out in accordance with the directions of the consigner or consignee. In practice, transport of waste nuclear licensed sites is likely to be carried out under exclusive use conditions.

As well as meeting some general design requirements covering such matters as lifting attachments, collection and retention of water, and package closures, both IP-2 and IP-3 packages, if subject to specified impact and stacking tests to represent normal conditions of transport, must prevent:

• loss or dispersal of the radioactive contents; and
• loss of shielding integrity that would result in more than a 20% increase in the radiation level at any external surface of the package.

The specified impact and stacking tests are:

• the package is to be dropped from a height that is dependant upon the package weight (e.g., 1.2 m high drop for a package weighing less than 5 te, but from 0.3 m for a package weighing more than 15 te) on to an unyielding target; and
• for a period of 24 hours, the package is to be subject to a compressive load of the greater of 13 kPa multiplied by the vertically projected area of the package or the equivalent of 5-times the mass of the actual package.

There are additional requirements for IP-3 packages, most of which are relatively straightforward design issues, such as the minimum package dimension and security seal, but there are some additional test requirements, and any tie-down attachments on the package must not lead to impairment of the compliance of the package with the requirements of the regulations during either normal or accident conditions of transport. The additional test requirements comprise:

• a water spray test; and

• a penetration test, with the latter involving a 3.2 cm diameter 6 kg ball being dropped on to the package from a height of 1 m.

The acceptance criteria from these tests are those identified above relating to loss or dispersal of the radioactive contents and the loss of shielding integrity.

Freight containers may be used as IP-2 or IP-3 packages, provided that when the freight container tests are carried out, as well as meeting the specified acceptance criteria for freight containers, the acceptance criteria identified above relating to loss or dispersal of the radioactive contents and the loss of shielding integrity are also met.

Industrial packages do not require competent authority approval, unless they are carrying fissile material.

Type A Packages

Type A packages are permitted to carry contents up to a total activity of one A_2. All the general design requirements for industrial packages, both for IP-2 and IP-3 packages, apply to Type A packages.

The test requirements are that Type A packages must be subject to specified water spray, impact, stacking, and penetration tests, i.e., those tests that apply to IP-3 packages, must prevent:

• loss or dispersal of the radioactive contents; and
• loss of shielding integrity which would result in more than a 20% increase in the radiation level at any external surface of the package.

For Type A packages carrying liquids or gases, apart from tritium gas or noble gases, the package must be subject to the following tests:

• a free drop test on to an unyielding target from 9 m; and
• a penetration test with a 3.2 cm diameter 6 kg ball being dropped on to the package from a height of 1.7 m.

For packages carrying liquids, the liquid must be contained after these tests, either by means of absorbent material or by an outer secondary containment. For packages carrying gases, there is to be no loss or dispersal of the radioactive contents.

Type A packages do not require competent authority approval, unless they are carrying fissile material.

Type B Packages

There is no specified contents limit for Type B packages, although contents limits will need to be set for each specific package design, such that the relevant requirements of the regulations are met.

All the design and test requirements for Type A packages also apply to Type B packages, apart from the requirement for there to be no loss or dispersal of the radioactive contents after the specified tests representing normal conditions of transport. Additionally, the specified tests for Type A packages carrying liquids or gases do not apply to Type B packages.

Additional design requirements are laid down for Type B packages, including limits on the maximum normal operating pressure, permitted temperatures on accessible surfaces, and specified design ambient temperatures and solar insulation. Tests representing normal and accident conditions of transport and the associated acceptance criteria are also laid down.

The tests representing normal conditions of transport comprise:

- the water spray test;
- the free drop test;
- the package is to be dropped from a height that is dependent upon the package weight (e.g., 1.2 m high drop for a package weighing less than 5 te, but from 0.3 m for a package weighing more than 15 te) on to an unyielding target; and
- for a period of 24 hours, the package is to be subject to a compressive load of the greater of 13 kPa multiplied by the vertically projected area of the package or the equivalent of 5-times the mass of the actual package.

After completion of these tests, the following must be met:

- the loss of radioactive contents must not exceed $10^{-6} A_2$/hour; and
- there must be no more than a 20% increase in the radiation level at any external surface of the package.

The tests representing accident conditions of transport comprise:

- the package being dropped from 9 m on to an unyielding target;
- the package being dropped from 1 m onto a solid steel bar of 150 mm diameter and at least 200 mm length;
- the package in its damaged condition following the above two drop tests being subjected to a fully engulfing fire with an average temperature of at least 800°C for a period of 30 minutes; and
- immersion of the package under a head of water of at least 15 m for a period of not less than 8 hours.

After completion of these tests, the following must be met:

- the loss of radioactive contents must not exceed $10 A_2$ of krypton-85 and not more than $1 A_2$ of all other radionuclides; and
- the radiation level at 1 m from the external surface of the package must not exceed 10 mSv/h.

Additionally, any package for radioactive contents with an activity greater than $10^5 A_2$ is required to be subject to a water immersion test at a depth of 200 m for at least 1 hour, with there being no rupture of the containment system.

Type B packages are subdivided into Type B(M) and Type B(U) packages. All Type B packages require competent authority approval, but Type B(M) packages require multilateral approval, that is approval by the competent authority of every country in which the package is to be used, whereas Type B(U) packages only require unilateral approval, that is approval by the competent authority of the country of the origin of the package design.

Certain of the requirements for Type B packages need not be met for Type B(M) packages, subject to acceptance of this by the relevant competent authorities. The most common area where this is used is where a narrower range of design ambient conditions is used, where that is appropriate for the countries where the package is being operated.

Where a Type B package contains fissile material, the package type becomes Type B(U)F or Type B(M)F, as appropriate. Multilateral approval of a Type B package containing fissile material is required, irrespective of whether it is a Type B(U)F or Type B(M)F package.

Packages Containing Fissile Material

Four provisions are given in the regulations that enable packages containing fissile material to be exempt from the requirements for packages containing fissile material. The detail of these provisions is not included here, although it should be noted that if a package contains less than 15 g of fissile material, it should be exempt from the requirements for packages containing fissile material.

Additionally, only certain radionuclides are considered by the regulations to be fissile material. These radionuclides are:

- uranium-233;
- uranium-235;
- plutonium-238 (note that this radionuclide is not included in the definition of fissile material in the 1996 edition of the IAEA Transport Regulations);
- plutonium-239;
- plutonium-241; and
- any combination of these radionuclides.

Subcriticality must be demonstrated for:

- a package in isolation;
- package arrays under normal conditions of transport; and
- package arrays under accident conditions of transport.

The tests representing normal and accident conditions of transport are similar to those for Type B packages.

22-2-4. *Mode-Specific Requirements*

The IAEA Transport Regulations are generally based on the approach that safety is primarily invested in the transport package and, therefore, there needs to be only limited requirements that relate to the transport mode.

There are some mode-specific requirements, such as the total activity limit for a conveyance, which is dependent upon whether or not the transport package is being carried in an inland water craft. There are also different limits on the total sum of transport indexes in a single freight container or aboard a conveyance, which depend upon the type of freight container or conveyance being used. Other minor differences between the different transport modes, such as those applying to labeling, are also set out in the IAEA Transport Regulations.

When these regulations are applied in national legislation and by international organisations, some additional mode-specific requirements are introduced. For example, the International Maritime Organisation (IMO) requires in its dangerous goods code (IMDG) that ships carrying transport packages containing irradiated fuel must meet certain requirements, such as having double hulls.

22-2-5. *Operational Requirements*

Various operational requirements are imposed by the IAEA Transport Regulations. These include:

- labeling to indicate the radioactive nature of the contents, the transport index (which is related to the external dose rate), and the criticality safety index;
- predespatch requirements;
- consigner's responsibility; and
- storage in transit.

22-2-6. *Special Arrangements*

A "special arrangement" is where a consignment of radioactive material does not meet all the relevant requirements of the regulations, but provisions are put in place to ensure that the overall level of safety in transport is at least equivalent to that which would be provided if all the applicable requirements of the Regulations had been met. Such shipments require competent authority approval.

Para. 238.1 of the Advisory Material [2] to the 1996 Edition (Revised) of the Regulations states that "this type of shipment is intended for those situations where the normal requirements of the Regulations cannot be met." It goes on to say that an example of this would be "the

disposal of old equipment containing radioactive material where there is no reasonable way to ship the radioactive material in an approved package." Furthermore, "the hazard associated with repackaging and handling the radioactive material could outweigh the advantage of using an approved package, assuming a suitable package is available" and "reliance on administrative measures should be minimised in establishing the compensating measures."

Amongst the types of transport where the special arrangement approach is potentially applicable is decommissioning waste, where it is impractical to design and manufacture a package that meets all the regulatory requirements and where size reduction to enable the waste to fit within an approved package would involve significant dose uptake.

The DfT Applicants Guide [6] is what they require to assess before they would give approval to a special arrangement transport operation. As well as details of the package, conveyance, transport mode and route, the following is required:

- "state in which respect, and justify, the reasons why the consignment cannot be made in full accordance with the applicable requirements of the regulations;"
- "identify and justify what compensatory safety measures, or controls, are proposed to compensate for failure to meet the requirements of the Regulations;" and
- "demonstrate how the appropriate regulatory standard of safety will be achieved and how these will be put into effect."

It is also stated that "it is in the interest of the applicant to demonstrate that all alternative options have been fully explored" and "such applications should only be sought on a short-term basis or to cover minor shortfalls in some regulatory requirements."

It can be seen from the above that, in principle, the special arrangement approach is of potential use in the transport of large decommissioning items. However, where the radioactive nature of a large decommissioning item is such that it would require an IP-2 or IP-3 package, it is difficult to see how it could be applicable as the key performance requirement is for a free drop from 0.3 m (packages heavier than 15 te) with no loss or dispersal of the contents and an increase in external dose rate of no greater than 20%, and any argument for not being able to meet this is going to need to be extremely robust and probably difficult to sustain.

22-3. **Examples of Waste Transport Packages**

Table 22-5 includes some examples of packages that are either currently in use or which are being developed for

Table 22-5. Some Examples of UK Transport Packages

Organisation and designation	Dimensions and weights	Use	Specific features
BNFL, Chapelcross ILW flasks	Type A package 105 mm minimum shielding thickness. 540 × 905 mm cavity. Type B package 250 mm minimum shielding thickness. 300 × 676 mm cavity.	Transport of a variety of irradiated ILW components from the Chapelcross reactor core(s).	• Disposable liners. • Forged carbon steel. • Type B package max. activity carried up to 75 TBq and 30 A_2 and max. 100 W heat output.
BNFL, On-site ILW flasks (variety)		Sellafield site ILW movements.	
BNFL, Sellafield Waste Transport Container (SWTC) (under development)	Type B package shielding thicknesses of 70, 150, and 285 mm. Unladen weight range 16–53 te and max. gross laden weight range 28–65 te over shielding thickness range.	For transport of ILW to a possible future repository if located away from the Sellafield area.	Max. leakage rates for containment to be Standard Leakage Rate (SLR) of 10^{-3} bar cm^3/s under normal conditions of transport and 10^{-2} bar cm^3/s SLR under accident conditions of transport. Single bolted lid.
UKAEA, Modular flask	Type B package 230 mm lead shielding and 18 te unladen weight. 210 × 2134 or 2794 mm cavity.	Wastes and irradiated fuel movements from Winfrith to Harwell.	
RWE Nukem, Transactive-20 container	Shown to meet Type B(U)F package performance. 6187 mm long × 2442 mm wide × 2716 mm high compliant with ISO standards for normal road transport.	Transport of drummed PCM.	Outer structure (for impact and thermal protection) with internally mounted stainless steel containment vessel (5.5 m long × 1.9 m diameter).
RWE Nukem, NUPAK 200	4 × 200 l PCM drums forming an overall Type B package. 2.2 × 2.18 × 1.64 m high.	Transport of drummed PCM.	Special frame such that five packages may be transported on a 12 m flatbed road trailer.
Croft Associates, Reusable full-height ISO container	Large volume reusable IP-2 package. 6058 mm long × 2438 mm wide × 2591 mm high (for standard 20' freight container). Unladen 5 te, fully laden 25 te.	Primarily for road transport of bulk quantities of large items of radioactive LSA or SCO material.	Mild steel with large end door and double seal system.
Nirex, Reusable Shielded Transport Container (RSTC) (under development)	Type B(M) package with range of shielding thicknesses (70–285 mm) & to meet road vehicle weight limit of 38 te or UK rail loading gauge. Unladen weight range 16 to 48 te and max. gross laden weight range 28–60 te over shielding thickness range.	For packaged ILW in UK Nirex standard waste containers (4 × 500 l drums, 1 × 3 m^3 drum or 1 × 3 m^3 ☐) from waste arising/store to future ILW repository.	Purge/vent valve. Nitrogen pressurised.
Nirex, Industrial Package Transport Container (IPTC) (under development)	IP-2 package requirements.	To carry those LSA and SCO wastes that do not require Type B packages.	A lightweight reusable transport container made up from two components — a lid assembly and the body. Bolted lid.

Continued

Table 22-5. Continued

Organisation and designation	Dimensions and weights	Use	Specific features
Nirex, 4 m ILW box	Nonfissile IP-2. 4 m long and max. gross weight of 65 te. Concrete shielding in range 100–300 mm.	For transport by rail, road, and sea on standard arrangements.	Prototype tested to ISO 1496/1 freight container requirements. A skeletal frame, comprising the corner posts, top, and bottom rails, withstands all the test forces applied to the container, without assistance from the wall panels, lid, or concrete shielding.

use in the UK for radioactive waste transport. Waste transport packages used outside the UK include the Trupact, Cogema Logistics LR56, and Cogema Gemini, BNFL Vitrified HLW Return Flask, etc.

22-4. Transport of Large Items of Decommissioning Waste

22-4-1. *Application of the Regulations to Large Items*

There are two key drivers in considering the application of the IAEA Transport Regulations to the potential transport of large decommissioning items. First, the package type that would be required for the transport, which in turn is determined by the nature and activity of the radioactive material that is associated with the large item. Secondly, whether the large item, either on its own or with minor modifications, can act itself as the packaging, i.e., that no specific packaging needs to be provided within which the large item would be transported.

The benefits of this are that the packaging costs and the handling activities associated with loading the large item into the packaging are avoided.

The performance requirements for an IP-2 (suitable under exclusive use for LSA-III material or SCO-II) are less than those for an IP-3 or Type A package, although the differences, which comprise a water spray test and a penetration test involving a 6 kg bar being dropped onto the package from a height of 1 m, would be expected to be readily met for a large decommissioning item forming its own package.

However, the performance requirements for a Type B package which include tests to represent accident conditions of transport, are much more demanding. Furthermore, impact tests using a scale model are typically used to demonstrate the packages performance. For a one-off move of a large decommissioning item, it would be a significant cost to manufacture and drop test a scale model. An alternative approach would be to demonstrate the impact performance by means of finite element dynamic analysis, but this may not be favored by the relevant competent authority due to uncertainties of this approach.

If large decommissioning items are to be transported such that they themselves form the package without any additional packaging, it is clearly preferable to be able to transport them as IP-2 packages, as the performance requirements are less demanding. However, there would still be a need to demonstrate performance for the free drop, which would be between 0.3–1.2 m, depending upon the package weight. Demonstration of this by analysis is likely to be more acceptable to the DfT than the 9 m drop test for Type B packages.

The use of special arrangement shipments would be anticipated to be the most appropriate approach for some large decommissioning items, although there may be significant effort involved in obtaining approval from the relevant competent authority for such a shipment.

22-4-2. *General Requirements*

Road

A load being transported by road in the UK is considered to be abnormal if the combination of the road vehicle and the load exceeds 18.65 m in length, 2.90 m in width, 40,000 kg gross weight, or has an overhang to the rear of more than 3.05 m. There are no legal requirements regarding height, although when the load is over 4.70 m high, it is advisable to check the route with the relevant Highway and Bridge Authorities.

There are three categories within the Special Types General Order (STGO) [7], which classifies vehicles according to the total laden weight of the vehicle, with the weight ranges being 38–46, 46–80, and 80–150 te. The particular category determines specific requirements relating to axle numbers, spacing and weights, speed limits, and notification to the Police, Highways, and Bridge Authorities.

Rail

There are three key drivers that determine whether rail would be a practical option for the transport of large items of decommissioning waste:

- the overall dimensions of the waste item;
- the weight of the waste item; and
- the availability of rail routes between the despatching and receipt sites.

The loading gauge in the UK is relatively small, with the loading gauge on most rail networks in Continental Europe being larger, and that in the US larger still.

An object of 2 m width and 15 m length would be capable of being readily transported by rail, subject to a suitable rail wagon being available to carry it, but larger objects would need careful examination to ascertain the practicality of rail transport, including ascertaining the permissible loading gauge on the actual planned route.

The allowable weight of a waste item being transported by rail is primarily dictated by the permitted axle weight on the applicable rail route and the rail wagon to be used. Most of the UK rail controlled infrastructure is able to take rail vehicles with an axle loading of 22.5 te, although some stretches of line have a lower allowable axle loading, perhaps down to 12 te. A four axle rail wagon would give an allowable weight of the load of 60–65 te.

The rail network is operated on a tightly timetabled basis, and one-off movements are discriminated against in favor of regular, preferably daily, timetabled trains. This makes it very difficult to make the necessary arrangements to move one-off large items by rail.

Rail is, therefore, of limited potential for the transport of large loads in the UK.

Water

The key issues in respect of sea transport are the availability of:

- a suitable ship;
- loading and unloading facilities commensurate with the load and the ship; and
- access routes between the despatching site and the ship loading facility, and between the ship unloading facility and the despatching site.

Sea transport provides a practical option for the transport of large items, provided that there are readily available facilities for loading the large item onto a ship. This means that sea transport is most suitable for despatching and receiving sites that are near to the sea or to large estuaries.

22-4-3. *Examples of the Transport of Large Decommissioning Items*

Transport and disposal of complete heat exchangers or reactor pressure vessels or other large items as whole units, rather than cutting them up prior to transport and disposal, has been carried out in a number of instances. The main reasons for this approach have been:

- lower cost;
- lower overall dose uptake; and
- the overall dose uptake being ALARP.

Such transport can be a valuable, integral aspect of the decommissioning, waste management, and final disposal plans for shut-down nuclear facilities, particularly where they can save significant operator dose uptake by the avoidance of size reduction.

Significant achievements in this area have included:

(a) WAGR Heat Exchanger (Cumbria, UK) — achieved 1995.
(b) Trojan Reactor Pressure Vessel (Oregon, USA) — achieved 1999.

WAGR Heat Exchangers

A number of options were considered for disposal of the heat exchangers, including cutting up prior to disposal and disposal of each heat exchanger as an entire unit. The latter approach was the selected option.

The main heat exchanger pressure vessels comprised a carbon steel cylinder with dished ends, a diameter of approximately 3.5 m, an overall height of 20.6 m, and weight of nearly 190 te each. Radiation surveys were carried out on each boiler, and the results of these showed that the material of the boilers was essentially nonactivated, but the extent of surface contamination was such that the boilers could be classified as SCO-I. Additionally, the dose rate at 1 m was an important controlling criterion. SCO-I can be transported in an IP-1 transport package, which has no package performance requirements. Consideration of the boiler configuration showed that, subject to design of the lifting and tie-down arrangements, the boiler could be shown to meet the requirements for IP-1 packages.

The 6 km road journey from the reactor to the disposal site was achieved using a four module two-by-two arrangement of 24-wheeled trailers, which gave a total deck area of 16.8 m long × 5.3 m wide. Two concrete saddles were located on this platform to support the heat exchanger, with steel mat load spreaders under each saddle. The heat exchanger was secured to the transporter by steel wire ropes tensioned by turnbuckles.

The total mass of the heat exchanger, transporter, tie-down, and ancillary equipment was 314 te. This load was spread over 96 wheels, resulting in a load of 1.63 te on each wheel. The maximum speed was 6 km/hr, and each of the 48 axles was capable of being hydraulically rotated to steer the transporter.

Transport of each of the four heat exchangers took place at night to reduce the disruption to other road traffic. Detailed surveys of the route were carried out ahead of the movements, with key parameters being to ensure that it could take the axle loads and that there was sufficient clearance for the load to buildings, road furniture, and telephone and power lines. It proved necessary to carry out temporary removal of a central road island and associated traffic lights, to lay temporary wooden boards to protect road kerbs and verges, and to raise telephone wires.

The typical journey time was about 8 hours, with all four heat exchangers being successfully transported to Drigg over a period of just under 3 weeks.

Trojan Reactor Pressure Vessel

The Trojan 1178 MWe PWR Power Plant in Prescott, Oregon, USA operated from 1976 to 1992, and was permanently closed in January 1993. All fuel was removed from the Trojan reactor and placed in an on-site storage pool in 1993.

Some large low-level radioactive components from the plant, including four steam generators and a pressuriser, had previously been shipped to the commercial low-level disposal facility at the Hanford reservation in Washington State, USA.

The remaining large component was the reactor vessel, and four options were considered for this:

• shipping the entire reactor vessel, complete with its internals, to a disposal site;
• storage of the reactor vessel on site;
• disposal of the reactor vessel in one piece, with certain internals left inside; and
• separate disposal of the reactor vessel and the internals.

Assessment of the options resulted in the first option being selected as the preferred option. The key reasons for this were:

• overall lower costs;
• practicable transport route available (barge on the Columbia river); and
• total dose uptake was ALARP.

The actual shipment was successfully carried out in August 1999.

The reactor vessel is a carbon steel cylindrical shell with an integral lower head and a removable upper head. All reactor vessel penetrations are closed with welded plates. The overall dimensions are 13 m in length and 5.2 m diameter (excluding the nozzles).

Prior to transport, the reactor pressure vessel was filled with low-density concrete to prevent movement of radioactive material within the reactor vessel, closures were welded over the reactor nozzles to provide containment, and steel shielding was installed on the exterior surface of the reactor vessel to reduce the external dose rates. Impact limiters were installed to minimise reactor vessel stresses in the event of an accident involving an impact of the reactor vessel.

The resultant transport package, designated the Trojan Reactor Vessel Package (TRVP), weighed approximately 950 te without the impact limiters. Prior to transport, the TRVP was rotated to a horizontal position, loaded and tied down onto a specially designed transporter, which was then moved onto a specially selected barge, which was grounded for this activity, and secured using an engineered tie-down system. Following barge transport, a heavy-haul mover was connected to the transporter and moved it off the barge and overland to Hanford, where the TRVP was off-loaded at the disposal facility.

The specific activity of the activated material in the TRVP was in excess of the LSA material limits, and so the TRVP had to be transported as a Type B package. However, it was not possible for the TRVP to meet all the requirements for a Type B transport package, and exemption was granted from three of these requirements:

• a drop height of 3.3 m instead of 9 m;
• exemption from the 0.3 m drop test for orientations other than horizontal; and
• a minimum ambient temperature of 45°F (7°C) was used.

This approach, which included the use of stringent operational and administrative controls, ensured that the probability of the TRVP encountering accident conditions beyond those for which it has been analysed is low.

Although the TRVP is approved as a Type B package in the US, it would not be able to be approved as such within the UK. This is because the US Federal Regulations allow the principle of equivalence of safety with the use of environmental and test conditions different from those specified for normal and accident conditions of transport provided suitable controls are exercised during the shipping.

Within the UK, such an approach would require consideration as a special arrangement.

22-5. Regulatory Considerations in the UK

22-5-1. *DfT (Department for Transport)*

The remit of the DfT, as competent authority for the transport of radioactive material in the UK, is concentrated upon assuring compliance with the regulations for the transport of radioactive material, and thereby assuring safety. This is achieved by issuing package and shipment approval certificates, where these are required and where packages comply with the regulations, and by auditing the activities of organisations involved in radioactive material transport.

DfT does not take account of issues of whether specific material should be transported or retained in its current location.

If waste were to be moved from one UK nuclear licensed site to another one, whether in raw, partly conditioned, or completely conditioned states, DfT would only be interested in whether the transport of this radioactive material complied with the regulations, which is essentially focused upon the package, make up of the packaging and its contents.

22-5-2. *NII*

The Nuclear Installations Inspectorate (NII) are responsible for regulating safety on licensed nuclear sites, and this applies to any facilities needed for the handling, treatment, packaging, or storage of radioactive waste. Plant safety cases will have to be produced in order that the NII may issue consent under NIA65 [8] as part of the site licence conditions.

Where such facilities receive waste from other sites for processing and storage, the operations involved in unloading the waste from the incoming transport package will need to be covered in the safety cases for the facility for acceptance by the NII.

A Memorandum of Understanding exists between the Health and Safety Executive (HSE) and DfT covering the transport of radioactive materials. The extent of this MoU is such that the HSE, of which the NII is a part, would not make any input to any decision by the DfT as to the acceptability or otherwise of a proposal for the transport of radioactive waste.

22-5-3. *Environmental Agencies*

Environmental agencies are responsible for the regulation of waste movements and disposals, including any discharges from a nuclear site. These agencies are the Environment Agency (EA), which covers England and Wales, and the Scottish Environmental Protection Agency (SEPA), which covers Scotland.

RSA 93 [9] requires authorisation for disposal of radioactive wastes. It should be noted that Article 47 of RSA 93 defines disposal, in relation to waste, as including its removal from a licensed site. Removal has been interpreted by the EA as covering the transfer of waste

from the despatching site until it reaches an appropriate destination, in an analogous way to waste being sent for ultimate disposal. Therefore, separate authorisations in the form of an Intersite Transfer Authorisation (ISTA) will be needed from the EA/SEPA for each waste transport operation, even if the waste is not going directly to a disposal site. It should also be noted that the authorisations, when granted, specify limits on waste quantities that may be removed.

22-6. Waste Transport Planning

The key issues that need to be considered when planning the movement of waste between sites in the public domain essentially fall into three areas:

- the availability of a suitable package for the waste transport;
- the availability of a suitable shipment route; and
- the obtaining of the necessary approvals.

In order for waste to be transported safely and legally, irrespective of whether the waste is in a raw, partially conditioned, or fully conditioned form, a suitable packaging must be used whereby the combination of the packaging and its contents meets the regulatory requirements.

A shipment route from the despatching site to the receiving site must be available, taking account of whether the package is to be transported by road or a combination of road and rail, noting that many nuclear licensed sites do not have a direct rail connection, but would need to utilise a railhead away from the site for trans-shipment of transport packages from a road vehicle to a rail wagon.

Approvals must be obtained for the following before any shipment takes place:

- design approval of the package from the DfT, or approval may be issued by another organisation if the package does not require competent authority approval;
- shipment approval from the DfT, where required; and
- EA/SEPA for the disposal (including removal and transfer to another site) of waste from a licensed nuclear site.

22-7. References

1. International Atomic Energy Agency. *Regulations for the Safe Transport of Radioactive Material, 1996 Edition (Revised)*, IAEA Safety Series No. TS-R-1, IAEA, Vienna (1996).
2. International Atomic Energy Agency. *Advisory Material for the IAEA Regulations for the Safe Transport of Radioactive Material*, IAEA Safety Standards Series TS-G-1.1 (ST-2), IAEA, Vienna (2002).
3. International Atomic Energy Agency. *Planning and Preparing for Emergency Response to Transport Accidents Involving*

Radioactive Material, IAEA Safety Standards Series No. TS-G-1.2, IAEA, Vienna (2002).

4. International Atomic Energy Agency. *IAEA QA document*, IAEA, Vienna (2002).

5. International Atomic Energy Agency. *IAEA Compliance Assurance Document*, IAEA, Vienna (2002).

6. International Atomic Energy Agency. *Department of the Environment, Transport and the Regions, Guide to an Application for UK Competent Authority Approval of*

Radioactive Material in Transport, IAEA 1996 Regulations, DETR/RMTD/0003 (January 2001).

7. Her Majesty's Stationery Office. *Motor Vehicles (Authorisation of Special Types) General Order*, SI 1198/79, as amended, HMSO, London (1979).

8. Her Majesty's Stationery Office. *Nuclear Installations Act*, HMSO, London (1965).

9. Her Majesty's Stationery Office. *Radioactive Substances Act*, HMSO, London (1993).

Chapter 23
Site Remediation — Principles and Regulatory Aspects

23-1. Introduction

The UK has a large number of industries and sites on which radioactive materials are processed or used. In addition to commercial power reactors, there are supporting fuel fabrication and processing plants, waste disposal sites, and research facilities. The UK also has a nuclear weapons production capability. However, there is no uranium mining or weapons testing sites. Hence, the problems of radioactive contaminated land in the UK are generally much smaller than those experienced in the US, the former Soviet Union, and many other eastern European countries.

There is a wide range of sites, apart from licensed nuclear sites, which can be contaminated with radioactive materials, mainly due to processes which cause an enhancement in concentration of natural radioactivity. These include sites which have processed the following materials:

- radium for luminising,
- thorium and rare earths for alloying,
- catalysts (e.g., in gas works),
- gas mantles,
- electron emitters in filaments,
- refractory bricks,
- phosphates, fertilisers, and detergents,
- heavy metal mining and smelting (e.g., lead, bismuth, tin, and zinc),
- descaling equipment from oil and gas production,
- uranium for glazing ceramics,
- mineral sands processing, and
- fuel and fly ash from coal-fired power stations.

In many cases, the processing and contamination occurred before such materials were regulated as radioactive materials, and the organisations responsible for the contamination have long since disappeared.

Some nuclear licensed sites also have an historical legacy from previous use, typically as airfields or ordnance manufacturing plants during World War II,

where a range of chemical substances were used, some of which may have caused contamination of the land.

The problems of radioactive contamination have many parallels with nonradioactive chemical contamination of the kind associated with heavy industry, coal mining, gas works, etc., which are much more widespread than radioactive contamination. Contaminated land can be defined as any land which appears to be in such a condition, by reasons of substances in, on or under the land, that significant harm is being caused or is likely to be caused.

Harm is determined using a risk-based approach which requires a pollutant linkage between a source, pathway, and receptor. Note that groundwater and other controlled waters can be both pathways and receptors.

23-2. Delicensing

The ultimate end-point for decommissioning a nuclear licensed site is the termination of the licence and release of the site for unrestricted use. However, in some cases, this is neither economic or practicable. Therefore, as a preliminary step in the development of the decommissioning and environmental restoration strategy for a site or facility, the proposed "end-point" should first be determined. Table 23-1 gives proposed definitions for "green field" and "brown field" end-points for sites.

In the UK, a site (or part of a site) can only be delicensed if the regulator is satisfied that there is "no danger" from ionising radiations from anything on the site or that part of the site to be delicensed. The term "no danger" is not defined in legislation. The NII are currently seeking public responses to a consultation paper on how delicensing should be carried out in practice. The approach which UKAEA has adopted involves addressing three questions:

- Is there contamination above the background level for the area concerned?
- If so, is the risk from the contamination $< 10^{-6}$ per year?

Table 23-1. Proposed Green Field and Brown Field "End-Point" Definitions for the Environmental Restoration of Sites

Green Field	This describes an end-point which allows a site to be released from institutional control, i.e., the nuclear site licence is terminated. Decontaminated structures will be demolished and removed to a depth of 1 meter below grade. Contaminated foundations, drains, and earth will be removed. The site will be back-filled with clean material, which may be recycled crushed concrete and masonry so long as it meets the Substances of Low Activity Exemption Criterion (SoLA) under the Radioactive Substances Act 1993 (RSA 1993) (<0.4 Bq/g) or meets a "no danger" criterion of $< 10^{-6}$ per annum risk to human health (whichever is the lower). In some circumstances, massive foundations and deep drains may be left *in situ*, provided that they meet the above free release/delicensing criteria.
	Unless the site is required for other purposes, it would ultimately be allowed to return to its natural state, perhaps with suitable landscaping appropriate to the location. The site would be deemed to be suitable for release from regulatory control. Records of the work done and the final radiological characterisation should be sufficiently robust for future delicensing.
Brown Field	This describes an end-point which falls short of the greenfield description, such that the site would remain under institutional control for the foreseeable future. Typically, building foundations and other underground structures may be left *in situ* unless they are contaminated in excess of 40 Bq/g $\beta\gamma$ or 1 Bq/g α. Underground piped services should be physically isolated and cleaned, but not necessarily removed. Underground electrical services should also be isolated, but not necessarily removed.
	The end-point should comply with the ALARP and BPEO principles in relation to both radiological and other contamination and/or hazards. Some formal care and custody measures may be required, such as fencing to control access or monitoring of groundwater. The site would not be suitable for release from regulatory control without further characterisation and possibly further clean-up. Records should be kept of any contamination or remaining structures or services.

- Are the contamination levels below the SoLA exemption level of 0.4 Bq/g?

If the answer is either negative to the first question or positive to the second two, then UKAEA considers that the site is delicensable without further cleanup and makes a submission to the regulator accordingly.

This, in turn, is related to a dose to individuals in the order of 10 µSv per year. The Environment Agency may grant authorised disposals on nuclear licensed sites. Paradoxically, the presence of such an authorised disposal may prevent the delicensing of the area containing the disposal.

In the US, licence termination may be for unrestricted or restricted release of the site. For unrestricted release, the licensee must demonstrate that the dose from residual radioactivity (excluding background radiation) does not exceed 250 µSv per year to an average member of the critical group over a 1000 year period. This is a Federal limit; some States require a more restrictive limit. In the case of a restricted licence termination request, the licensee must propose institutional controls which will ensure that the criteria for unrestricted release are complied with.

23-3. Chemically Contaminated Land

In the UK, the regulatory framework for managing chemically contaminated land on a site which is also potentially contaminated with radioactivity falls within the scope of the (chemically) Contaminated Land Regulations (Part IIA, Section 57 of the Environmental Protection Act 1990). Local Authorities have a duty to identify potentially contaminated land in their areas. If a nuclear licensed site (or one which has a history of using radioactive substances) could also potentially contain chemical contamination that could cause harm, the Local Authority will designate it as a Special Site and pass on regulatory responsibility to the relevant Environmental Agency. The EA (or SEPA) also has powers under this act to investigate a site on its own initiative.

Both the Local Authority and the Environment Agency have powers to serve remediation notices after 3 months of notification of the person deemed to have caused the pollution or, if that person cannot be identified, the current owner or occupier of the land. Liability is determined on the basis of the polluter pays principal. The LA or EA may carry out remediation itself if necessary and recover the cost from the polluter or appropriate person. A site is deemed to be an "orphan site" if no appropriate person can be identified (it is then the responsibility of the relevant authority to undertake remediation). Complex rules defining liability make it possible for several individuals to be jointly liable.

Contaminated materials, substances, or products resulting from remedial action with respect to land are defined as waste when discarded.

Land remediation in the UK has traditionally relied on simple landfill disposal. Landfill was generally seen as simple and cost effective. Under a recent Landfill Directive from the European Commission, waste sent to landfill must be pretreated so as to minimise the amount of material disposed. The directive also places a ban on the co-disposal of hazardous and nonhazardous waste. This will pose an additional problem related to hazardous waste disposal routes as it is generally expected that few hazardous landfills will be licensed and that those will be far apart, making it necessary to dispose of waste far from its point of origin. This is a contravention of the Proximity Principle, which states that waste should be disposed of as near as possible to its point of origin.

23-4. Radioactively Contaminated Land

EPA 90 does not cover harm or pollution caused by radioactivity, and there is no specific definition of radioactively contaminated ground. The current legislative control of land contaminated by radioactivity (other than on nuclear licensed sites) comes under the Radioactive Substances Act 1993 (RSA 93), which requires that users of radioactive substances must register with the Environment Agencies (EA or SEPA) in order to keep, use, or dispose of radioactive substances. Exemption from registration applies to Nuclear Licensed Sites, as they are regulated under NIA 65.

Disposal of radioactive substances may only be made under authorisation granted by the Environment Agencies, and this includes licensed sites. This includes both disposal of solid wastes and the discharge of liquid and gaseous wastes to the environment. The EAs require a demonstration that the disposal is regulated by best practicable means. There is a clear area where HSE and EA/SEPA interests can overlap. Memoranda of Understanding exist between the Agencies and HSE defining their regulatory responsibilities.

With regard to the nuclear licensed sites which it regulates, the HSE considers land and materials contaminated by radioactivity to constitute storage of bulk quantities of radioactive waste. Accidental leaks are not authorised disposals. Contaminated land on licensed sites requires inclusion in site-specific strategies for radioactive waste management and is subject to safety justification.

Double jeopardy may arise from authorised discharges, e.g., from stacks, which have contaminated surrounding area. Because the discharge was authorised, operators are deemed not to be liable for related contamination adjacent to licensed sites. However, authorised gaseous discharge of, say, tritium leading to fallout onto the licensed site could give rise to a further discharge via the surface water drainage system which could be deemed unauthorised.

The Substances of Low Activity (SoLA) Exemption Order made under RSA 93 provides a general threshold at 0.4 Bq/g for solids that are substantially insoluble in water or organic liquids which are radioactive solely due to ^{14}C and/or ^{3}H. Specific thresholds are given for lead of 0.74 Bq/g, for thorium of 2.59 Bq/g and for uranium of 11.1 Bq/g. The exemption order criteria apply when the material becomes waste. These levels correspond broadly to a dose limit of 20 μSv per year or a risk target of 10^{-6} per year.

In the case of natural radon gas, the UK Government uses as "action level" of 200 Bq per m^3 (roughly equivalent to 10 mSv per year), above which it advises householders to take remedial action.

In February 1998, the DETR issued a draft consultation paper on a regime parallel to the EPA 90 Part IIA legislation for chemical contamination to cover sites (other than nuclear licensed sites) which are contaminated by radioactive material. In effect, the Environment Agency would enforce the regime in the same way as "special site" under the terms of the main contaminated land legislation.

23-5. Principles for Management of Contaminated Land

The SAFEGROUNDS project was established by a number of interested organisations in the UK to prepare best practice guidance about the management of contaminated land on nuclear licensed and defence sites.

Five key principles have been established through a consultative process, which included representatives of a variety of stakeholder groups. The principles are non-overlapping and complementary and should be applied together. The key principles are as follows:

- *Principle 1: Protection of People and the Environment.* The fundamental objective of managing contaminated land on nuclear-licensed sites and defence sites should be to achieve a high level of protection of people and the environment, now and in the future.
- *Principle 2: Stakeholder Involvement.* Site owners/ operators should develop and use stakeholder involvement strategies in the management of contaminated land. In general, a broad range of stakeholders should be invited to participate in decision-making. The level of Stakeholder involvement is related to the scale of the problem.
- *Principle 3: Identifying the Preferred Land Management Option.* Site owners/operators should identify their preferred management option (or options) for contaminated land by carrying out a comprehensive, systematic, and consultative assessment of all possible options. The assessment should be based on a range of factors that are of concern to stakeholders, including health, safety, and environmental impacts, and various technical, social, and financial factors.

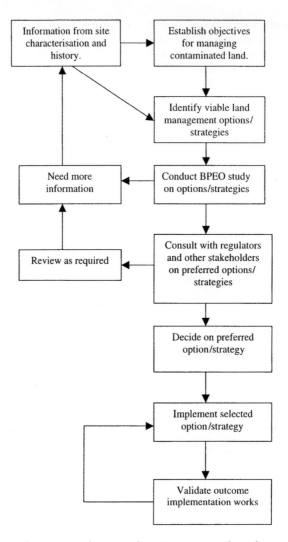

Figure 23-1. The Structured Approach to Contaminated Land Management.

- *Principle 4: Immediate Action.* Site owners/operators should take measures immediately to monitor and control all known (or suspected) contamination and continue such measures until an acceptable management option has been identified and implemented.
- *Principle 5: Record-Keeping.* Site owners/operators should make comprehensive records of the nature and extent of contamination, the process of deciding on the management option for the contaminated land, and the findings during the implementation and validation of the option. All records should be kept and updated as necessary.

Figure 23-1 shows the basic steps in a structured approach to putting the principles into practice. The principles operate throughout the process. In some cases, the process may be iterative, but in others it will be once through. The extent to which stakeholders other than the site licensee and the regulators will be involved will vary from site-to-site.

23-6. Best Practicable Environmental Option

Key Principle 3 involves identifying the preferred management option. It is suggested that the Best Practicable Environmental Option Study (BPEO) method is used for this process. BPEO is defined as:

the outcome of a systematic consultative and decision-making procedure which emphasises the protection and conservation of the environment across land, air, and water. The BPEO procedure establishes, for a given set of objectives, the option that provides the most benefit or least damage to the environment as a whole at acceptable cost, in the long-term as well as the short-term.

BPEO is a decision-making tool which gives structure to the process by which decisions are made. Provided proper records are kept in the form of an audit trail, the process can be subjected to independent review if the decision is subsequently challenged. The essential steps in a BPEO process are shown in Table 23-2.

The assessment should be carried out by a team with a range of expertise to ensure a balanced decision. The criteria and weightings should be agreed at the outset, in order to avoid personal bias influencing the decision. The assessment should be presented in a clear and transparent way. Sensitivity analysis should be applied to check the robustness of the decision. Finally, an independent review of the process is advisable.

23-7. Summary

- Residual contamination on a nuclear licensed site must be removed in order to allow the site to be delicensed.
- The "no danger" criterion for delicensing under the UK Nuclear Installations Act is not defined in law, but pragmatic approaches are being developed.
- Land on a nuclear licensed site (or one which has a history of use of radioactivity) that is also potentially contaminated comes under the (chemically) Contaminated Land Regulations (Part IIA, Section 57 of the Environmental Protection Act 1990).
- The SAFEGROUNDS project sets out principles for the management of contaminated land.

Table 23-2. Essential Steps in the Best Practical Environmental Option (BPEO) Approach

Step	Comment
Define the objectives clearly	The objective(s) must be established in terms which do not preclude the means by which the objective is to be achieved.
Identify all the options	All options should be included at the outset. Options must meet the objective.
Assess against constraints	Allows nonfeasible options to be rejected at an early stage.
Identify and agree criteria	All relevant criteria should be included and should avoid personal interpretation.
Assessment Methodology	Ranking/scoring system, e.g., Keppner-Tregoe. Weightings can be applied to criteria to reflect relative importance.
Evaluation of Options	Must have data on each of the options. Data may be quantitative or qualitative.

Chapter 24

Characterisation of Contaminated Land

24-1. Introduction

Regulators require that site licence holders manage radiologically contaminated ground safely to protect the public and the environment. The licence holder must demonstrate that contamination is not migrating through groundwater or the air from its point of source and that direct exposure to radiation from contamination is acceptably low.

Site characterisation involves collecting and collating the information about a site which will allow an understanding of the extent of actual or potential ground contamination to be built up. This information should enable an assessment of the extent to which contamination might impact on receptors, either by migration or direct exposure. An understanding of the geology, hydrogeology, and hydrology of the site is necessary, so that contaminant transport processes can be understood. The assessment will lead to a decision on the most appropriate actions which need to be taken to minimise hazard and risk in the long-term.

24-2. Desk Studies

The first stage of site characterisation should involve a desk study to collate known information about the ground. An initial desk study will save time and money in the long-term. This information can be obtained from a wide variety of sources, including:

- site records,
- previous site investigation reports,
- interviews with current and past personnel,
- incident records,
- water resource and catchment records, and
- public records offices.

The information obtained may relate to any of the following:

- history of the site and building use, including any incidents which may have given rise to contamination,

- infrastructure, e.g., drains, waste disposal sites, etc.,
- geological and hydrogeological data,
- contaminant data (both radioactivity and hazardous chemicals), and
- adjacent land use and adjacent receptors.

Building up a knowledge of the history of the site enables the licensee to determine previous occupants, their activities, and the potential contaminants. Maps showing service locations and previous site activities are important in terms of both health and safety and the siting of boreholes and trial pits.

Geological and hydrogeological data can be obtained from nationally held information sources (e.g., British Geological Survey), previous borehole logs, and reports. Such information allows the operator to construct an investigation program that will fill in the data gaps and link into existing features like borehole monitoring locations. Even limited information on groundwater flow directions can be invaluable when constructing the site investigation program. Gaining some previous knowledge of ground conditions can also influence the choice of intrusive and nonintrusive site investigation techniques.

Evidence of contaminant data may also be present in previous reports, either from borehole sampling or health physics monitoring. This is another data set which can help to pinpoint the areas requiring further investigation. Information on previous building use can help to highlight known contamination or the potential for contamination through knowledge of the operations which were undertaken. Site Incident Reports may also describe incidents where accidents and spills led to ground contamination. Such reports are forwarded to the regulators as part of the legislative process.

24-3. Walk Over Surveys

It is important to visually inspect the site and to add to the information gleaned from the desk study. The walk over survey records visible evidence of possible surface and shallow subsurface contamination and also provides a recent record of site activities undertaken within and around designated areas. The following list highlights examples of data that can be gleaned from the walk over survey:

- ground conditions, e.g., boggy, dry, burrows, uneven ground;
- access arrangements;
- condition of building structures and surface covers;
- presence and layout of below ground services and locations of below ground structures;
- description of area including operations and land use;
- vegetation type and visible signs of distress;
- visible evidence of contamination, i.e., disturbed ground, discolored soil/water, subsidence, above ground deposits, accidental/uncontrolled releases;
- presence of significant odors;
- presence of surface water including presence of ponding and direction of run off flow; and
- potential access constraints, e.g., overhead cables/services, machinery, site operations, vegetation, heavy duty fencing with no gates.

All this information can be collated onto a reconnaissance observation sheet, and sketches should be made to record the location of the observations.

24-4. Planning the Characterisation Program

Once the desk study and walk over survey are complete, it is possible to specify the requirements for data measurements and develop a costed program for the work required in order to meet the objectives.

If the work is being performed to underpin known or suspected ground contamination, most site investigation effort will be on the area around, and down the hydraulic gradient from, the suspect contamination. The desk study may highlight that most contamination is expected to be contained within the top few meters of the ground. In this instance, there will be a need for a more focused investigation on the near surface soil, while still determining the potential for contaminant migration.

If the budget and time constraints allow, it is best to put together a phased site investigation program, especially when the existing data is scarce. Before carrying out intrusive surveys, such as sinking boreholes, it is usually very cost-effective to monitor the area using a global positioning satellite linked scanning technique such as Groundhog™, a proprietary system marketed by RWE Nukem in the UK. This will identify areas of enhanced radiation and help to target intrusive sampling, which will provide more detailed information.

The data acquired from surveys needs to be appraised to build up the assessment of the site in a continuous and interactive manner. When pre-existing data has helped to focus the site investigation, the drilling and sampling is used to verify the original assessment and quantify the levels of contamination. Additionally, careful design of the characterisation program and ongoing appraisal of the information obtained is necessary to avoid the possibility that the boreholes can allow contaminants to migrate either vertically or laterally within the sediments and, therefore, make the situation worse.

This type of technical program is likely to require the combination of a number of drilling and geophysical techniques. A clearly-defined sampling regime will be set up so that samples can be taken and sent through to the laboratories for analysis. Samples are usually taken every meter and at a noticeable change of lithology, or where visible evidence of contamination (through discoloration or odor) can be seen.

The technical program, once formulated, needs to be configured into a Bill of Quantities, so that accurate costings can be made. Depending on company policy, it may be necessary to undergo a tendering process for some components of the work. These costings need to cover the following:

- nonintrusive surveys;
- intrusive sampling and laboratory analysis;
- health and safety support;
- field- and office-based personnel working on
  - logging and databasing information,
  - interpretation,
  - modeling,
  - meeting attendance,
  - travel and subsistence expenses, and
  - quality assurance;
- equipment and consumables;
- peer review; and
- risk contingency.

24-5. Health, Safety, and Logistical Issues

Prior to undertaking site investigation work, it is important to ensure that a health and safety plan is developed. Health and safety plans on nuclear licensed sites need to cover the potential of radioactive exposure as well as chemical exposure and conventional safety. Wearing the appropriate level of protective equipment (PPE) and gaining the advice of a radiological advisor and health physics monitor may be necessary. If the potential of contamination is significant, controlled area working may be enforced, whereby contractors and other personnel change out of their own clothes into personal protective equipment. It may be necessary to monitor samples prior to transport offsite (under the appropriate regulations for transport of radioactive materials) and monitor the equipment such as drilling rigs prior to moving to the next location. The health and safety plan, therefore, should determine all the procedures that need to be adhered to when undertaking the site investigation.

Risk assessments for personnel may also be produced which highlight all the potential risks associated with a particular job and how they should be mitigated.

Plant modification proposals (PMP) are sometimes required on nuclear licensed sites for site investigation work and are often utilised to highlight a change that may be made to the vicinity of the work which may have a health and safety implication. An example may be that the site investigation work blocks an access route, so the PMP will show a new route to be used while the work is being carried out.

The management regimes on such sites can sometimes be quite onerous, and it is necessary to ensure that all personnel are aware of their responsibilities and that the chain of command is clear. Linking into the roles and responsibilities is the necessity for personnel to have received the correct training. Relevant training courses include those covered under COSHH (Control of Substances Hazardous to Health Regulations (1999)) and IOSH (Institute of Occupational Safety and Health). Most sites require personnel to undergo a site induction course, where they learn about the site emergency instructions and about general site operations.

One of the areas to consider when putting together the site investigation is the issue of access and services. This again has safety implications. For example, working in confined spaces will need special permits and procedures. Services like electric cables, water mains, and drain lines need to be highlighted from the relevant plans and avoided at all costs. Scanning equipment to determine locations of services should be used in conjunction with hand-dug inspection pits prior to undertaking intrusive work.

Because material being brought to the surface from drilling or trial pitting may be contaminated, it will be necessary to identify disposal routes. This can, in itself, be quite a complex process because there may be mixed contamination or there may be different levels of contamination. Only authorised disposal locations can be utilised, and they themselves may create financial implications. Different companies have slightly varying authorised disposal routes available to them.

Many nuclear licensed sites are a haven for wild animals and plants. Some of these may be protected species. It may be necessary to carry out environmental impact assessments to identify habitats, etc., prior to site investigation work, and possibly seek specialist advice as to where intrusion should be avoided.

24-6. Nonintrusive Surveys

24-6-1. *Radiological Surveys*

Radiological surveys in the field can be divided into two types: scanning surveys or direct (point) measurements. Scanning surveys are undertaken on foot or with a vehicle using portable radiation detection equipment that rapidly responds to the presence of primarily gamma emitting radionuclide contamination on or very close to the ground surface. The results of the measurements are generally presented in terms of "counts per second," and give an indication of the relative levels of radioactivity across the site. The measuring equipment may be linked to a global positioning satellite system to give an accurate positional reference to the data, which can automatically be plotted on a map by a computer. Direct measurements on the site are used to determine absolute values for certain parameters or to provide a better understanding of which radionuclides are present.

24-6-2. *Geophysical Surveys*

Geophysical techniques provide an indirect means of characterising a site prior to or in conjunction with intrusive work. A number of geophysical techniques are commonly used for the investigation of contaminated land. These include electrical, magnetic, and ground penetrating radar. Such methods focus on the near surface sediments and can help to detect buried objects, areas of disturbed ground, and services. Other methods like seismic profiling and gravitational surveys tend to map deeper ground and help to understand the geology on a more regional basis. It is the near surface techniques that are of more interest for contaminated land appraisal.

Electromagnetic surveying can often identify buried objects, disturbed ground, and metallic services. Anomalous readings can be given if the survey is carried out adjacent to buildings and fences. Resistivity profiling provides a cross-section of ground resistivity and can highlight distinct lithological changes within the subsurface strata as well as buried metallic objects.

Magnetic methods are primarily used to detect buried metallic objects such as cables, drums, pipes, or waste materials. While the resolution of the method decreases with depth, the technique can be used to estimate both the depth and mass of an object.

Ground Penetrating Radar (GPR) systems are used to detect both metallic and nonmetallic objects like pipes, void spaces, drums, and concrete. The depth of penetration varies upon the electrical properties of the soil at any particular location.

24-7. Intrusive Surveys

Most site investigations involve some form of intrusive survey on the lines of drilling or digging trial pits. The choice of intrusive technique links into understanding the overall objectives of the site investigation and the type of data required. The two most common drilling techniques are rotary and cable percussion (also known as shell and auger).

Truck-mounted rotary drilling produces a high quality of core and samples. This technique is extremely beneficial if the investigation is required to drill to depths greater than 30 m and if continuous relatively undisturbed core recovery is one of the prerequisites. It is, however, comparatively expensive and usually requires a larger working area than cable percussion rigs. There is a potential that the drilling flush, usually water, polymer, or bentonite, can contaminate the surrounding strata and waters that are themselves required for chemical analysis. The use of an air flush can sometimes help to mitigate this, but cross-contamination between aquifers is a possibility.

Cable percussion rigs on the other hand are small enough to be towed behind a 4 × 4 utility vehicle and require a smaller operating footprint area than the rotary rigs. The samples retrieved may be disturbed, which makes it harder to determine the exact vertical location of the retrieved sample. Fine material can sometimes be lost, so it is important to empty the core sample onto a tray which allows logging and sampling to be carefully undertaken. Cable percussion has the ability in certain materials like soft clays to retrieve about 25 m of core/sample. However, in dense gravels or where large cobbles/boulders are present, progress may be restricted to less than 1 meter a day due to standing time and chiseling. A driller will normally add water to help progress the borehole, but if the program needs to highlight perched water bodies it will be necessary to drill without adding water until the regional aquifer is encountered. This type of drilling technique can be quite messy and there is always the potential that contamination can be spread from one horizon to another.

A number of different variants of shallow surface probes exist. These are often used in areas where access is restricted or where a quick near surface sampling campaign is required. For this technique to be successful there is generally a dependency on soft soil or subsurface material being present.

Trial pits or trenches are a very quick and economical way of investigating the near surface soil. Using a mechanical excavator, it is usually possible to dig a trench or pit about 4 meters deep. This allows a cross-section of the geology to be seen, which is especially useful in glacial soils as the operator can spot water seepages, micro-faulting and subtle changes in lithology. If subsequent analysis shows the material to be contaminated above certain levels, it is not permissible to backfill the pit with this spoil, and inert material must be used.

Most, if not all, drilling techniques are suitable for the installation of permanent groundwater or gas installations. In order to facilitate long-term sampling, the borehole is designed so that a screen is set adjacent to the specific horizon being targeted.

Down-hole geophysical logging is often used within the oil industry and for acquiring geotechnical data, but it can be used in a cost effective manner for hydrogeological and environmental investigations. Natural gamma logs, for example, can produce information on the clay content of strata or the variations in sandstone matrices. This can assist in the overall approximations of hydraulic conductivity and allow data to be cross-correlated with other field data.

Choosing the correct intrusive techniques for a specific site investigation is not always an easy task. Different techniques have their own positive and negative points which will depend on the site geology, contaminants, access, and overall objectives of the program. There is always a trade off between the amount and quality of information acquired and the time and costs associated with the drilling program.

24-8. Logging, Sampling, and Analysis

All boreholes and trial pits should be logged to an appropriate recognised standard (e.g., British Standard BS5930) and, additionally, the following information should be recorded:

- depth and results of any *in situ* radiological monitoring,
- depths and depth ranges and type of any samples collected for radiochemical analysis, and
- depths of any man-made features.

While all field data should be transferred to a field note book, it is often useful to support this by taking photographs of the work. A digital camera can facilitate the quick transfer of images into the computer for report or presentation production.

Some geotechnical testing could be advantageous to the site investigation, in that useful information on hydraulic conductivity of the various sediments can be gleaned. Some simple tests can be carried out in the field, while others *require undisturbed* samples to be sent to the laboratory.

Probably the most important component of a contaminated land assessment is the collection of groundwater samples. Groundwater samples are usually taken by two methods, pump sampling or bail sampling. Bail sampling is usually undertaken in trial pits and trenches. Pump sampling is the preferred method from boreholes, because a large volume of water can be withdrawn prior to sampling, thus ensuring that the sample is representative of the location being targeted. As mentioned previously, an authorised disposal route for the water needs to be considered prior to commencing the work. The selection of suitable containers and preservation techniques is imperative, especially when sampling and testing for organic compounds which can degrade over short periods of time.

The transportation of radiologically contaminated material needs to be undertaken within the regime set out in the Radioactive Substances (Road Transport) Regulations. On-site screening or on-site laboratory analysis may need to be undertaken to determine the levels of activity, as certain laboratories may have upper levels that they are licensed to handle.

The chosen laboratory needs to be competent for the analysis required and should be able to demonstrate an appropriate accreditation (e.g., UKAS), and that quality management systems in line with ISO9001 are adhered to.

24-9. Interpretation and Modeling

Once the data has returned from both the field and the laboratory, interpretation, and modeling will commence. Modeling involves creating a mathematical description of the movement of contamination from a source through migration pathways to a receptor, so that appropriate management and remediation measures can be devised and instigated. The output from the modeling work will then feed into the production of environmental risk assessments.

The modeling work may require the use of computer codes. Some of these are readily available off the shelf, while others may need to be developed in house in order to model site specifics.

24-10. Databasing and GIS

Having collected all the data, it is imperative that it is clearly documented. The field data is immediately logged onto field sheets, but these then need to be transferred into an electronic medium as well as being kept for QA purposes. Proprietary software such as Microsoft Access or Excel are ideal for tabulating the field information. The data can easily be downloaded onto a larger relational database. There are various ways of ensuring that the data is correctly transposed electronically, and duplicate entry and checking can help to minimise errors.

All data, whether new or previously acquired, should be given a quality tag. It is not best practice to ignore any data without having a transparent audit trail documenting the reasons why.

Computerised Geographical Information Systems (GIS) linked into an underpinning database are frequently used to store, link, and view a wide variety of information. Examples of data to be held and facilitated include historical site maps, service plans, photographs, building footprints, and site investigation data.

UKAEA has developed an application tool called IMAGES (Information Management and Geographical Evaluation System) which consists of three main elements:

- a database system for storage, searching, and reporting functionality,
- a data management system to capture data from a number of sources in a controlled and auditable way, and
- a Geographical Information System (GIS) for geographical analysis and reporting.

The integrated package enables:

- quality controls on data and data capture, e.g., identification of data custodians, data quality marking, revision control, updateability, archivability, traceability;
- data selection through querying, filtering, and searching within the database;
- integration with GIS, e.g., bringing selected data into GIS, recording file locations for themes, views, projects;
- full functionality of the GIS;
- recording GIS generated data in a database; and
- control of modifications and distribution of data and GIS inputs and outputs.

Information within IMAGES features a "relationship manager" which enables the user to create dependencies between allowable classes of information. Thus, information can be organised in a number of ways — by locality, by building, by area or zone. In addition, a "workflow" procedure enables management of information, e.g., preliminary input, checking, and issuing (see Figures 24-1 and 24-2).

Once captured, the information can be searched and queried and then delivered to the GIS. For example, the query might relate to searching for buildings where oils were used or stored. The resulting data set produced from this search can then be viewed in the GIS in order to build up a picture of areas where oil contamination may be present in the ground.

24-11. Guidance on Site Investigation

In the UK, guidance on site investigation can be found from the SAFEGROUNDS Learning Network, which is a website managed by the Construction Industry Research

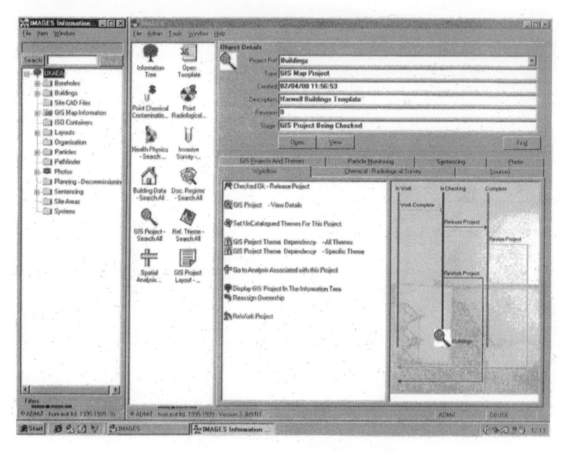

Figure 24-1. Example of Workflow.

and Information Association (CIRIA), WS Atkins, and The Environment Council on behalf of the main UK nuclear licensed site owners (BNFL, UKAEA, AWE, and British Energy) and the Ministry of Defence. It is focused on the management of contaminated land on nuclear licensed sites and defense sites. It was set up to identify and disseminate best practice in the health, safety, and environmental aspects of managing contaminated land, chemically (nonradioactively) contaminated land, and land with mixtures of radioactive and nonradioactive contamination.

Comprehensive guidance relating to US requirements can be found in MARSIMM (Multi-Agency Radiation Survey and Site Investigation Manual). This is sponsored jointly by the Department of Defense, Department of Energy, Environmental Protection Agency and the Nuclear Regulatory Commission. MARSIMM provides "information on planning, conducting, evaluating, and documenting building and surface soil final status radiological surveys for demonstrating compliance with dose or risk-based regulations or standards." It can be accessed electronically at NRC's Public Electronic Reading Room at www.nrc.gov/NRC/ADAMS/index.html. Other useful references and web-sites are included here [1–14].

24-12. References

1. British Standards Institution, *Code of Practice for Site Investigations*, BS5930, British Standards Institution (1999). BSI Publications, London.
2. CIRIA website, www.ciria.org.uk. CIRIA, London.
3. CIRIA. *SAFEGROUNDS Learning Network. Managing contaminated land on nuclear licensed sites. Best practice guidance for site characterisation*, CIRIA (September 2000). CIRIA, London.
4. CLAIRE website, www.claire.co.uk, CL:AIRE, London.
5. Clark and Lewis. *The field guide to water wells and boreholes*, Geological Society of London. John Wiley & Sons (1992).
6. DETR website, www.detr.gov.uk.
7. DTI website, www.dti.gov.uk. *Department Environment, Transport and the Regions, UK.*

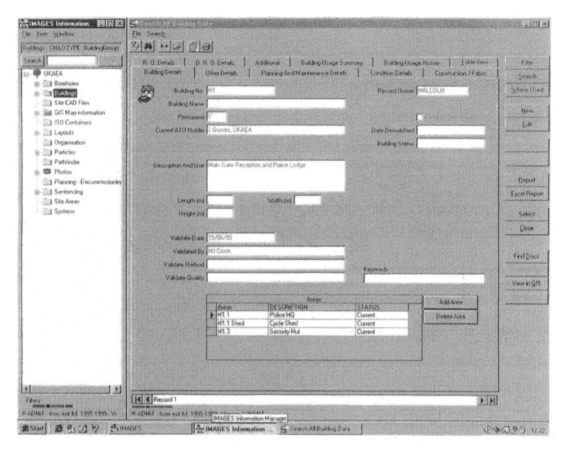

Figure 24-2. Illustration of Part of the Standard Input Form.

8. Environmental Agency website, www.environment-agency. gov.uk.

9. HSE website, www.open.gov.uk/hse. Health and Safety Executive, UK.

10. Health and Safety Executive. *Avoiding Danger from Underground Services*, Health and Safety Executive document HS(G)47, HSE (1988) HSE Books, Suffolk.

11. International Standardisation Organisation. *Guidance on Sampling Groundwaters from Potentially Contaminated Sites*, DIS/ISO 5667-18, ISO (May 2000). ISO, Switzerland.

12. DETR. *Model procedures for the management of contaminated land*, DETR Contaminated Land Research report CLR11, DETR (2000). HMSO, Norwich.

13. Nuclear Installations Inspectorate website, www.hse.gov.uk/ nsdhome/ Health and Safety Executive, UK.

14. SAFEGROUNDS Learning Network website, www. safegrounds.com. CIRIA, London.

Chapter 25

Technologies for Remediating Contaminated Land

25-1. Introduction

This chapter gives an overview of technologies that can be used for the remediation of radioactively contaminated ground. There is a wide variety of techniques available. Many of them were originally developed for remediating chemically contaminated ground and have only recently been adapted for clean-up of radioactive contamination.

Remediation techniques for radioactively contaminated ground involve either:

- removal of the contamination and transfer to a controlled disposal facility such as the UK national LLW repository at Drigg;
- immobilisation, solidification, and stabilisation *in situ* where the physical nature of the soil is changed, or an "agent" is added to the soil, to reduce the migration of the contaminants; and
- on-site containment of the contamination where barriers or hydraulic control measures are placed adjacent to the contaminated soil to reduce migration of the contamination and control potential detrimental effects to human health.

Where removal of the contamination is considered the most appropriate option, key factors to be taken into account should include:

- the clean-up target required for the ground and the method of validating that this level has been reached;
- the cost of disposal of the contaminated material; and
- waste minimisation methods which might reduce the disposal costs.

Immobilisation, solidification, and stabilisation methods and on-site containment systems are used to control the spread of contamination, perhaps as an interim measure pending future removal. These techniques may be appropriate to circumstances where:

- contamination has leaked under structures such as buildings, where removal could be delayed until the

final decommissioning and demolition of the building; and
- the contaminants have a relatively short half-life which will decay to clearance levels during a period of institutional care.

A general overview of waste minimisation, immobilisation, and containment systems is presented in the following sections. Appendix 3 gives a practical example of a particular site remediation project.

25-2. Waste Minimisation

Remediation of contaminated land can generate very large volumes of waste unless measures are taken to minimise waste by careful monitoring and segregation. Processes available for waste minimisation can be either *ex situ* or *in situ*:

- *ex situ*: the soil is excavated and subsequently processed to reduce the volume of radioactive waste; and
- *in situ*: the soil is treated to remove the radioactive contaminants in the ground.

Ex situ processes include:

- Detector-based segregation;
- Soil washing by particle separation;
- Soil washing with chemical leaching agents; and
- *Ex situ* electroremediation.

In situ processes include:

- Electroremediation; and
- Phytoremediation.

Table 25-1 describes the *ex situ* processes and Table 25-2 describes the *in situ* processes.

Detector-based segregation contaminated soil may be segregated from uncontaminated soil by excavating it and measuring the radiation from a predefined volume of excavated material (typically an excavator bucket or a

Table 25-1. *Ex situ* Waste Minimisation Systems for Radioactively Contaminated Soil

Technology	Description
Detector-based segregation	This is the most commonly used waste minimisation process for radioactively contaminated soil. It is based on real time measurements of the radioactivity levels where "zones" of contamination are detected and physically removed or segregated from soil which is relatively contaminant-free. This type of process is a "dry" process and has been mainly used where the contaminant is a gamma emitter or where a fingerprint can be used to reference a gamma emitter to alpha and beta activity. Simple systems involve excavating soil in layers and using hand held instrumentation to manually scan over the surface of soil and soil in the excavator bucket. Simple systems are generally labor intensive. Sophisticated versions of this type of system can have high throughputs and have a more auditable record of the contamination. They use conveyors, a detector counting "chamber," and microprocessor controlled segregation gates. Some presizing of the feed to the conveyor is required to remove large rubble and boulders. There are potential problems with materials handling with respect to feeding and segregating cohesive soils. The most reliable methods using detector based segregation are based on measurements of thin layers of the soil in order to take into account attenuation of the activity caused by self-absorption. This technique is particularly successful where the contaminants are in the form of discrete particles.
Soil washing by particle separation	Generally a waste minimisation process in which the particular soil particles which contain the contaminants (e.g., clays, carbonaceous matter) are removed from the contaminant-free bulk particles. The equipment used is commonly found in the mineral/coal processing industry for concentrating minerals from the low grade ore. Separation devices are based on exploiting differences in particle size, settling velocity, specific gravity, surface chemical properties (using froth flotation), and magnetic properties. These devices operate with the particles suspended in water — but it is generally found that the contaminants adsorbed on the soil particles are insoluble. Soils with >30% of the particles <0.063 mm are not usually economically treated by particle separation-based soil washing alone. Soil washing by particle separation can be combined with upstream detector-based segregation and downstream chemical leaching systems.
Soil washing by chemical leaching and extraction	A form of soil washing where leaching agents such as complexants, acids, and alkalis are used to transfer the contaminants from the soil into aqueous solution. The contaminants are then removed from solution using precipitation, adsorption, or ion exchange.
Batch *ex situ* electroremediation	A batch version of *in situ* electroremediation. *Ex situ* electroremediation involves treatment in treatment cells.

Table 25-2. *In situ* Waste Minimisation Systems for Radioactively Contaminated Soil

Technology	Description
Electroremediation	A three stage process involving the desorption of the contaminants from the soil, the movement of the contaminants through the soil pore water to buried electrodes, and the capture and removal of the contaminants at the electrodes. In the process, an array of electrode assemblies are inserted into the contaminated soil and a DC current is applied. The contaminants are desorbed from the soil either by the "acid" front created by the electrolysis of water at the anode, or by the controlled movement of complex ions from solutions added to the electrode housings. These then move to the electrode housings by electromigration or electroosmosis and are either captured in the ground on a solid sorbant (which can be treated later or disposed of) or are pumped in solution to an above ground treatment facility. Electroremediation is one of the few technologies that can be effective with clay-rich soils. Electroremediation can also be used as a containment system where the electrodes form an "electrokinetic fence."
Phytoremediation	Phytoremediation is a developing technology where plants are used to accumulate metal contaminants in an harvestable biomass, or are able to stimulate biodegradation by protecting and supporting microbial communities. Phytoremediation can also be used as a containment system where the plants are used to form a "biobarrier."

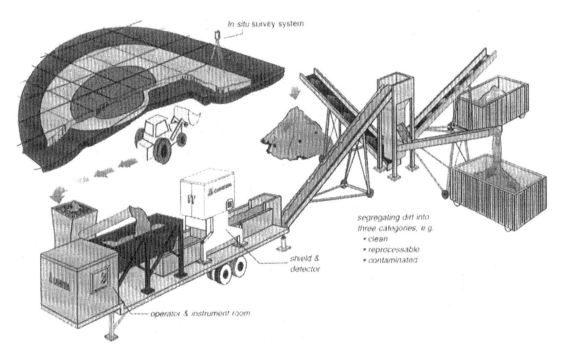

Figure 25-1. Typical System for Segregation of Contamination (Source: Canberra-Harwell).

small skip) and sentencing it into the appropriate waste stream. This is traditionally done in a batch-wise process. A case study in the appendix to this chapter describes this methodology for one particular project — remediation of the Dounreay Castle. For larger scale projects, an automated system may be used to increase the throughput (tens of tonnes/hour). This involves transferring the excavated soil to a conveyor belt and measuring its radioactivity as it passes underneath a detector array. The equipment automatically separates the portion exceeding the cleanup standards. Figure 25-1 illustrates the sort of process which can be used. The use of segregation for volume reduction is limited by the proportion of the soil which is contaminated and can be separated from the bulk of the soil, and by the costs associated with the disposal (or further treatment) of these fractions.

A commercially-available system for soil segregation is available from Eberline Services Inc. (www.eberlineservices.com). This is known as the Segmented Gate System (SGS). Table 25-3 summarises the results which have been obtained with this system on various remediation projects.

Soil segregation has the following limitations:

- it requires a fingerprint reference to gamma emitters, as alpha and soft beta radiation cannot be measured directly;

- the contamination should occur in discrete zones which can be separated from those zones containing no contamination;
- hand held/field monitoring systems are subject to operator inaccuracies and can cause a slowing of excavator operations (unless several operators are used);
- coarse rubble will need to be manually removed and crushed where it is suspected that contamination may be internal; and
- for systems which measure activity in a thin layer of soil on a conveyor belt, there can be considerable problems in the materials handling of the soil — e.g., crushing, but also soil which contains a high clay content or is moist, can also cause problems as the soil will tend to hang in hopper feeders.

Processing can be carried out either on the contaminated site using mobile, trailer mounted units, or off-site in fixed, centralised plants where the contaminated soil is brought to the plant. Treatment using mobile equipment has the advantage that transport of large volumes of contaminated soil off-site can be avoided. Conversely, fixed centralised plants avoid large commissioning and dismantling costs and are, thus, more able to treat smaller batches of soil economically.

Soil washing is a well established remediation technology in The Netherlands, Germany, Belgium, and

Table 25-3. Results Obtained from Remediation Using the Eberline Segmented Gate System

Site	Type of contamination	Volume of contaminated soil treated (m^3)	Treatment target	Volume reduction
At Johnston Atoll	Pu	76,000 at 1250 m^3 per week	<0.5 Bq/g for the majority and to <5 Bq/g for hot spots	90%
Savannah River Site	Cs^{137}	960	<0.15 Bq/g	99%
Los Alamos National Laboratory	Natural and depleted uranium	160	No details	97%
FUSRAP New Brunswick	Radium	3,000	<0.19 Bq/g	60%

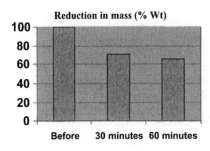
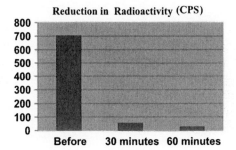

Figure 25-2. Results of Rock Scrubbing Trials.

Switzerland. It is also used in the US and Canada. It is a volume reduction/waste minimisation process based on particle separation and leaching techniques used in the mineral processing industry. In soil washing, contaminated particles are segregated by physical processes from the relatively uncontaminated bulk, or contamination is chemically leached and then recovered from solution in a concentrated form.

To be successfully applied to contaminated soils, residual levels of contamination in the bulk of the soil which remains after removal of the concentrate should be lower than a threshold value (e.g., guideline value, risk based clean up target) or waste class.

The bulk (clean) fraction can then be:

* recycled on the contaminated site as backfill;
* used on other sites as a relatively inert building fill; and
* disposed of as material less hazardous than the original soil and having a lower associated disposal cost.

The cost of the processing is offset by the significantly reduced volumes (and, hence, costs) for disposal of radioactive waste.

Soil washing is appropriate where the contamination is associated with a particular group of particles in the soil matrix which can be separated by physical processes, or the contamination can be selectively removed from the

soil by leaching processes (and recovered from solution in a concentrated form). Physical separation processes are generally cheaper than those that involve chemical leaching.

Laboratory tests are required to determine what type of soil washing processes might be suitable in given circumstances, if any. Although many contaminated soils display a differential distribution of the contaminants to different types of particles, the fraction containing the least contamination may not meet the required limit for clean-up or be in a sufficient proportion to justify application. Where leaching solutions are used, the cost effectiveness of the process is determined not only by the extraction efficiency of the leaching solution, but also by its selectivity, the ability to recover the contaminants from solution, and the ability to recover and regenerate the reagent used for leaching.

Figure 25-2 shows the results of some scrubbing trials using contaminated rock from the Dounreay site. The process involved tumbling the rocks in a rotating drum with water for up to 60 minutes. The action of the rocks against one another resulted in the surface layers being eroded. This generated fines in which the radioactivity was concentrated, leaving the bulk of the rocks relatively clean. The graphs compare the loss in mass of the rocks with the reduction in contamination.

Electroremediation is defined as "the redistribution of contaminants in soil using an electric field." To date, it has not been used commercially, but extensive trials have been carried out, notably in Russia and Uzbekistan. In one trial, the level of Cs-137 in an area of 15 m^2 and a depth of 2–3 m was reduced by 56% after treatment for 111 days. The method is believed to be particularly suited to treating low permeability soils, such as clay, which are not responsive to other techniques. It is strongly dependent on the soil chemistry, particularly the cation exchange capacity and the soil pH buffering capacity (e.g., the presence of calcium carbonate). Buried metal objects also impair performance. Some contaminants (e.g., Cs) have been found to be particularly difficult to treat due to problems in desorbing them from the soil particles prior to electromigration.

Phytoremediation is a term used to describe the use of plants in remediating contaminated ground. Plants can extract and concentrate a range of chemical and radioactive contaminants from the soil and incorporate them into their leaves, stems, and roots (Phytoextraction). These can then be harvested and subsequently treated (by incineration, composting, or anaerobic digestion) to concentrate and recover the pollutants. It is a relatively new development which has been applied experimentally, mainly in eastern Europe and the US. In a variation of the process, known as Rhizofiltration, the plants are used to contain the contaminants by precipitating them in their root system. The main problems with phytoremediation processes are:

- release of contamination to air through the foliage;
- migration up the food chain (by herbivores);
- the long timescales involved; and
- the disposal of secondary waste arisings.

25-3. Immobilisation, Stabilisation, and Solidification

Immobilisation, stabilisation, and solidification systems aim to reduce the mobility of the contaminants by:

- forming chemically immobile compounds of the contaminant — stabilisation;
- binding the soil together to form a monolithic block to prevent access by external mobilising agents such as wind, rain, and groundwater — solidification; and
- melting and rapidly cooling the soil so that the contaminants are immobilised in a glassy matrix — vitrification.

These methods are likely to be most acceptable in dry or desert regions (e.g., Idaho and Hanford in the US), but less successful in areas of high rainfall where leaching is a problem. Both *ex situ* and *in situ* processes exist (Table 25-4). Solidification processes add up to 30–130% to the soil volume, whereas vitrification processes can reduce the volume by 25–40% due to losses in pore space, moisture removal, and volatile emissions.

Stabilisation and solidification processes do not destroy the contaminants, and the stability of solidified masses has only been tested over periods up to 20 years. Therefore, it is important to include long-term monitoring arrangements to ensure continued integrity.

Other limitations affecting stabilisation and solidification processes, particularly for *in situ* treatment, include those relating to site-specific conditions such as difficulties in mobilising large construction equipment on uneven ground and problems of mixing soil which contains buried large boulders.

As an example of application and solidification, stabilisation systems are being assessed on waste pits at INEEL

Table 25-4. Solidification/Stabilisation and Immobilisation System

	Technology	Description
Ex situ	Stabilisation and Solidification	*Ex situ* stabilisation/solidification involves mixing the contaminated soil with a chemical immobilising agent and/or binding agent (e.g., Portland cement, lime, fly ash, silicates, polymers, bitumen, asphalt, and other proprietary agents). On curing, the contaminants in the resulting material are less mobile than in the original soil.
	Vitrification	*Ex situ* vitrification involves mixing the soil with a fluxing agent and heating the soil to high temperatures until the material melts. Radioactive contaminants are incorporated in a melt, which on cooling forms a leach resistant glassy matrix.
In situ	Stabilisation and Solidification	*In situ* processes involve incorporating a stabilisation/solidification mixture directly into the soil using hollow stem augers or pressure injection. Treatment proceeds as a series of overlapping treatment columns. The solidified soil can have superior civil engineering properties.
	Vitrification	In *in situ* vitrification, an array of four electrodes are inserted into the contaminated ground and an electric potential is applied between the electrodes to melt the soil by Joule heating. Temperatures typically range 1600–2000°C. Radioactive contaminants are incorporated in a melt which cools to form a leach resistant solidified glassy mass. The melt from an *in situ* vitrification process takes months to fully solidify [1].

in the US. Here, the technology is seen as a "holding" solution to the problem of migration of contaminants from these pits. Thus, the tests include not only assessments of injection methods, solidification formulations, and long-term integrity, but also ease of retrieval of the solidified mass (which may be necessary at some time in the future).

With vitrification processes, the composition of the soil (particularly the alkali, chloride, and SiO_2, Al_2O_3, CaO, MgO, Fe_2O_3 contents) need to be ascertained and if necessary adjusted (by mixing with other soils/wastes or by adding fluxing materials) to ensure that the material will melt at the temperatures achievable with the system.

In situ vitrification is most suited to contaminated ground between 2–6 m deep. Where the contamination is relatively shallow, vitrification is best performed as an *ex situ* process or using *in situ* vitrification equipment operating on heaps of excavated contaminated soil [1]. With *in situ* vitrification, the moisture content of the soil will have a significant effect on the cost of treatment, and may also cause problems with steam build-up in the melt.

25-4. Containment Systems and Hydraulic Measures

Containment systems and hydraulic measures are listed in Table 25-5. These systems and measures are used to control the migration of the contamination and to reduce its toxic effects. Because the main function of the system is to contain the contaminants and not to remove or immobilise them, it is necessary to ensure that the system remains functional. This, therefore, necessitates long-term monitoring to confirm integrity and effectiveness.

In some instances, containment systems can be part of the remediation process, for example:

- as a safeguard against migration whilst the contaminants naturally decay over a period of time; or
- as a control measure for long-term remediation such as during electroremediation; or
- integrated with a long-term passive treatment system such as in the use of containment walls and permeable

Table 25-5. Containment Systems and Hydraulic Measures

Technology	Description
Cover systems	A multilayer construction placed over contaminated soil to reduce the harmful effects of the contaminants at the surface, minimise water infiltration through the contaminants by rain, prevent upward migration of groundwater by capillary rise, prevent airborne migration of the contaminants, and where appropriate control gas migration. The optimum combination of the layers, in terms of composition, thickness, and sequence of materials is based on an assessment of the physical and chemical properties of the entire system (e.g., chemical resistance, physical resistance to climatic conditions, and ground conditions such as cracking and channeling due to drought, freeze/thaw, settlement), construction aspects, consideration of the reduction in environmental risk of the underlying contaminated land and cost.
Vertical barriers	Vertical barriers are installed adjacent to contaminated ground (i) to prevent the off-site lateral migration of contaminated groundwater, (ii) to divert clean groundwater away from contaminated ground, and (iii) to reduce the extraction rates of contaminated groundwater from hydraulic control measures. They can be used to funnel groundwater to an in-ground treatment center (so called funnel and gate — see permeable active barriers in Section 25.5) and also be used to cut-off the underground migration of gases. To be effective, vertical barriers are normally tied into a natural low permeable layer at depth (e.g., a clay layer) or to an in-ground horizontal barrier. There are three common types of vertical barriers (i) displacement systems (e.g., sheet piling, membrane walls), (ii) excavated barriers (e.g., shallow cut-off walls, slurry trench walls, secant walls), and (iii) injection barriers (e.g., chemical grouting, auger mixing, jet grouting).
Horizontal barriers	In-ground horizontal barriers are installed below the contaminated ground to prevent vertical migration. They can be used in combination with vertical barriers to isolate potentially mobile contamination. Horizontal barriers are generally formed by injection of cement slurries at depth. The quality and integrity of the construction is difficult to guarantee and remedy where there are deficiencies.
Hydraulic control measures	Hydraulic control measures are used to adjust the groundwater flow around a contaminated area so that no further spread of contamination takes place. This can involve preventing or reducing the contact of the groundwater with the contaminated mass (e.g., lowering the water table), reducing, intercepting or containing a plume of contaminated groundwater, supporting other remediation methods such as in-ground barriers, or being part of groundwater remediation operations [2]. Hydraulic control measures are commonly carried out by pumping out groundwater from a number of wells, or using diversion or interceptor trenches. Where the groundwater that is pumped is contaminated, consideration has to be given to its treatment before it can be returned to a water course.

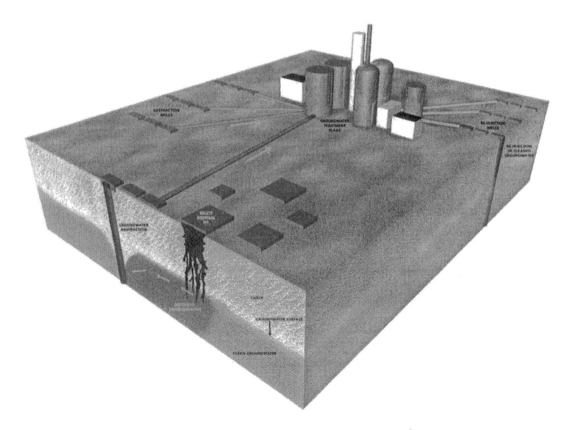

Figure 25-3. Schematic Diagram of a Groundwater Pump-and-Treat System.

active walls in a funnel and gate system to treat contaminated groundwater.

Potential constraints with containment systems are related to their long-term effectiveness. With hydraulic control measures, because water has to be continuously pumped out of the ground over an indefinite period of time, pumps need to be regularly maintained and checked.

With some containment systems, such as cover, consideration may need to be given to the civil engineering and geotechnical properties of the materials as it may be necessary for the cover to support structures and services.

25-5. Treatment of Contaminated Groundwater

An integral part of contaminated land remediation is the treatment of contaminated groundwater. On a working site, and where the flow is not excessive, it may be appropriate to intercept the groundwater and pump it to an existing low level effluent treatment works. On other sites, it may be necessary to construct an above-ground facility specifically for the treatment of contaminated

groundwater. Figure 25-3 is a schematic diagram of a pump-and-treat system installed for the remediation of chemically contaminated groundwater at Harwell in Oxfordshire, England. Over a period of 7 years, this plant has treated a total of 2.6 million m^3 of groundwater and removed 5.1 tonnes of volatile organic chemicals (chlorinated hydrocarbon).

An alternative solution being developed in the US is the use of in-ground permeable reactive barriers. These are vertical barriers which contain sorptive or reactive constituents which capture and/or destroy the contaminants as the groundwater flows through them. The constituents used in these barriers include iron filings, zeolites, and peat.

25-6. Best Practicable Environmental Option

Although many technologies are potentially applicable, their application to specific contaminated site remediation is dependent on a number of factors and related to detailed site characterisation studies. Development trials can be conducted to determine what performance might be expected from a given technology in specific circumstances. This information can be used to inform a Best Practicable Environmental Option (BPEO) study.

The factors considered of particular importance in a BPEO study are:

- the clean-up target (determined by regulatory requirements and/or the future use envisaged for the land);
- technical feasibility relative to the particular site, soil and contaminant characteristics, and timeframe;
- site infrastructure arrangements and needs, the working life of the site, and the duration of institutional care;
- long-term monitoring arrangements for slow remedial techniques or for immobilisation and containment techniques;
- validation of the remediation;
- health and safety aspects;
- regulator and public acceptance; and
- cost.

See useful references for further information [3–11].

25-7. References

1. www.ierp.org.
2. International Atomic Energy Agency. *Establishing a National System for Radioactive Waste Management*, IAEA Safety Series No. 111-S-1, IAEA, Vienna (1995).
3. Pearl, M. and P. A. Wood. *Review of Pilot and Full Scale Soil Washing Plants*, Warren Spring Laboratory Report LR 1018 prepared for the UK Department of the Environment. Available from the AEA Technology National Environmental Technology Centre, Culham (1994).
4. United States Environmental Protection Agency. *Remediation Technologies Screening Matrix and Reference Guide*, US EPA (1993) Available through www.clu-in.org/remed1.cfm #tech_desc.
5. Construction Industry Research and Information Association. *Remedial Treatment for Contaminated Land Special Publication 108*, Volumes 1–12. Available from CIRIA, 6 Storey's Gate, Westminster, London SW1P 3AU (1995).
6. United States Environmental Protection Agency. *Approaches for the Remediation of Federal Facility Sites Contaminated with Explosives or Radioactive Wastes*, US EPA/635/R-93/013 (1993).
7. Selentec. *Presentation at INEEL Soil Restoration Meeting*, United States Department of Energy Tech Co Programme, Idaho Falls (November 1996).
8. Tech Con. *Presentation at INEEL Soil Restoration Meeting*, United States Department of Energy Tech Co Programme, Idaho Falls (November 1996).
9. Department of Energy. *FY-95 Technology Catalog Technology Development for Buried Waste Remediation*, DOE Environmental Management, Office of Technology Development DOE/ID-10513 (1995).
10. Geosafe Corp. *Presentation at INEEL Soil Restoration Meeting*, United States Department of Energy Tech Co Programme, Idaho Falls (November 1996).

Appendix 1
Country Specific Examples of Radioactive Waste Management Programs

A1-1. Belgium

How Radioactive Waste Management is Organised in Belgium

ONDRAF/NIRAS is the Belgian radioactive waste management organisation. It is responsible for waste minimisation, identification, processing, interim storage, long-term management, and transport. It also takes on tasks related to the decommissioning of facilities and the maintenance of an inventory of enriched fissile materials.

ONDRAF/NIRAS, as a public agency, reports regularly to the Minister of Energy within the Ministry of Economy and annually to Parliament.

ONDRAF/NIRAS operates on a nonprofit basis, taking into account the costs that will have to be incurred in the future, in particular for long-term management. These costs are then passed onto the producers of radioactive waste, in line with the "polluter pays" principle.

The Origin of the Wastes

Eighty percent of Belgian radioactive waste arises from activities related to nuclear power production:

- nuclear power stations,
- nuclear fuel manufacture,
- spent fuel reprocessing by Cogema of France, and
- nuclear research.

Other sources include medical, industrial, and agricultural applications, and the decommissioning of nuclear facilities. This results in a very diverse range of wastes. In 1998, the Belgian government announced a moratorium on further spent fuel reprocessing. Upto then, 650 tonnes of spent fuel had been reprocessed by COGEMA.

Waste Classification

Belgium divides its radioactive waste into three categories:

- *Type A or low-level waste.* This contains low levels of short-lived (half-life less than 30 years) radionuclides. It mostly arises from power station operation and during the use or manufacture of radionuclides.
- *Type B or intermediate-level waste.* This contains some long-lived radionuclides. It arises during the manufacture of nuclear fuel, in research on irradiated nuclear fuel, and during fuel reprocessing.
- *Type C or high-level waste.* This contains significant quantities of short, medium, and long-lived radionuclides. It mostly originates from research on irradiated nuclear fuel and from fuel reprocessing. It is estimated that by 2060 there will be 60,000 m^3 of Category A waste, including decommissioning waste. With no further reprocessing, there will be 7800 m^3 of category B and 2100 m^3 of category C. There would also be 4000–5000 tonnes of spent fuel.

Strategies for Radioactive Waste Management in Belgium

All categories of radioactive waste are currently held in interim storage. However, in 1998, the Belgian government rejected storage as a long-term (as opposed to an interim) management solution. It also announced a moratorium on further spent fuel reprocessing. In 2001, the government asked ONDRAF/NIRAS to take part in initiatives to study the possibilities of developing a regional disposal concept in parallel with its own national efforts.

- *Short-lived, low-level waste.* A disposal solution for short-lived, low-level waste must provide protection for up to 300 years. There are two options for this type of waste: surface disposal or underground disposal and ONDRAF/NIRAS is carrying out research to evaluate these alternatives.
- *Long-lived waste.* Long-lived waste will require deep disposal. In partnership with SCK-CEN, ONDRAF/NIRAS is continuing to conduct research into the possibility of building a repository for long-lived waste in the deep clay that covers much of Belgium. Much of this research is performed at the underground laboratory in the Tertiary Boom Clay layer below the Mol-Dessel nuclear site in northern Belgium.

Future Program

- *Short-lived, low-level waste.* Following an unsuccessful attempt to site a near-surface facility in the mid-1990s, since 1999 ONDRAF/NIRAS has concentrated its activities on the development of local "partnerships" to facilitate project proposals in sites showing an interest to host a facility. Such local partnerships involve ONDRAF/NIRAS working through independent (University research-based) mediators with local stakeholders in the development of a proposal for a disposal project which is seen as an integrated part of local development. The project is intended to satisfy both technical/scientific and socio-economic criteria before being proposed to government.
- *Long-lived waste.* The underground laboratory constructed at the Mol-Dessel site came into operation in 1983, having been selected for a detailed site investigation program. The laboratory was extended in 1987. Since 1983, a wide range of experiments has been performed to demonstrate technical feasibility and long-term safety on a site-specific basis.

Detailed studies for a deep facility in clay are planned to start around 2015, with construction commencing about 5 years later. It is intended that, by 2035, the disposal of Type B wastes will begin; disposal of Type C wastes and/or spent fuel will start around 2050. With the most recent government announcement, this research will be accompanied by collaboration with international partners for a regional (i.e., trans-national) disposal concept (see Figure A1-1).

Figure A1-1. Hades Underground Laboratory at Mol, Belgium. (photo courtesy of SCK-CEN).

A1-2. Canada

How Radioactive Waste Management is Organised in Canada

Historically, Atomic Energy of Canada Ltd (AECL), a Government funded corporation for nuclear research and development, has been responsible for developing disposal concepts for radioactive waste.

More recently, the Government has said that it expects the waste producers and owners to form a Waste Management Organisation and to establish a segregated fund to finance nuclear waste management. The Government is also considering how best to regulate these activities. The federal regulator is the Canadian Nuclear Safety Commission.

The Origin of the Wastes

In 2000, nuclear electricity accounted for 13.5% of total Canadian electricity production. Canada has an indigenous reactor design in the CANDU reactor, which allows the use of natural uranium (i.e., nonenriched) fuel. A number of smaller reactors are licensed for research and medical isotope production.

Canada is also a major uranium producer, which leads to significant volumes of mine and mill tailings.

Waste Classification

Canada does not reprocess spent fuel, and radioactive waste is grouped into three categories: nuclear fuel waste, low-level radioactive waste, and uranium mine and mill tailings.

- Nuclear fuel waste (spent fuel) is expected to amount to about 65,000 tonnes ($14,500\,m^3$) by 2035.
- Low-level waste is expected to amount to 2.1 million m^3 by 2035. Of this, 1.5 million m^3 is "historic" waste. This dates back to the 1930s, when radium was extracted for medical uses at a refinery in Port Hope, Ontario. The remainder "ongoing" waste is an operational by-product of Canada's nuclear reactors, nuclear fuel processing and fabrication facilities, and medical, research, and industrial uses of isotopes.
- Uranium mine and mill tailings are expected to amount to 248 million tonnes by 2035. These wastes are subject to onsite decommissioning. They are not discussed further here.

Strategies for Radioactive Waste Management in Canada

- *Spent fuel.* Between 1981 and 1994, AECL developed a disposal concept for spent fuel that envisaged a deep geological repository in the Canadian Shield. The concept was examined in public hearings in 1996/1997 led by a Government appointed Independent Assessment Panel. The Panel's report, issued in March 1998, judged that the concept appeared to be technically sound, but that public support had not been demonstrated. In response to the Panel's findings, the Government published a policy framework for radioactive waste in 1996. This expects the waste producers and owners to form a Waste Management Organisation to follow up the Panel's recommendations. It also expects the waste producers and owners to establish a segregated fund to finance nuclear waste management. A new Nuclear Fuel Waste Act is to be enacted. This will require the waste management organisation to carry out an options study into the long-term management of spent fuel. This is to be completed within 3 years of the Act being brought into force. This options study is expected to lead to new Government decisions on the long-term management and regulation of spent fuel in Canada.
- *Ongoing low-level wastes.* These (short-lived) wastes will continue to be kept in purpose-built stores pending the development of a disposal facility. They remain the responsibility of the producer. For financial planning purposes, it is assumed that this could be in operation as early as 2015.
- *"Historic" low-level wastes.* Low-level "historic" wastes are the responsibility of the Federal Government. They are currently stored at or close to the production sites: principally at Clarington and the Municipality of Port Hope in Southern Ontario. The preferred long-term management option is a locally developed solution to place the wastes in newly constructed aboveground engineered mounds, designed to last for at least 500 years with minimal maintenance costs. Legal agreements with the host communities were completed early in 2002, and environmental assessments are now underway on two projects. These are the Port Hope long-term low-level radioactive waste management project (with a number of components, including a Port Hope facility, a facility in the former Township of Hope

and interim waste management measures) and the Port Granby (Clarington) long-term LLRWM project. The two projects are expected to take around 10 years to complete.

A1-3. Finland

How Radioactive Waste Management is Organised in Finland

Posiva Oy (Posiva) is the company responsible for research and development into final disposal of spent nuclear fuel and, ultimately, for the construction and operation of a final repository in Finland.

Posiva was established in 1995 and is owned by Teollisuuden Voima Oy (TVO) and Fortum Power and Heat Oy (Fortum), Finland's two nuclear power companies. They have reactors at Olkiluoto and Loviisa.

The two power companies retain responsibility for the treatment and final disposal of low and intermediate-level waste and for plant decommissioning.

Posiva compiles the operating plan and reports annually on nuclear waste management at its owners' power plants where spent nuclear fuel is currently in interim storage in water filled pools.

The Origin of the Wastes

The wastes consist of spent nuclear fuel and operational wastes from the nuclear power stations. These produced 27% of Finnish electricity in 2000. Decommissioning wastes will also occur in the future.

Waste Classification

Nuclear waste is divided into three categories:

- operating waste (low and medium-level),
- decommissioning waste (low and medium-level), and
- spent nuclear fuel (high-level).

Strategies for Radioactive Waste Management in Finland

The aims and schedules relating to implementation of nuclear waste management and the associated research and planning were defined in the Council of State's decision in the principle of 1983. This specifies that research into spent fuel disposal should aim to choose and study a suitable disposal site to allow selection by 2000.

- *Spent fuel.* In May 1999, Posiva applied for a Decision in Principle on the construction of a final disposal facility at Olkiluoto. This is to be for spent nuclear fuel generated by the existing Finnish nuclear power plants. The application was based on the results of research over 20 years, including a site characterisation program. The Finnish Parliament agreed to this in May 2001, having first established that the community of Olkiluoto consents to a repository being built there. The Decision in Principle upholds the view that the construction of a final disposal facility is for the overall good of the society.
- *Operational waste.* In addition to spent fuel, operation of the Olkiluoto and Loviisa power plants also produce intermediate and low-level operational waste such as ion exchange resins and miscellaneous maintenance waste. Resins are packaged into 200 liter drums using bitumen as an immobilisation matrix. Metallic wastes may be compacted and placed into drums or other containers. These wastes are stored and then disposed of in purpose-built repositories in the bedrock about 70–100 m below the site. The Olkiluoto repository has been in operation since 1992 and, at the end of 2000, 3500 m^3 of this waste had been disposed of there. The Loviisa repository went into operation in 1998, and the equivalent figure is 700 m^3. The Olkiluoto repository also accepts waste from small producers such as hospitals.
- *Decommissioning waste.* Both repositories will be expanded to take the NPP decommissioning short and intermediate level waste. Used reactor internals such as control rods and core instruments from the power plants are classed as decommissioning wastes. It is currently planned that these will be disposed of when the power plants themselves are decommissioned.

Future Program

The Finnish Parliament's agreement to the Decision in Principle to construct a spent fuel repository near Olkiluoto makes it possible for Posiva to construct an underground rock characterisation facility, ONKALO. This will allow the geohydraulic, geochemical, and mechanical properties of the Olkiluoto bedrock to be studied in detail. This will provide the information needed to design the repository and give an opportunity to test disposal technology in realistic conditions.

Construction of ONKALO is expected to start in 2003–2004, allowing investigations at final disposal depth to commence around 2006.

The Government approved schedule of 1983 envisages operation of a final disposal facility starting in 2020. This suggests that construction of the facility should start after 2010. Separate licences will be required to permit both construction and operation of the final disposal facility (see Figures A1-2 and A1-3).

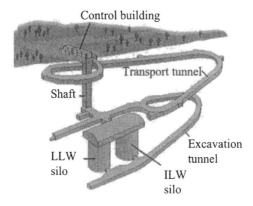

Figure A1-2. The Olkiluoto Repository for Low- and Intermediate-Level Wastes.

Figure A1-3. Copper Canister at SKB's Encapsulation Laboratory, Sweden. Finland Will Have a Similar Disposal Concept to Sweden.

A1-4. France

How Radioactive Waste Management is Organised in France

The National Agency for Radioactive Waste Management (ANDRA), financed by the waste producers, was established in 1979; it was reconstituted as a "public industrial and commercial entity" in 1991. ANDRA is responsible to the Government for designing, siting, constructing, and operating long-term disposal facilities and underground laboratories.

ANDRA's tasks also include compilation of the waste inventory, the specification of waste packaging, and research into long-term safety.

The Origin of the Wastes

French radioactive wastes arise from extensive use of nuclear technology: for energy production, defense, medicine, research, and industrial uses. In 2000, nuclear power stations produced about 75% of French electricity. French involvement with all stages of the nuclear fuel cycle, from uranium mining to spent nuclear fuel reprocessing, has led to the production of significant quantities of all types of radioactive waste with diverse chemical and physical properties.

Waste Classification

Radioactive waste in France is divided into four categories: very low, low, intermediate, and high-level wastes (VLLW, LLW, ILW, and HLW, respectively). The last three of these are further subdivided into short and long-lived wastes (based on a 30 year half-life).

Short-lived LLLW and ILW wastes represent 90% of the volume of French radioactive wastes. They typically consist of contaminated clothing, filters, water treatment resins, etc. They contain only trace quantities of long-lived radioactivity.

Long-lived waste includes heat generating HLW, either as vitrified waste or spent fuel, and ILW that comes mainly from spent fuel reprocessing. Long-lived LLW is mostly (historical) radium contaminated material or graphite from the now decommissioned French gas graphite reactors.

It is expected that, by 2040, 7600 m^3 of vitrified HLW, 25,000 m^3 of spent fuel and 80,000 m^3 of other long-lived wastes will be stockpiled.

Strategies for Radioactive Waste Management in France

- *Short-lived waste.* Final disposal was carried out in the near-surface repository at Centre de la Manche near the La Hague reprocessing plant from 1969 until it reached its capacity of 0.5 million m^3 in 1994. Here, waste that needed some shielding was encapsulated in concrete structures known as monoliths. Less hazardous waste was disposed of in surface structures known as tumuli; these were then covered in clay and topsoil and allowed to return to nature. A new surface disposal facility with a capacity of 1 million m^3 opened in 1992 at Centre de l'Aube, about 50 km east of the city of Troyes. Here, the waste is being disposed of in engineered concrete vaults which, when full, will be covered with a concrete slab and sealed. A final cap of clay, bitumen, and seeded topsoil will then be placed over the structures.
- *Long-lived waste.* For long-lived waste, legislation adopted in December 1991 provides for a three-pronged approach.

    (1) Research on partitioning and transmutation of long-lived radioactive elements in the waste, carried out by the French Atomic Energy Commission (CEA).
    (2) Evaluation of options for retrievable or nonretrievable disposal in deep geologic formations; particularly through the creation of underground laboratories. This work is carried out by ANDRA.
    (3) Study of waste conditioning processes and long-term surface storage techniques for the waste, carried out by CEA.

By the end of 2006, the Government is to submit an overall assessment of the above research to Parliament. This will be accompanied by (if appropriate) a draft law authorising the construction of a disposal facility for high-level long-lived radioactive waste at one of the laboratory locations.

Future Program

The French Atomic Energy Commission (CEA) consider that the separation of minor actinides and fission products contained in long-lived radioactive waste and their subsequent transformation into shorter-lived isotopes is scientifically feasible. However, whether the required technologies can be successfully developed and whether such technologies will be commercially viable remains to be seen.

Following a volunteer siting program started in 1992, at the end of 1998, the French government decided that the country would have two underground research laboratories.

The first was to be in the clay formations of the Paris Basin, at Bure, a commune straddling the Haute-Marne and Meuse departments. ANDRA was authorised to construct this laboratory in August 1999. At the same time, a consultative mission began to select one or more sites for a second, granitic laboratory.

Research into conditioning and storage of long-lived wastes has examined industrial processes for waste treatment and state-of-the-art storage techniques. Designs are also being developed for surface or underground storage facilities that would ensure the containment and the retrievability of packages over several centuries.

Work on these three approaches will continue until 2006, when the French Parliament will decide on the best solution or solutions for the long-term management of long-lived high-level waste.

A1-5. Germany

How Radioactive Waste Management is Organised in Germany

The Federal Office for Radiation Protection (BfS) controls radioactive waste management under the Atomic Energy Act, and BfS has authorised DBE, the German Company for the Construction and Operation of Waste Repositories, to plan, construct, and operate facilities for the final disposal of radioactive waste.

However, in forming the Federal Government that came into office in 1998, a Coalition Agreement was drawn up between the coalition partners. Amongst other things, this seeks to amend the Atomic Energy Act to allow the phasing out of nuclear power and to develop a national plan for the orderly disposal of all nuclear waste.

The phasing out of nuclear power passed into law in December 2001. This and other measures will change the way radioactive waste management is organised in Germany.

The Origin of the Wastes

Nuclear electricity produces around one third of all German electricity supplies. Until 1994, government policy was that spent fuel should be reprocessed and fuel was sent to France or Britain for this purpose. A 1994 amendment to the Atomic Energy Act then made direct disposal of spent fuel a possibility, and the Consensus Agreement now requires transportation for reprocessing to end by mid-2005 at the latest. After that, the disposal of radioactive waste from the operation of nuclear power stations will be limited to direct final disposal.

Waste Classification

Wastes are classified into heat generating and negligible heat generating wastes, according to their likely impact on the temperature of a repository.

Heat generating wastes are spent fuel and the vitrified waste (HLW) resulting from reprocessing.

Wastes with negligible heat generating capacity include power station operating wastes and decommissioning wastes.

Strategies for Radioactive Waste Management in Germany

In addition to the phasing out of nuclear power, the Coalition Agreement includes a range of political aims relevant to radioactive waste management:

- a national plan for the orderly disposal of all kinds of nuclear waste;
- one single, deep, final repository for the disposal of all radioactive wastes;

- a target date for 2030 for the disposal of highly radioactive wastes;
- a pause in the investigations at the Gorleben salt dome to allow other sites to be explored;
- cessation of waste disposal at Morsleben followed by decommissioning; and
- power station wastes to be kept, so far as possible, in onsite interim storage until a disposal route is agreed.

Future Program

The Federal Government has initiated amendments to the Atomic Energy Act to pursue the "one repository" aim and to allow the suitability of further sites in different host formations to be investigated. The Federal Minister of the Environment has established a 15-member panel to recommend a site selection methodology. The current situation at four sites where investigations have previously taken place is described below.

- *Gorleben.* Until 2000, the Gorleben salt dome was under investigation as a potential repository, most notably for high-level waste and spent fuel. Issues were raised, however, in relation to safety related issues such as gas evolution, waste retrievability, and criticality. Under the Consensus Agreement (June 2000), the Federal Government and the utility companies therefore agreed to suspend investigations at Gorleben for between 3 and 10 years pending clarification of these issues. In the meantime, the Federal Environment Ministry will develop procedures and new criteria for a suitability assessment.
- *Konrad.* Konrad is a former iron ore mine at Saltzgitter that was planned to be a final repository for nonheat generating wastes. The licensing procedure for disposal at Konrad is at an advanced stage, having been ongoing with the licensing authority (the Ministry for the Environment of the Federal State of Lower Saxony) since 1983. In July 2000, BfS withdrew its licence application to allow the application to be decided by a court of law.
- *Morsleben.* Prior to German reunification, low and intermediate-level waste with low alpha emitters had been disposed at the disused Bartensleben salt mine near Morsleben in the German Democratic Republic. After reunification, BfS became responsible for the Morsleben repository. Operations were handed over to DBE who, until 1998, continued to dispose of low-level wastes there at depths of around 500 m. In September 1998, the Magdeburg Higher Administrative Court decided that storage of radioactive waste at Morsleben should stop. Emplacement operations are not expected to resume, but the waste that has already been emplaced there will be allowed to remain. A new licence will be required to seal and close the facility which it is estimated will require 6,000,000 m^3 of backfill.
- *Asse mine.* From 1965 to 1978, 141,000 drums of LLW and short-lived ILW were disposed of at the Asse salt mine in Lower Saxony. This mine is now used as a research and development facility (see Figure A1-4).

A1-6. Japan

How Radioactive Waste Management is Organised in Japan

In June 2000, the Japanese parliament approved a new framework for the underground disposal of HLW. This allowed the formation, in October 2000, of a new organisation, NUMO. NUMO is responsible for the identification of a site for HLW disposal and the subsequent construction, operation, maintenance, closure, and postclosure institutional control of a repository. A 15-year period is envisaged during which a fund will be accumulated to pay for the disposal of existing HLW.

Research into the disposal of long-lived intermediate level (TRU) wastes is carried out under the direction of the Japan Nuclear Cycle Development Institute (JNC) and the Federation of Electric Power Companies.

JNFL (Japan Nuclear Fuel Company) owns and operates a facility at Rokkasho-Mura for the disposal of low and intermediate-level waste.

The Origin of the Wastes

Nuclear generated electricity accounted for 34% of all electricity produced in Japan in 2000. Nuclear capacity is planned to increase in the period to 2010.

Most spent fuel is shipped to Europe for reprocessing with the first vitrified waste canisters being returned in 1995. Spent fuel can also be reprocessed at a small plant at Tokai. A larger one is being built at Rokkasho in Aomori Prefecture, where spent fuel is being stored pending the start of reprocessing in 2005.

(a)   (b)

(c)

Figure A1-4. (a) Morsleben disposal vault (photo courtesy of DBE, Germany); (b) Gorleben exploratory mine (photo courtesy of DBE, Germany); (c) A transportation cask undergoing a drop test (photo courtesy of BAM, Berlin, Germany).

Waste Classification

Radioactive waste is classified according to origin, and the type and strength of the radioactivity it emits. In terms of origin, it can be broadly classified into: operational waste generated by the nuclear power plants; fuel cycle waste (including waste returned from overseas reprocessing plants); and other wastes arising from decommissioning, research, and radioisotopes. Fuel cycle waste is further divided into (vitrified) high-level waste, long-lived intermediate-level waste (known as transuranic or TRU waste in Japan and some other countries), and uranium-bearing waste.

Strategies for Radioactive Waste Management in Japan

- *Interim storage of spent fuel and HLW.* Spent fuel is first kept in cooling ponds at the reactor sites before being sent for reprocessing. HLW is held at the Vitrified Waste Storage Centre at Rokkasho operated by JNFL.
- *Operational wastes generated by nuclear power plants.* Low-level wastes are emplaced in JNFL's shallow disposal facility at Rokkasho. Typically, drummed wastes are emplaced in concrete lined trenches. Wastes with higher amounts of gamma and beta emitting radionuclides are currently in interim storage. It has been suggested that a repository at 50–100 meters depth could be constructed for these wastes.

- *Fuel cycle wastes.* Low level wastes from reprocessing and MOX fuel fabrication are disposed of at Rokkasho. Uranium-bearing wastes are currently in storage; most are thought to be suitable for shallow disposal. Wastes with higher uranium content will be treated as TRU waste. TRU wastes will require deep disposal, and a range of concepts is being considered. For repository design purposes, the total volume of conditioned waste is estimated at $56,000 \, m^3$. Deep disposal concepts for HLW are being developed by NUMO. For example, canisters could be placed in a thick carbon steel overpack, surrounded by a clay-based buffer material, and placed within a vertical or horizontal disposal cell leading off from a repository vault. It is estimated that, by 2030, the total number of HLW canisters (each with a volume of 150 liters) will approach 50,000.
- *Other wastes.* Radioactive wastes generated by decommissioning, research reactors, laboratories, and radioisotope facilities are to be disposed of together with nuclear power plant waste, TRU waste, or uranium-bearing waste as appropriate.

Future Program

NUMO intends to commence a stepwise program of site selection and repository development that will lead to the operation of a deep HLW repository no later than the mid-2040s.

In support of this program, underground rock laboratories have been in operation at Tono in Gifu Prefecture (sandstone) and at Kamaishi in Iwate Prefecture (in granite). The latter ceased operation in March 1998, and will be replaced by a new deep facility at Mizunami close to Tono. A facility in sedimentary rocks is also planned at Honorobe in Hokkaido Prefecture (see Figure A1-5).

A1-7. The Netherlands

How Radioactive Waste Management is Organised in The Netherlands

The Central Organisation for Radioactive Waste, COVRA, was established in 1982 to carry out all aspects of radioactive waste management in The Netherlands. COVRA is owned by the owners of the two nuclear power plants (Dodewaard and Borsele), The Netherlands Energy Research Foundation at Petten (ECN), and the State. However, because it has now been decided to phase out nuclear power entirely, COVRA is in the process of being taken into state ownership.

COVRA has developed a system of fees for charging waste producers for the waste that is transferred to COVRA.

All activities relative to the import, transport, use, storage, disposal, and export of radioactive material are subject to the provisions of the Nuclear Energy Act 1963, last revised in 1994. Enforcement of the Act mostly falls to the Ministry of Housing, Spatial Planning, and the Environment.

The Origin of the Wastes

In December 1994, the Government decided that the country's two nuclear power plants should be closed at the end of 2003. Consequently, the Boiling Water Reactor at Dodewaard ceased operation in March 1997. However, the operators of the second plant successfully contested the Government's decision in the courts. Therefore, in 2001, The Netherlands had one operating power reactor, the Pressurised Water Reactor at Borssele, which made a contribution of 4% to national electricity supplies in 2001.

Fuel from both nuclear power plants is reprocessed either in France or the UK. Dismantling of the Dodewaard plant is to be postponed for up to 50 years to allow radioactivity to decay. There are also two operating research reactors in The Netherlands: the JRC in Petten (owned by the European Commission), and IRI at Delft.

Waste Classification

There are three categories of radioactive waste in The Netherlands:

- *Spent fuel and high-level waste.* Spent fuel from the two nuclear power plants is to be reprocessed and will be returned as vitrified high-level waste.
- *Low and intermediate-level waste (L/ILW).* L/ILW originates from Borssele and Dodewaard, from the two research reactors, from a uranium enrichment facility at Almelo, and from minor users.

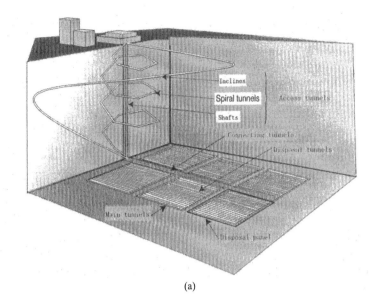

(a)

(b)   (c)

Figure A1-5. (a) Japan HLW Disposal Concept (photo courtesy of JNC); (b) Low Level Waste Disposal at Rokkasho-Mura (photo courtesy of JNFL); (c) Spent Fuel Storage Facility at Rokkasho-Mura (photo courtesy of JNFL).

- *Very low-level waste (VLLW).* Around $1000 \, m^3$ of VLLW are produced annually by the ore processing industry. A dedicated storage building for this waste was constructed in 2000.

Strategies for Radioactive Waste Management in The Netherlands

- *Waste storage.* In 1984, the Dutch Government decided to store all existing wastes and future arisings in one central facility pending final decisions on disposal methods and sites. The capacity of this facility (HABOG) is to be large enough to allow interim storage of all the wastes arising over the next 50 to 100 years. In 1988, a site near to the Borssele plant was selected for the storage facility and construction started in 1990 and was nearing completion in 2001. The facility has capacity for $200,000 \, m^3$ of LLW and ILW, $3000 \, m^3$ of vitrified HLW and 5000 tonnes of spent fuel, of which little is likely to be needed because current policy is for spent fuel to be reprocessed.
- *Waste disposal.* It was decided in 1984 that any disposal option should accommodate all types of radioactive waste in one facility. A long-term research program was established to investigate disposal options. The first to be investigated was disposal in Dutch salt domes. Subsequently, the Government decided that, to comply with sustainable development, any wastes disposed of must be retrievable. Consequently, concepts are now being developed for

retrievable disposal. It is considered that the benefits of this policy outweigh the disadvantages of burdening future generations with the responsibility of deciding when to close the repository. A Commission on Radwaste Disposal (CORA) ran from 1995 to 2000. Its main task was to compare the retrievability of waste for options that included above ground storage, and disposal in clay and salt formations.

Future Program

An issue still to be resolved is the length of time during which waste retrieval would be feasible. Suggestions vary between 25 and 200 years.

A1-8. Spain

How Radioactive Waste Management is Organised in Spain

The National Radioactive Waste Management Company (ENRESA), established in 1984 by Royal Decree, is responsible for radioactive waste management in Spain. Its tasks include the conditioning of some radioactive waste and uranium mine tailings, the identification of disposal sites and their subsequent operation, the decommissioning of nuclear facilities, and the transport of radioactive material.

Waste producers are responsible for financing radioactive waste management. ENRESA supervises the funding arrangements and charges the direct cost of services to small producers while the electricity utilities pay fees based on their electricity sales.

ENRESA is a limited liability company whose shareholders are the Technological, Energy and Environmental Research Centre (CIEMAT), and the National Institute of Industry (INI). The Nuclear Safety Council regulates all parts of the nuclear industry in Spain.

The Origin of the Wastes

In 2000, Spain had nine nuclear power units supplying 28% of its electricity. Spent fuel from currently operating plant is not reprocessed. Fuel from the early gas cooled nuclear power plant, Vandellós I (now being dismantled), was reprocessed.

Spain has significant uranium deposits. Since 1974, a state owned company, ENUSA Industrias Avanzadas SA, has been responsible for uranium prospecting, mining, concentrate production, enrichment, and fuel element manufacture. However, production of uranium concentrates ceased at the end of 2000.

Waste Classification

In Spain, radioactive waste is divided into two categories:

- Low and intermediate-level wastes (L/ILW) contain low radioactivity concentrations, short-lived radionuclides and limited quantities of long-lived alpha emitters.
- High-level waste consists of spent fuel and a limited amount of vitrified waste from the reprocessing of Vandellós fuel.

Strategies for Radioactive Waste Management in Spain

- *Low and intermediate-level waste (L/ILW)*. ENRESA operates a near surface disposal facility for L/ILW at El Cabril, the site of a former uranium mine 100 km from Córdoba. Conditioned waste is placed in preconstructed concrete vaults using a crane. El Cabril has been in operation since 1992, and will be capable of taking 40,000 m^3 of waste. It is estimated that 193,000 m^3 of L/ILW will arise during the lifetime of the current nuclear power plants.
- *High-level waste and spent fuel*. Some 20,000 spent fuel assemblies (equivalent to 6750 tonnes of uranium metal) are expected to accumulate during the lifetime of the current nuclear power plants. In addition, there will be 80 m^3

of vitrified HLW from reprocessing of Vandellós fuel). The 5th General Radioactive Waste Plan, approved by the Government in July 1999 states that no decision on the final solution for spent fuel and HLW will be taken before 2010. A Centralised Interim Storage facility will be available by 2010, to store both spent fuel and other wastes and materials not amenable to disposal at the El Cabril facility. These include HLW and fissile materials returned from reprocessing and other long-lived high-level wastes from different sources, in particular the dismantling of nuclear power plants. Until the Centralised Interim Storage facility is available, spent fuel will continue to be stored at the power plants.

Future Program

By December 2000, 15,400 m^3 of L/ILW had been disposed of at El Cabril. The facility is expected to be capable of taking all the L/ILW generated until 2016.

For HLW and spent fuel, ENRESA has developed plans and carried out generic safety assessments for disposal facilities in three candidate geologies: granite, clay, and salt. However, no siting or further design studies will be carried out in the period leading up to a decision on the long-term waste management strategy in 2010. This period will be used to assess the possible impact of waste retrievability and new technologies such as partitioning and transmutation (see Figure A1-6).

(a)   (b)

(c)

Figure A1-6. (a) Dismantling of Vandellós I NPP; (b) Waste Emplacement Operations at El Cabril; (c) Disposal Vaults at El Cabril (Photographs courtesy of ENRESA).

A1-9. Sweden

How Radioactive Waste Management is Organised in Sweden

The Swedish Nuclear Fuel and Waste Management Company (SKB) is owned by the four Swedish nuclear electricity generation companies. It is financed through a levy on the price of nuclear electricity, which accounts for about 50% of all electricity generated in Sweden. SKB started its research into radioactive waste management in the mid 1970s. SKB is responsible for the handling, transportation, storage, and ultimate disposal of all Swedish radioactive waste.

The Origin of the Wastes

Radioactive waste in Sweden arises mostly from the generation of nuclear electricity. Radioactive waste is also produced by research activities at Studsvik and from the recovery and collection of radioactive materials used for industrial and medical applications.

Sweden does not reprocess its spent nuclear fuel but, instead, classifies this as waste that requires interim storage followed by disposal.

Waste Classification

Radioactive waste in Sweden is divided into three categories:

- spent nuclear fuel (high-level, long-lived),
- operational waste (intermediate and low-level, short-lived), and
- decommissioning waste (intermediate and low-level, short-lived).

Assuming that all the Swedish reactors operate for 25 years, the estimated stockpiles of waste will be:

- $13,000 \, m^3$ spent fuel,
- $2,000 \, m^3$ LILW from Studsvik,
- $10,000 \, m^3$ reactor internals,
- $80,000 \, m^3$ operational waste, and
- $155,000 \, m^3$ decommissioning waste.

Strategies for Radioactive Waste Management in Sweden

- *Operational and decommissioning wastes.* The Swedish Final Repository (SFR) for the final disposal of low and intermediate-level waste is located near the Forsmark nuclear power plant. It is built in the bedrock beneath the Baltic Seabed at a depth of about 60 meters. Two parallel 1-km long tunnels run from the surface down to the repository area, which consists of various rock caverns, designed according to the different activities of the waste. Intermediate-level waste is housed in a concrete silo surrounded by bentonite clay. The SFR could be expanded to accommodate decommissioning wastes, but this would require a new licence from the Government. When SFR has been filled, the entrance tunnels will be sealed with concrete to isolate the caverns and tunnels to prevent future access. There will be no requirement for further monitoring following sealing of the repository.
- *Spent fuel.* Spent nuclear fuel is currently stored underground at the Central Interim Storage Facility for Spent Fuel (CLAB). This is located next to the nuclear power plant at Oskarshamm, and has been in operation since 1985. The intention is that spent fuel will be stored for 3040 years before encapsulation and final disposal to a site that has yet to be chosen. It is planned that spent fuel will be deposited in sealed copper canisters with inner steel containers in a repository about 500 meters underground. The canisters will be surrounded by highly compacted bentonite clay and the tunnels backfilled, thus providing a number of barriers to prevent radionuclides from reaching the biosphere in harmful concentrations. SKB has been conducting research on the disposal of radioactive waste since the mid 1970s. An important part of the spent fuel research program has been the construction of the deep Hard Rock Laboratory ("HRL") at Äspö near Oskarshamm. The HRL enables tests to be carried out at large scale to demonstrate that repository technology works in a realistic environment.

Future Program

Since 1993, feasibility studies have been carried out in eight municipalities to see whether they might be suitable to host a deep repository for spent fuel. The feasibility studies consider social factors as well as infrastructure and geology.

The next stage is to carry out site investigations. These will include a program of deep drilling to obtain more knowledge about the rock at depth. The intention is to conduct these investigations in three Swedish municipalities: Oskarshamm, Östhammer, and Tierp. All these sites lie on the Baltic Sea.

It is expected that these investigations will enable SKB to propose a site for a deep repository around 2008 (see Figure A1-7).

A1-10. Switzerland

How Radioactive Waste Management is Organised in Switzerland

The National Co-operative for the Storage of Radioactive Waste Nagra (German) or Cedra (French) was established in 1972 to take responsibility for all research and development work related to the final disposal of radioactive waste. Nagra is owned jointly by the owners of the nuclear power plants and the Swiss Federal Government, which takes responsibility for all the waste from medicine, industry, and research.

The utilities are responsible for waste conditioning and for interim storage of the wastes. A centralised facility, ZWILAG, has been constructed for the conditioning and storage of intermediate-level waste at Würenlingen. This came into operation in 2001.

The Origin of the Wastes

In 2000, nuclear electricity accounted for about 38% of total Swiss electricity production. There were five reactors at four different sites.

Nuclear power accounts for most of the radioactive waste produced in Switzerland but, as in other countries, waste also arises from industrial, medical, and research applications of radioactive material.

Waste Classification

Switzerland operates a spent nuclear fuel reprocessing policy, with spent fuel being sent to Sellafield or La Hague for this purpose. However, in February 2001, the Federal Government decided that, once the existing contracts had been fulfilled, this practice would cease.

Radioactive waste therefore falls into three broad categories:

- High level waste consisting of vitrified reprocessing waste and spent fuel. It is estimated that the current nuclear program will produce about 700 (150-liter) flasks of HLW along with 2000 tonnes of spent fuel.
- Long-lived intermediate level waste consisting of wastes produced during the reprocessing of spent fuel. This is estimated at $700 \, m^3$.
- Low-level and short-lived intermediate-level waste amounting to $100,000 \, m^3$ over the life of the current nuclear program.

Strategies for Radioactive Waste Management in Switzerland

- *Interim storage.* Spent fuel is first kept for 1–10 years in cooling ponds at the reactor sites. It is then sent for reprocessing in France or the UK. Reprocessing wastes, which have yet to be returned to Switzerland, will be stored at the ZWILAG central waste storage facility. This facility will also provide storage for any spent fuel not sent for reprocessing.
- *Disposal.* Switzerland plans on having two repositories: one for low and intermediate-level wastes, and another for high-level waste, spent fuel and long-lived intermediate-level waste. In both cases, the repository designs are

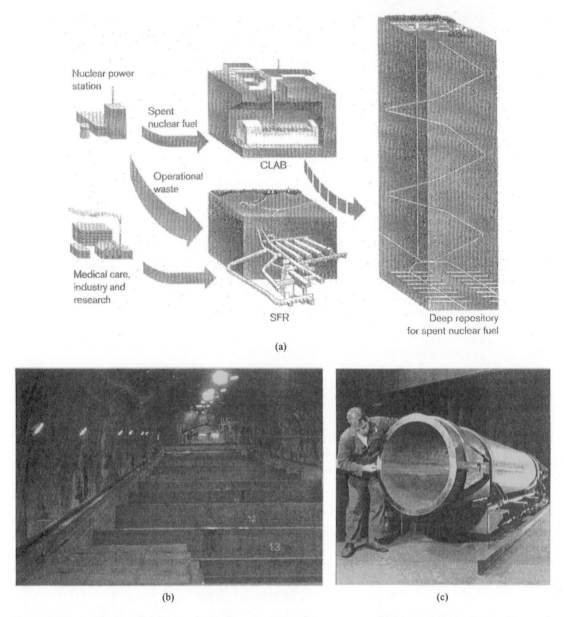

Figure A1-7. (a) The Swedish System for Radioactive Waste Management; (b) Disposal Operations at Forsmark (GU); (c) Copper Canister at SKB's Encapsulation Laboratory. Photos supplied courtesy of SKB, Sweden.

such that a decision on repository closure could be indefinitely postponed, giving future generations the option of whether to retrieve the waste or not.

- *Low and intermediate-level wastes.* Four sites were investigated for a repository for low and intermediate-level wastes. This process resulted in the selection, in June 1993, of a preferred site at Wellenberg in central Switzerland. Federal Government experts confirmed this. In June 1994, the local community voted in favor, but this was rejected by a cantonal referendum in 1995. This impasse requires resolution on a political level. A Wellenberg repository would consist of a mined cavern system in a low permeability sedimentary host rock (marl). This would be accessed from a horizontal tunnel into a valley side.

• *High-level waste.* High-level and long-lived intermediate-level waste (ILW) will be disposed to a deep geological repository. High-level waste and spent fuel (HLW/SF) would be emplaced in tunnels and long-lived ILW in silos or caverns. Two host rocks are under consideration: the crystalline basement and the Opalinus Clay, both in Northern Switzerland. For both rock options, rock laboratories are available in Switzerland: the Grimsel test site in crystalline rock and the Mont Terri rock laboratory project. HLW/SF will be kept in interim storage for at least 40 years to allow radiogenic heat to fall to an acceptable level. The start of repository operations is likely to be around the middle of the century.

A1-11. The United Kingdom

How Radioactive Waste Management is Organised in the UK

Radioactive waste management in the UK is currently subject to a wide ranging Government consultation that is expected to continue to 2007. Existing responsibilities for radioactive waste management are split between a number of organisations.

Disposal facilities for low-level waste are managed by British Nuclear Fuels (BNFL) and the United Kingdom Atomic Energy Authority (UKAEA).

Nirex, a company wholly owned and financed by the main waste producers, is responsible for the disposal of intermediate-level wastes and some low-level wastes that are unsuitable for near-surface disposal. However, given the ongoing review, this disposal option is not being actively progressed.

BNFL and UKAEA also manage the interim storage of vitrified high level waste (HLW). Management of intermediate-level wastes also falls to these two organisations and to the nuclear power utilities.

Spent nuclear fuel is stored at nuclear power plants until it has cooled sufficiently to allow the fuel to be sent for reprocessing.

The Origin of the Wastes

Most UK wastes are historical in nature, reflecting the UKs early involvement in the large scale use of nuclear technology. In particular, the reprocessing of spent nuclear fuel, initially for production of materials for weapons later as part of the nuclear fuel cycle, has produced vitrified high level waste and a wide range of intermediate level wastes.

In 2000, there were 33 nuclear power units generating 24% of the UKs total electricity supply.

Other wastes arise from the medical, industrial, and research uses of radionuclides.

Waste Classification

UK wastes are classified as very low, low, intermediate, and high level wastes (VLLW, LLW, ILW, and HLW, respectively). VLLW and LLW are defined by the concentration and type of radionuclide that they contain. HLW is defined by its heat output and is an end product of reprocessing that consists of vitrified fission products. ILW is defined as anything other than these.

Whether spent fuel and other nuclear materials such as plutonium and uranium should be classified as waste is one of the questions included in the Government consultation.

Strategies for Radioactive Waste Management in the UK

• *Low-level waste.* LLW is disposed of in near-surface, concrete lined trenches at BNFL's Drigg site in Cumbria. LLW produced at Dounreay is similarly disposed of on that site.
• *Interim storage of radioactive waste and spent fuel.* Spent fuel is first kept in cooling ponds at the reactor sites before being sent for reprocessing at Sellafield. The resulting HLW is stored at Sellafield. UKAEA also stores some HLW at Dounreay as a result of reprocessing activities carried out there. Intermediate level wastes are stored at various licensed sites around the country, Sellafield being the most notable.

- *Disposal of ILW and HLW.* HLW was to be stored for 50 years to allow the heat output to decay. It was then to be placed in a deep geological repository. ILW and any LLW not suitable for Drigg were to be disposed to a "Nirex repository."

Future Program

A Nuclear Decommissioning Authority (NDA) is to be established. This will co-ordinate the decommissioning of facilities and the interim storage of radioactive waste.

The Government's consultation paper on radioactive waste envisages a five-stage process with approximate timescales as show:

- Stage 1: Consult on the proposed program — September 01 to March 02;
- Stage 2: Research and public debate to examine the different options and recommend the best option (or combination) — 2002–2004;
- Stage 3: Further consultation seeking public views on the proposed option — 2005;
- Stage 4: Announcement of the chosen option seeking public views on how this should be implemented — 2006; and
- Stage 5: Legislation if needed — 2007.

A1-12. The United States of America

How Radioactive Waste Management is Organised in the USA

Management of all long-lived wastes, both civilian and military, is the responsibility of the US Department of Energy (DOE). The development of a disposal system for spent fuel and high-level radioactive waste is also a matter for the DOE, through the Office of Civilian Radioactive Waste Management (OCRWM), based in Washington DC. OCRWM integrates all aspects of HLW management from transportation through to the construction and operation of a deep repository. It is regulated by other Federal agencies, particularly the Nuclear Regulatory Commission, the Department of Transportation, and the Environmental Protection Agency. OCRWM gets its finance from the Nuclear Waste Fund. The fund collects fees from the electricity generating companies at a rate of 0.1¢ per kWh.

Defence-generated low-level waste is the responsibility of the DOE, but civilian low-level waste rests with the State in which the waste arises. Many States have grouped together to form "compacts," to reduce the number of disposal sites.

The Origin of the Wastes

The USA has a very wide range of radioactive wastes, the consequence of research, development, and exploitation of nuclear technology for military use and power generation since the 1940s. The following map shows the locations of the principal DoE sites. The total volume of radioactive waste being managed by the DOE amounts to 36 million m^3.

Nuclear energy amounts to about 20% of all US electricity production and, although national policy now favors a once-through fuel cycle, significant quantities of heat generating wastes exist from reprocessing of nuclear fuel carried out in the past.

Waste Classification

In the USA, four main categories of waste are defined:

- low-level waste (LLW), where the radioactive content is low and short-lived;
- transuranic waste (TRU), which contains significant quantities of long-lived, alpha emitting isotopes of uranium, neptunium plutonium, etc.;
- high level waste (HLW), which is heat generating waste resulting from reprocessing; and
- spent fuel.

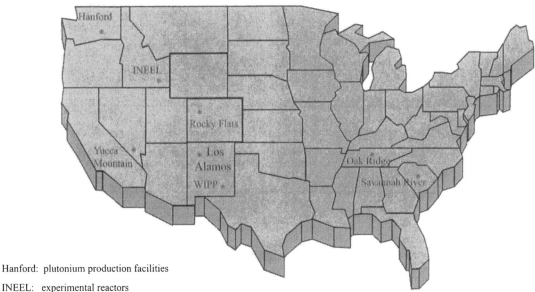

Hanford: plutonium production facilities

INEEL: experimental reactors

Los Alamos: research facilities

Rocky Flats: plutonium metal fabrication

Oak Ridge: uranium enrichment and research

Savannah River: nuclear weapons manufacture

WIPP: transuranic waste disposal site

Yucca Mountain: high level waste and spent fuel disposal site

Location and Function of Principal US DoE Sites

Strategies for Radioactive Waste Management in the USA

- *High-level waste and spent fuel.* The 1982 Nuclear Waste Policy Act gave the go-ahead for the search for a number of sites for a deep repository to take HLW and spent fuel. In 1987, amending legislation directed the DOE to evaluate only the Yucca Mountain site on the Nevada nuclear weapons test site. The technical rationale for the selection of Yucca Mountain in 1987 for detailed characterisation is that at this site, which is an unsaturated tuff rock formation and has minor groundwater movements, there is little risk of groundwater bringing radionuclides to the surface. Extensive site characterisation work has been carried out including the construction of an 8-km long underground Exploratory Studies Facility. The current repository concept for Yucca Mountain contains spent fuel and HLW equivalent to up to 70,000 tonnes of uranium metal. Repository closure could occur anytime between 50 and 300 years after the final waste emplacement.
- *Transuranic waste (TRU).* The strategy for managing military TRU is to dispose of it in a geologic repository built in salt deposits. The Waste Isolation Pilot Plant (WIPP) is a series of chambers carved into salt beds 645 meters underground. It is located about 30 miles east of Carlsbad in New Mexico. After two decades of development, WIPP opened for disposal operations in March 1999. The disposal capacity of WIPP is about 175,000 m^3 and will take 30 years to fill. By September 2001, the WIPP repository had received and emplaced 352 waste shipments (about 2100 m^3) from five DOE sites, filling one room of the repository.
- *Low-level waste (LLW).* At present, shallow land burial repositories for civilian LLW are operating at Barnwell, South Carolina; Clive, Utah; and Hanford in Washington State. Seven other facilities have already been filled in and three further sites are being evaluated. Approximately 1,000,000 m^3 of defence related LLW will require disposal in the period to 2020. The DOE currently operates six surface disposal sites for these wastes.

(a)

(b)

(c)

Figure A1-8. (a) Yucca Mountain Aerial View; (b) Waste Isolation Pilot Plant Aerial View; (c) Waste Emplaced Underground at WIPP. Photos supplied courtesy of US DoE.

Future Program

A Viability Assessment for a HLW/SF repository at Yucca Mountain was published in 1998 followed by a final environmental impact statement in 2001. In January 2002, the Secretary of State for Energy announced his intention to recommend the site to the President as scientifically sound and suitable for development as the nation's long-term geological repository for nuclear waste (see Figure A1-8).

A1-13. Central and Eastern European Countries

Introduction

It is more than 10 years since the events of 1989 led to the liberalisation of the countries of Central and Eastern Europe.

That same period has seen enlargement of the European Union. This is now extending membership to the Applicant Countries of Central and Eastern Europe (CEE). However, before full membership is attained, certain economic and environmental requirements have to be met. In addition, nuclear safety has to be addressed. Implicit in this is the safe management of radioactive waste, including provisioning for future liabilities.

Institutional Arrangements

The reorganisation of the nuclear sector has not necessarily been a priority in the CEE countries. However, the "first round" Applicant Countries have had more incentive to harmonise their legal and institutional systems with those of the EU. Some have already created separate waste management organisations: PURAM in Hungary, Agency RAO in Slovenia, RAWRA in the Czech Republic, RATA in Lithuania, and RAPA in Estonia. The creation of RAWRA was assisted by the European Union's PHARE program, undertaken by the Cassiopee consortium of EU radwaste agencies (see below).

During 2001, Cassiopee provided advice to the Bulgarian Government to help set up a new WMO there.

Other Applicant Countries do not have significant radioactive waste arisings, and so the urgency to establish separate new organisations has not been as great.

EU Assistance

The EU has allocated large sums of money to finance studies and safety improvements in nuclear safety and radioactive waste management. A clear picture of what needs to be done to attain an acceptable level of safety and environmental protection has, thus, been achieved.

Cassiopee was established in February 1993 to assist countries of Eastern Europe in developing radioactive waste management systems within a framework of the European Union's assistance programs PHARE and TACIS. Its membership comprises ENRESA of Spain, Andra of France, DBE of Germany, Nirex of the UK, ONDRAF/NIRSAS of Belgium, and COVRA of The Netherlands.

The creation of the consortium marked an important step forward in international cooperation on radioactive waste management. Building upon existing relationships between the radioactive agencies of the European Union, the consortium provides a vehicle for specialists in Western European countries to combine capabilities and share experiences with their counterparts in Eastern Europe.

The countries of Eastern Europe with nuclear power programs face a challenge in ensuring the safe management of radioactive waste, and it is in the interest of all involved that the West shares its experience in this field. Much effort is devoted to reactor safety issues in the Eastern Countries, but it is vital that similar effort is devoted to waste management if unnecessary problems are to be avoided in the years to come.

One of the first tasks undertaken by the consortium in 1993 was a 1-year long project of major importance to the Eastern European countries. Working under contract to the European Union, teams from the consortium went to Bulgaria, the Czech Republic, Hungary, Lithuania, Poland, Romania, and the Slovak Republic to discover at first hand the radioactive waste management situation in those countries.

A report identifying the issues and priorities was presented to the Commission in June 1994. Since that time, Cassiopee has been asked by the Commission to follow through its earlier work and draw up terms of reference for specific projects. Cassiopee considers that hardware and engineering projects are of value only if they form part of a coherent strategy which takes account of the institutional, financial, and legal aspects of disposal.

Waste Management Situation: Spent Nuclear Fuel Management

- *Power reactors.* The management of spent fuel from nuclear power plants became a crucial issue in many applicant countries following the collapse of "take-back" agreements with the USSR. Such agreements allowed for the return of the spent fuel, with reprocessing wastes remaining in the Soviet Union. All operating VVER and RBMK reactors are affected by problems of spent fuel storage. Most if not all countries operating NPPs have still to decide on their long-term strategy regarding spent fuel, i.e., open versus closed fuel cycle (direct disposal versus reprocessing of spent fuel).
- *Research reactors.* In the case of Soviet-designed research reactors, some countries made regular returns of spent fuel to the Soviet supplier, but these arrangements broke down in the late 1980s. Consequently, there are accumulations of spent fuel at all sites in Bulgaria, the Czech Republic, Hungary, Latvia, Poland, and Romania. In comparison, the TRIGA research reactors in Romania and Slovenia still benefit from agreements with the US, allowing return of spent fuel until May 2006.
- *Treatment and conditioning.* Before 1990, operational waste was simply stored on-site with very little treatment, and all decisions relating to volume reduction, conditioning, long-term storage, and disposal were postponed until the

time of NPP decommissioning. However, new treatment facilities are being commissioned or planned, including a new waste treatment centre at Bohunice in Slovakia, treatment facilities at Kozloduy NPP in Bulgaria, and additional facilities planned at Cernavoda NPP in Romania.

- *Storage and disposal sites.* Only the Czech Republic has a licensed and operating disposal facility for NPP operational waste, though a new facility in Slovakia is currently in the licensing phase. In all other countries with operating NPPs, operational waste is being stored on-site at the power plant. Concerning institutional waste, there are operating repositories in several of the Applicant Countries. Most of these existing disposal sites were constructed in the 1960s or 1970s and were also used for military waste. Many are considered to be of unsuitable construction and contain inappropriate waste packages with unknown radionuclide inventories. Some sites have now been closed with the intention of retrieving and repackaging the waste (e.g., Tammiku in Estonia, Maišiagala in Lithuania), while others have been closed pending upgrading (e.g., Novi Han in Bulgaria). Some are still operating as storage sites pending further safety assessments or the availability of alternative disposal sites (e.g., Baldone in Latvia, Rozan in Poland). At Ignalina NPP in Lithuania, changes in regulations have meant that what was originally intended as a disposal facility can only be used as an interim store for operational and institutional waste. Other disposal facilities are operational, but very close to full capacity (e.g., Püspökszilágy in Hungary), and alternative sites need to be found soon. Finally, some sites are in operation, but upgrading is acknowledged to be necessary (e.g., Baita Bihor in Romania). Siting programs for low and intermediate level waste (LILW) disposal are on-going in some countries, though they suffer from the same problems of public acceptance experienced in the West.
- *Geological disposal.* Only the Czech Republic, Hungary and Slovakia have begun siting investigations for a deep repository, though these are at a very preliminary stage.
- *International assistance — Siting.* There has been international assistance in the field of site selection (e.g., Hungary, Slovenia) and in topics such as safety assessments of existing repositories (Rozan in Poland, Novi Han in Bulgaria, Maišiagala in Lithuania, others planned in the Czech Republic, Hungary, and Latvia).
- *International assistance — Decommissioning.* In several countries, detailed decommissioning plans often do not exist, and they have, until recently, made little or no financial provisions for the decommissioning. Perhaps the first reactors to be affected will be at Kozloduy NPP (Bulgaria) and Ignalina NPP (Lithuania). In both these countries, decommissioning funds have recently been created. International assistance programs have addressed some of the problems at sites such as the Bohunice A1 reactor in Slovakia and the Paldiski nuclear naval training center in Estonia. In the case of the Bohunice A1 reactor, decommissioning waste is likely to be disposed of at the new Mochovce repository. In Estonia, the waste must be stored until a national repository is available. In other countries, decommissioning waste may be accommodated by extensions to existing disposal sites, e.g., Baldone (Latvia) and Baita Bihor (Romania) for waste from decommissioning of research reactors at Salaspils and Magurele, respectively.
- *Spent sealed sources.* There can be little doubt that spent sealed radioactive sources pose a potentially serious threat to public health; of particular concern are the sources that have become "lost" and are no longer under any regulatory control. During the Soviet era, a large number of sources were in use and eventually were disposed of in boreholes at existing repositories. More recently, use of sources in these countries has declined and they also now return sources to the foreign suppliers, where appropriate.
- *Uranium mining and milling.* Such operations have been widespread in many of the applicant countries, though now most have ended for economic reasons. The only countries not effected are Latvia and Lithuania. The legacy is one of open pits, tailing ponds, and low-grade ore or waste heaps — all constituting a health or environmental hazard, either through radon emanation or contamination of water supplies. The worst affected countries are perhaps Bulgaria, the Czech Republic, Estonia, and Romania.

Appendix 2

An Example of a Project Sanction Case — Repacking of Harwell Legacy Intermediate Level Wastes

DTI PANEL
A. THOMAS
J. D. WILKINS

Executive Summary

A number of legacy wastes have been identified as requiring remote handling to bring them to a state acceptable for future storage. Option studies, undertaken to identify the preferred route for repackaging each of the legacy wastes prior to further storage and ultimate disposal, have concluded that B459 provides the most suitable facility for these operations.

When this waste repackaging work is complete, it is expected that B459 will be closed and promptly decommissioned, as there is no further work identified for this facility that will be economically viable.

The project estimate (50%) of the identified waste treatment operations is £6.7 million over 4.5 years commencing in 1999/2000, with completion during 2004; the sanction estimate (90%) is £7.3 million, with a completion date of late 2004. The required funds are as identified in the 1999/2000 Harwell strategy.

	Signature	Date

Prepared by: A. Thomas
Checked by: A. J. Inns
Approved for Issue: R. A. Simpson
Approved for Issue to DTI: J. D. Wilkins

A2-1. Introduction

This appendix presents a proposal to repackage a number of $\beta\gamma$ legacy waste items, currently held in buildings within the B462 complex at Harwell. The waste items have been identified during the Harwell ILW waste study, which is developing plans for the treatment and long-term storage of ILW in a form acceptable to NIREX. Option studies have been undertaken for each item, in order to determine the preferred treatment and packaging option. It should be noted that many of the lower active wastes can be handled in existing facilities in B462 and are not covered by this appendix. These studies have identified B459, from a number of alternative facilities, as the preferred location to size reduce, characterise, and repackage four of the legacy waste streams. A description of B459 is presented in Section A2-8. A summary of the option studies for these four wastes is presented in Section A2-9.

A2-2. Objective

The project objective is to size reduce, characterise, and repackage items of remote handled legacy ILW, such that they can be converted into standard waste streams and stored on site in accordance with modern standards and in a form acceptable to NIREX for ultimate disposal. The legacy wastes are Harwell miscellaneous wastes, FINGAL vessels, Sea Disposal Drums, and RIPPLE waste crates. The wastes are described in Section A2-9. The repackaged ILW will be returned to the B462 complex for long-term storage in the B462.27 Vault Store, until a national waste repository becomes available.

A2-3. Recommendation

It is recommended that a program of ILW size reduction, characterisation, and repackaging is approved for execution in B459 over a period of approximately 4 years at an approval estimate of £6.7 million (sanction estimate £7.3 million).

A2-4. Technical Appraisal of Options

The following key aspects have been taken into account as part of the justification for this work:

- The current waste packaging is in a poor state and may leak its radioactive contents;
- There is a high risk of regulatory action against UKAEA if the situation is not rectified;
- The waste is not suitably characterised for acceptance by Nirex in the present form;
- UKAEA is currently failing to minimise the risks associated with the wastes;
- B459 is the only facility on site at Harwell suitable to treat all wastes;
- Transport of wastes between sites is difficult;
- There is an experienced team currently available to do the work;
- Annual costs for maintaining B459 in operational readiness are high;
- Waste volume reduction will reduce the final waste disposal costs; and
- Waste treatment operations will not significantly increase the decommissioning liability of B459.

The options considered in this business case for repackaging of the identified waste items are discussed below.

A2-4-1. *Option 1: Repackage Wastes Immediately in B459*

The B459 facility and current staff have experience of the RIPPLE crates, FINGAL Flasks, and Harwell miscellaneous wastes from previous operations. The miscellaneous wastes can be handled through the Medium Active cells in B459, while the other waste items are being processed through the High Active cells. Although the facility was built in the 1950s, the current ventilation system was installed in the late 1980s and is, therefore, suitable for the containment of loose contamination as well as remote handling. B459 is operationally ready to implement these works, thereby mitigating regulatory concerns about the unsuitable condition of the stored wastes. B459 can also handle all the identified wastes, unlike any other currently available facility.

A2-4-2. *Option 2: Delay Waste Treatment*

Option 2a: Delay for 2 years while maintaining B459 in operational standby

Since the wastes will not be treated as soon as reasonably practicable and, thereby, increasing the contamination risk, this option incurs possible regulatory action. In addition, annual costs of maintaining the facility in a standby state (approximately £450k yr-1) will be added to the eventual costs of treating the waste, making this option more expensive than option 1 in discounted cost terms.

Option 2b: Delay for 10 years while maintaining B459 in care and maintenance

Since the wastes will not be treated as soon as reasonably practicable, thereby increasing the contamination risk, this option increases the risk of regulatory action. Any delay in using the facility for other than a short campaign of waste handling before closure would require certain plant items, such as the zinc bromide cell windows, to be replaced. In addition, it is unlikely that, after 10 years C&M, approval would be given to operate the facility without bringing it up to modern standards. This option, therefore, incurs the long-term C&M costs as well as significant refurbishment costs on top of the waste treatment costs. Furthermore, facility knowledge and operational expertise would diminish during the 10 year closure, requiring a substantial investment in retraining, with the added difficulty of demonstrating operational competence.

A2-4-3. Option 3: Repackage Wastes Elsewhere and Seek Prompt Decommissioning of B459

Option 3a: Treat waste in an alternative facility on site (B220.29 or B443.26)

B220.29 will only be able to handle a proportion of the NDS wastes and would have difficulty dealing with larger items of waste without major modification. Some of wastes identified for treatment present a significant potential to contaminate the facility. B220.29 is largely uncontaminated, and the introduction of contamination from the variety of radionuclides present in these items would greatly increase the B220.29 decommissioning liability.

B443.26 (operated by Nycomed Amersham (NA) under their own NII licence at Harwell) is being prepared to size reduce their $\beta\gamma$ SDD. NA has offered to size reduce UKAEA $\beta\gamma$ SDD under a liabilities swap arrangement, but have yet to make a formal proposal. A liabilities swap is regarded as highly attractive by UKAEA. There is uncertainty regarding when this facility will be available and it may not be suitable for dismantling drums containing Pu and HEU (present in some UKAEA $\beta\gamma$ SDD). This option is not, therefore, considered available at present. UKAEA will review this option should it become available and select the approach based on optimising safety, environmental, and economic considerations. However, because of the high cost of keeping B459 available, and as the only other credible option capable of handling the full range of wastes, B459 is adopted as the reference strategy at this time.

Option 3b: Treat waste in a new facility at Harwell

Although UKAEA at Harwell are considering the need for a small flexible waste handling facility as part of the long-term site plan, this facility will be primarily designed to handle smaller items of abnormal wastes retrieved from B462.2/.9. In addition, it will provide an active maintenance and inspection facility and emergency clean-up/recovery capability. The specification for this facility will be significantly enhanced if it becomes necessary to process wastes such as the sea disposal drums and RIPPLE crates. Furthermore, this facility will not be available for a number of years, so the case against option 2b is equally valid for this option.

Option 3c: Treat wastes in an alternative facility off-site

The reference assessment has been to carry the work out in A59 at Winfrith. The issues relating to other suitable facilities such as D2001 at Dounreay, B13 at Windscale, and Berkeley caves, have been considered and are similar or less attractive. The main difficulties associated with this option are the technical, cost, and public acceptance issues related to transporting the waste items in their current state and the return of the ILW for storage at Harwell. Some specific issues are discussed below.

- It will be impractical to transport the RIPPLE crates off-site because of their unsuitable packaging. No suitable transport container is currently available, and there is high risk of transferring significant Sr90 contamination to whatever facility is used;
- Transporting the FINGAL Vessels off-site will involve moving them to B459 to package them for transport and size reduction can be carried out for little extra cost;
- The Pu content of some SDD will require type B transport arrangements for all off-site movements and suitable transport containers are not readily available;

- B459 is significantly contaminated from previous operations over its 40-year life and additional contamination from waste treatment operations would not significantly increase the decommissioning liability. For all other facilities considered, the decommissioning liability would be increased; and
- In addition, no off-site facility has been identified that offers better facilities than those of B459.

A2-4-4. *Summary of Technical Issues*

Any delay to waste treatment is not considered acceptable because of the poor state of the current packaging and storage arrangements, which will attract regulatory attention.

Option 3 has no advantages above options 1 and 2. The technical and contractual aspects of option 3a preclude this option from further assessment, although the possibility of using the B443.26 will continue to be explored as part of a liability swap. Option 3b places a greater requirement on the new facility to handle difficult wastes, which present a significant potential for contamination, thereby increasing the requirements of the facility specification and ultimate decommissioning costs. Furthermore, this facility is not currently available, requiring the waste to remain in its current unsatisfactory state for the foreseeable future. Option 3c presents significant transport costs and the associated public scrutiny, as well as increasing the contamination burden of the chosen facility, resulting in higher decommissioning costs. No facility more suitable than B459 has been identified.

Option 3 (a, b, and c) is not, therefore, considered any further in this proposal. A summary of the advantages and disadvantages for each option is presented in Table A2-1.

A2-4-5. *Financial Appraisal of Options*

Costs for each of the B459 options are summarised in Table A2-2. A breakdown of the costs of option 1 are presented in Table A2-3. The costs of maintaining B459 in an operational standby or C&M state, plus the cost of bringing the facility back-up to operational status, significantly increases the cost of both delay options. More detailed analyses are contained in Figures A2-1–A2-5.

Since this work is technically understood, the major risk relates to extended operations. The sanction estimate (90%) for option 1 has assumed a 1-year extension of operations, compared to the base estimate; the project estimate (50%) reflects a 5-month extension.

The figures in Table A2-2 for delayed action show that the effect of extending waste treatment operations in B459 is to incur the ongoing costs of operational standby or care and maintenance, which outweigh the small reduction in annual cost achieved by discounting at 6%.

The operational costs shown in Table A2-3 are divided into:

- fixed charges, which relate to maintaining B459 in an operational condition, e.g., limited team, health physics, maintenance, facility services; and
- project costs, which relate to variable charges associated with performing the waste treatment work; they are primarily labor costs.

A2-4-6. *Sensitivity*

The recommendation is not sensitive to significant inaccuracies in estimates of ±25%. This is due to dominant transport costs for the off-site option and the high ongoing costs of maintaining B459 in an operational standby state.

A2-5. Implementation

A2-5-1. *Proposal*

The project aims are to size reduce and repack a number of remote handled legacy wastes such that they may be disposed into established waste storage and disposal routes. All the tasks described below will be undertaken in B459.

Table A2-1. Summary Review of Options

Option	Issues
(1) Undertake size reduction and repacking operations immediately in B459	*Advantages* • Early resolution of unsatisfactory waste arrangements • Only facility at Harwell suitable for the work • Facility currently available • No off-site transport of ILW required • No need for new transport packaging • Meets Regulator expectation • POCO in parallel with operations, marginal cost • Continuity of operation and staff *Disadvantages* • Continued use of B459 • Only tasks in B459
(2a) Delay action for 2 years	*Advantage* • Short-term saving in management and operation *Disadvantages* • Wastes remain in inappropriate storage and packaging • Possibility of waste packages leaking. • Costs of operational standby little less than operation • B459 deteriorating increased possibility of major cost • Break in continuity of operation • Threat of regulatory action
(2b) Delay action for 10 years	*Advantage* • Saving in staff costs for C&M period *Disadvantages* • Wastes remain in inappropriate storage and packaging • Possibility of waste packages leaking • Need to bring B459 to "modern standards" • B459 deteriorating • Loss of experienced management and operators • Threat of regulatory action
(3a) Decommission B459 immediately and repackage waste in an on-site facility	*Disadvantages* • B220.29 is configured to accept only a proportion of NDS wastes • No on-site facility can handle all wastes • Increased decommissioning liability of B220 • B443.26 is not demonstrated as a practical option (SDD only)
(3b) Decommission B459 immediately and repackage waste in a new facility at Harwell	*Advantage* • Close B459 and decommission (annual costs similar to operations) *Disadvantages* • Wastes remain in inappropriate storage and packaging • Possibility of waste packages leaking • Threat of regulatory action • B459 deteriorating • Loss of experienced management and operators • Significantly enhanced requirement for future facility • Contamination of new facility highly likely

Continued

Table A2-1. Continued

Option	Issues
(3c) Decommission B459 immediately and repackage waste immediately in an off-site facility	*Advantage* • Close B459 and decommission (annual costs similar to operations) *Disadvantages* • Off-site transport of poorly packaged waste items • Return transport of ILW • RIPPLE repackaging will result in facility contamination • Requires the use of B459 to repackage RIPPLE and FINGAL for transport

For the purposes of this summary, the alternative off-site facility comparison is based on A59 at Winfrith. Cells at Berkeley and Windscale have also been considered.

• *Facility Commissioning and Operational Readiness.* This will include agreeing a commercial framework for operation of the facility and preliminary commissioning activities. In addition, method statements will be prepared and any outstanding safety documentation completed.

• *Harwell Miscellaneous Wastes Including NDS Waste.* This work is concerned with the size reduction of sources from various applications across the country. The work involves removing outer packaging and placing into ILW cans in accordance with B462 acceptance criteria. The remaining shielding will be despatched as LLW or to landfill.

1.00 Years from 2000	0	1	2	3	4	
FIXED COSTS - Fully Operational						
Total Full Operational Fixed Costs	90	813	813	813	745	3274
VARIABLE COSTS - Fully Operational						
Total Project Operations Costs	125	395	395	385	450	1750
Specific Harwell Miscellaneous Wastes Costs	47	42	17	17	25	147
Specific FINGAL Vessel Costs	46	23	16	0	0	85
Specific Sea Disposal Drums Costs	148	186	52	79	10	475
Specific RIPPLE Crates Costs	10	41	20	45	120	236
Total Variable ILW Repackaging Costs	376	687	500	526	605	2693
TOTAL ILW REPACKAGING COSTS	466	1500	1313	1339	1350	5967
Discounted @ 6%	*466*	*1414*	*1168*	*1124*	*1069*	
Cumulative Discount	*466*	*1880*	*3048*	*4172*	*5242*	

SUMMARY	
Total fixed cost	3274
Total variable cost	2693
TOTAL COSTS UNDISCOUNTED	5967
TOTAL COSTS DISCOUNTED @ 6%	5242

Figure A2-1. Option 1: Undertake Size Reduction and Repacking Operations in B459, to be Followed by Prompt Decommissioning.

1.00 Years from 2000	0	1	2	3	4	5	
FIXED COSTS - Fully Operational							
Total Full Operational Fixed Costs	90	813	813	813	745	366	3640
VARIABLE COSTS - Fully Operational							
Total Project Operations Costs	125	470	395	385	385	254	2014
Specific Harwell Miscellaneous Wastes Costs	47	42	17	17	17	17	156
Specific FINGAL Vessel Costs	43	15	11	16	0	0	42
Specific Sea Disposal Drums Costs	148	286	52	79	6	6	577
Specific RIPPLE Crates Costs	10	20	51	45	71	121	318
Total Variable ILW Repackaging Costs	373	833	526	542	479	398	3149
TOTAL ILW REPACKAGING COSTS	463	1646	1339	1355	1224	764	6789
Discounted @ 6%	463	1552	1191	1138	969	570	
Cumlative Discount	463	2015	3206	4344	5313	5884	

SUMMARY	
Total fixed cost	3640
Total variable cost	3149
TOTAL COSTS UNDISCOUNTED	6789
TOTAL COSTS DISCOUNTED @ 6%	5884

Figure A2-2. Option 1: Undertake Size Reduction and Repacking Operations in B459, to be Followed by Prompt Decommissioning. Assumes Project Overrun by 5.4 months (50% estimate).

1.00 Years from 2000	0	1	2	3	4	5	
FIXED COSTS - Fully Operational							
Total Full Operational Fixed Costs	90	813	813	813	813	723	4065
VARIABLE COSTS - Fully Operational							
Total Project Operations Costs	125	375	395	385	385	460	2125
Specific Harwell Miscellaneous Wastes Costs	47	47	17	17	17	17	161
Specific FINGAL Vessel Costs	43	12	11	16	0	0	82
Specific Sea Disposal Drums Costs	148	286	52	79	6	6	577
Specific RIPPLE Crates Costs	10	5	61	45	71	121	313
Total Variable ILW Repackaging Costs	373	725	536	542	479	604	3258
TOTAL ILW REPACKAGING COSTS	463	1538	1349	1355	1292	1327	7323
Discounted @ 6%	463	1450	1200	1138	1023	991	
Cumlative Discount	463	1913	3113	4251	5274	6265	

SUMMARY	
Total fixed cost	4065
Total variable cost	3258
TOTAL COSTS UNDISCOUNTED	7323
TOTAL COSTS DISCOUNTED @ 6%	6265

Figure A2-3. Option 1: Undertake Size Reduction and Repacking Operations in B459, to be Followed by Prompt Decommissioning. Assumes Project Overrun by 1 year.

1.00 Years from 2000	NPV				Standby 1	Standby 2	Waste Ops 3	4	5	6	Total		
FIXED COSTS - Operational Standby													
Total Operational Standby Fixed Costs					435	435	813	813	813	813	4122		
VARIABLE COSTS - Fully Operational	NPV												
Total Project Operations Costs	500	395	385	460			500	395	385	460	1740		
Specific Harwell Miscellaneous Wastes Costs	84	17	17	30			84	17	17	30	147		
Specific FINGAL Vessel Costs	55	11	16	0			55	8	8	0	71		
Specific Sea Disposal Drums Costs	334	52	79	10			334	52	79	10	475		
Specific RIPPLE Crates Costs	10	61	45	120			10	61	45	120	236		
Total Variable ILW Repackaging Costs	983	536	542	620			983	533	534	620	2669		
TOTAL ILW REPACKAGING COSTS					435	435	1796	1346	1347	1433	6791	870	5921
Discounted @ 6%					435	410	1598	1130	1066	1070			
Cumulative Discount					435	845	2443	3573	4640	5710			

SUMMARY

Total fixed cost	4122
Total variable cost	2669
TOTAL COSTS UNDISCOUNTED	6791
TOTAL COSTS DISCOUNTED @ 6%	5710

Figure A2-4. Option 2a: Delay Action for 2 years.

1.00 Years from 2000	Operational Standby 1	2	3	4	5	6	7	8	9	10	Waste Ops 11	12	13	14	Total		
FIXED COSTS - Operational Standby																	
Total Operational Standby Fixed Costs	435	435	435	435	435	435	435	435	435	435	813	813	813	813	7602		
VARIABLE COSTS - Fully Operational							NPV										
Total Project Operations Costs							500	395	385	460	500	395	385	460	1740		
Specific Harwell Miscellaneous Wastes Costs							84	17	17	30	84	17	17	30	147		
Specific FINGAL Vessel Costs							55	11	16	0	55	8	8	0	71		
Specific Sea Disposal Drums Costs							334	52	79	10	334	52	79	10	475		
Specific RIPPLE Crates Costs							10	61	45	120	10	61	45	120	236		
Total Variable ILW Repackaging Costs							1055	971	977	1055	983	533	534	620	2669		
TOTAL ILW REPACKAGING COSTS	435	435	435	435	435	435	435	435	435	435	1796	1346	1347	1433	10271	4350	5921
Discounted @ 6%	435	410	387	365	345	325	307	289	273	258	1002	709	669	672			
Cumulative Discount	435	845	1232	1598	1942	2267	2574	2863	3136	3393	4395	5104	5774	6446			

SUMMARY

Total fixed cost	7602
Total variable cost	2669
TOTAL COSTS UNDISCOUNTED	10271
TOTAL COSTS DISCOUNTED @ 6%	6446

Figure A2-5. Option 2b: Delay Action for 10 years.

• *FINGAL vessel.* The FINGAL vessels will be posted directly into cells from the flask. The flask will then be taken away for inspection and maintenance either by HRS or B459 operators. The vessel may then be size reduced by cutting with a diamond abrasive wheel. The cut sections would then be packed into standard ILW cans and returned

Table A2-2. Summary Costs of Options

Option	Base Estimate		Project Est. £k @ 50% Conf. Undisc'd	Sanction Est. £k @ 90% Conf. Undisc'd
	Undisc'd	Disc'd @ 6%		
(1) Undertake size reduction and repacking operations, immediately followed by decommissioning of B459.	5,967	5,242	6,789	7,323
(2a) Delay 2 years	6,791	5,710	7,824	8,354
(2b) Delay 10 years	10,271	6,446	12,174	12,704

Table A2-3. Costs Breakdown for Option 1 Project Estimate

		99/00 £k	00/01 £k	01/02 £k	02/03 £k	03/04 £k	04/05 £k
Operational Costs							
Fixed costs		90	813	813	813	745	366
Project costs		373	833	526	542	479	398
	Total	463	1646	1339	1355	1224	764
Discounted at 6%		463	1552	1191	1138	969	570
Cumulative discounted figure		463	2015	3206	4344	5313	5884

for storage in B462.27. Flask handling operations will determine if subsequent vessels can be retrieved from B462.9 and size reduced in B459.

- *Sea Disposal Drums.* There are approximately 150 drums identified for size reduction and repacking that require the shielding of B459 cells for the operation. Drums will be pretreated by radial drilling and the use of expanding grout, or by sawing through the concrete jacket. These operations will be undertaken in the Modular Containment System (MCS) before the drum is transferred to the cell for opening and repacking. All drum contents that cannot be classified as LLW will be transferred to standard ILW cans and returned to B462.27 for long-term storage.
- *RIPPLE Waste Crates.* The crates will be delivered to the facility in a half-length ISO container and transferred into the MCS. The crates will then be assayed before opening. Items of waste that emit a high radioactive dose will be transferred to the adjacent remote handling cell for decontamination and/or size reduction. Decontamination operations will be undertaken in the MCS, and waste will be disposed into ILW(R), ILW(C), and LLW. All operations within the MCS will require the use of RPE and are likely to be pressurised suit operations in the early stages of the campaign. This waste will be dealt with last, as it has the greatest potential to contaminate the facility.
- *Decommissioning.* Once waste processing operations are complete, all equipment used in the campaign will be decommissioned and disposed to the established disposal route. This is regarded as an integral part of the project and the costs are included in the estimate. A Gantt chart outlining all the above operations is presented as Figure A2-6.

A2-5-2. *Deliverables*

The deliverables shown in Table A2-4 are needed to secure the final objective of this sanction paper.

A2-5-3. *Risk Management*

The most significant risk items associated with the proposal are presented in Table A2-5, along with the measures established to minimise the risk impact. The software program @RISK was used to model the time-based risks associated with the project. Histograms showing the probability of a certain task being complete were produced. The program risk dates shown in Figure A2-6 show a potential (at 90% probability) for a 1-year over-run. Since the fixed operational (staff) costs represent the majority of the project costs, the 90% probability cost is determined by the cost of operating the facility during the 1-year over-run.

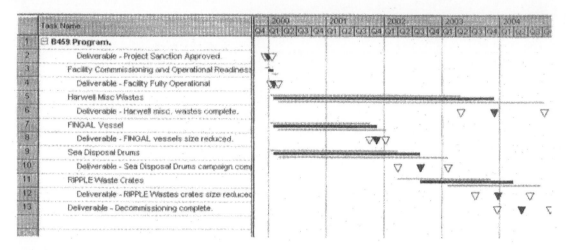

Figure A2-6. B459 Size Reduction and Repackaging Program — Summary Timescales at 10, 50, and 90% probability values.

Table A2-4. Project Deliverables and Target Dates

Task	50% Date	90% Date
Facility fully operational	02/02/00	29/02/00
Harwell miscellaneous wastes complete	25/11/03	08/10/04
FINGAL vessels size reduced	09/11/01	04/01/02
Sea Disposal Drums campaign complete	15/08/02	24/01/03
RIPPLE Wastes crates size reduced	24/12/03	07/06/04
Removal of waste treatment equipment complete	20/05/04	17/11/04

These projected timescales have been determined from the evaluation of project risks, and may be compared with the base estimate given in Figure A2-7.

A2-5-4. *Contract Strategy*

A review of the contract and commercial strategy has been carried out with Harwell Contracts Department. It is proposed that the operation will be carried out under two contracts.

(1) Preliminary work (starting around mid-2000 for 6 months) will be placed as a stand alone contract with NNC and will make use of labor which will become available from B393.6. This will be separate from the contract which provides for care and maintenance of B459 and decommissioning of B393.6 by NNC.
(2) The main operational contract will be competitively tendered and be in place from 2001. This contract will cover the provision of facility operators and supervisory support. A review has been carried out of the TUPE implications and it has been concluded that TUPE will not be applicable. The current Managing Agency (MA) contract will terminate in May 2000, but the direct employment of key members of the MA team has already strengthened the UKAEA management team, in readiness for controlling all future B459 operations.

A2-5-5. *Safety Management*

Contract staff, under the direct control and supervision of the UKAEA ATO holder, Mr T. Chambers and the UKAEA management team, will undertake the work. Four of this team of six staff already have experience of managing work within the facility. The UKAEA team will plan and design the work arrangements and direct operations. The work will be within the operational envelope specified within the revised Category 1 safety case, for continued operation of the facility, currently being peer-reviewed. It is expected that this safety case will be approved through UKAEA

Table A2-5. Project Risks

Risk	Risk Limiting Arrangement
(1) Unable to implement suitable safety and commercial arrangements (facility commissioning).	Arrangements are currently in place. However, the existing NNC operators contract is available for renewal mid 2001. The new contract will be used to establish satisfactory arrangements for the duration of this work.
(2) Detailed information on source construction not available. (Harwell miscellaneous wastes including NDS wastes.)	Where source construction details are unavailable, size reduction will proceed based on engineering judgment and an agreed local work instruction. Source size reduction will be abandoned if there is a possibility of breaching primary containment and the source will be packed for disposal in its container.
(3) Source corroded or sealed containment breached. (Harwell miscellaneous wastes including NDS wastes.)	Local containment and ventilation will be employed when the potential for a breach of primary containment is identified. Cell clean-up operations will be implemented should a breach occur.
(4) FINGAL vessel sticks in flask or posting port.	Detailed engineering drawings are being prepared for all flasking operations. Should the vessel stick, it will be withdrawn back into the flask. Suitable clearance will be established between vessel and flask wall.
(5) Unable to retrieve remaining FINGAL vessels.	Failure will not adversely affect the rest of the waste treatment program. Detailed engineering drawings will be prepared to minimise the risk, and trial operations will be undertaken once the current vessel is posted into B459 cell.
(6) Unable to detect base of inner drum (SDD).	Grout drilling will be undertaken in the ventilated modular containment. Breaking onto the inner drum will be undertaken in cells to manage dose and to ensure containment. Main risk is delay to program.
(7) Drum construction may not be in accordance with the specification (SDD).	Reinforcing bars may protrude below the base of the inner drum. If this occurs, these will be cut through with the drum in cell.
(8) Grout leaked into inner drum cementing contents (SDD).	Chisel attachments will be added to the master slave manipulators if necessary to remove waste that was accidentally grouted by poor manufacture.
(9) Waste drum contaminated externally (SDD).	Where it is necessary to place the LLW drum into the cell for loading, it will be wrapped in PVC to prevent external contamination.
(10) High operator doses (SDD and RIPPLE crates).	Where high operator doses are possible, remote handling methods and shielding will be employed. All such operations will be subject to Local Work Instructions.
(11) Commissioning problems associated with installed equipment and MCS requiring modifications.	B459 active workshop can undertake most engineering modifications. Ventilation system balancing may be required to ensure the correct air flow through the MCS, but tasks such as this can be undertaken by facility personnel.
(12) Crate containment breached (RIPPLE crates).	Emergency clean-up will be implemented. This procedure will be embodied into Local Work Instruction for handling the crates. Crates will not be handled without PVC secondary containment, will be constantly monitored for loose contamination, and will be handled in suitable ventilated areas.
(13) Possible spread of contamination and handling difficulties associated with RIPPLE crates.	RIPPLE crates will be double-wrapped in PVC and held in a half length ISO container. Once removed from the container, they will be handled using specially designed equipment in a fully ventilated facility. All lifting operations will be approved by the Harwell Appointed Person (lifting).
(14) Handling equipment failure.	All lifting equipment will be certified. All lifting will be undertaken by qualified slingers and approved by the Harwell Appointed Person (lifting).

and NII safety review processes in time for the work to start. However, the current safety case also encompasses the operational envelope of the work, and if necessary early tasks can be included through low category modification. Specific work on the Ripple Crates and the Sea Disposal Drums will require minor modifications to the safety case which are expected to be categorised no higher than C. These modifications will be reviewed by the Harwell Operations Safety Working Party before implementation. Control of contractors will comply with UKAEA/P/S310 requirements. All work carried out in the facilities will be covered by the facility Operating Instructions, Method Statements, and a Permit to Work system and approved by the ATO holder or his SQEP nominated representative. The ATO Holder or one of his Project Supervisors will supervise all tasks. Authority to Proceed (ATP) will only be issued for each task when satisfactory method statements have been received, reviewed, and are acceptable. Because of the potential high profile of this work, an experienced safety co-ordinator has been included in the UKAEA management team to review risk assessments and method statements before they are presented for approval.

Table A2-6. Waste Arisings

Waste	Original* ILW (m³)	Repacked ILW (m³)	Repacked LLW (m³)	Land fill
Harwell Miscellaneous Sources	8	0.5	1	8
FINGAL Vessels	2	2	0	0
Sea Disposal Drums	103	5	83	10
RIPPLE Waste Crates	7	2	3	0
Secondary Arisings	0	1	3	0
Total	120	10.5	90	18

*All waste items are currently regarded as ILW.

A2-5-6. *Waste Management and Environmental*

It is envisaged that the majority of radioactive waste arising from this project will be LLW that will be packaged into 200 l drums and dispatched to Drigg for disposal by HRS. LLW are expected to comprise waste packaging, ILW that has decayed to LLW, ILW that was decontaminated, and secondary arisings. Waste sentenced as ILW will be mainly remote handleable and will, therefore, be compacted, canned, and dispatched to B462.27 Head End Cell, where it will be assayed and transferred into a 500 liter drum for long-term storage. Provisional waste estimates are as shown in Table A2-6.

The waste volumes in Table A2-6 show the ILW volume reduction anticipated. Current estimates of long-term storage, cementation, and disposal costs (including a share of fixed costs) for 1 m³ of ILW are £570k. Size reduction, therefore, yields a substantial saving on waste disposal costs. In addition, the waste will be configured in a manner that will ensure its long-term isolation from the environment.

A2-5-7. *Project Management*

The Project Manager will be Mr T. Chambers, who has been involved in the management of the facility for 16 years. He will lead a UKAEA management team, as shown in Figure A2-7, and will be responsible for planning tasks and controlling all safety aspects of the work in the facility. Mr Chambers will report to the Harwell Projects Executive (HPE). Progress will be monitored (quarterly) by the Harwell Projects Executive through a Project Management Committee involving representatives from Harwell Radwaste Services, Contracts Department, and other appropriate UKAEA representatives. An external member will be chosen to provide an impartial view to the HPE. The project will also be monitored monthly through the Harwell Projects Progress Review, chaired by the HPE. Reporting of progress and financial performance will be carried out monthly, through the Harwell and SD projects reporting system.

A2-5-8. *Costs, Fundings, and Resources*

The annual project costs (project estimate) are shown in Table A2-7. Costs will be charged to the SAFER program, except for a proportion of the NDS wastes which were received between 1986–1994. The cost for repackaging some NDS wastes falls to the Waste Fund, because they were acquired as a liability through the previously operating "National Disposal Service" for radioactive materials and prepaid as part of the disposal agreement. The project costs are within the (1999) Harwell Strategy and Plans under SAFER PL 51310 (PIE Facilities).

Resources to manage the project are available within the planned continuation of the current B459/B393.6 decommissioning management team. An outline program of activities is presented in Figure A2-6. This program integrates with the Harwell site strategy and the HRS program of work.

A2-5-9. *Priority of Project*

In order to understand the relative priority of this project within the overall decommissioning program a project priority score is assigned in accordance with the principles described in Chapter 14. The project priority on this basis is 6.1.

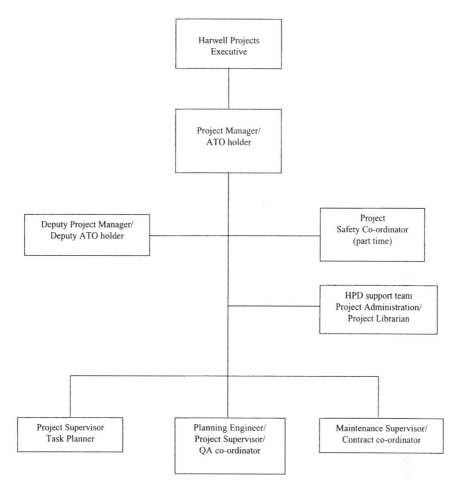

Figure A2-7. Planned Organisation for B459 Management Team.

Table A2-7. Spend Profile

	1999/2000 £k	2000/2001 £k	2001/2002 £k	2002/2003 £k	2003/2004 £k	2004/2005 £k
Waste Fund contribution*	87.75	190.5	178.3	155.5	236	98.1
SAFER Contribution	375.25	1455.5	1195.7	1271.5	1069	732.9
Total Project Cost	463	1646	1374	1427	1305	831

*85%, by volume, of the Harwell miscellaneous wastes are wastes originating from the NDS. Funding for the treatment of these wastes is, therefore, taken from the Waste Fund. The figure is calculated by adding a pro rata contribution from the facility fixed and operational costs to the NDS waste treatment costs.

A2-5-10. *Control of Contingencies*

The HPE, in consultation with the Harwell Finance Manager, the Harwell Planning Manager, and the PPED representative, will control variations in expenditure within the Harwell Projects Department budget; waste costs are specifically excluded. If significant additional funds are required, either within 1 year, or for the overall project, the Head of Site Harwell and SD Director will be consulted.

A2-6. Public Relations

Execution of this project provides tangible evidence to the public (and regulators) that UKAEA are actively pursuing a policy of clean-up and remediation in line with the UKAEA stated mission. The public relations aspect to this project is, therefore, highly positive. The potential adverse publicity, which would arise from public awareness of continued poor ILW storage, will be avoided.

A2-7. Conclusions

- The Harwell Miscellaneous wastes; RIPPLE Crates; FINGAL vessels; and the Sea Disposal Drums are currently packaged in a manner unsuitable for long-term storage or disposal. Continuing storage may result in these wastes leaking from their current packaging, contaminating the unventilated storage facility and the environment. UKAEA face the threat of regulatory action if these wastes are allowed to remain in their current unsatisfactory storage arrangements.
- B459 is the only facility on site that can treat all the wastes identified. Public concern, cost, and risk associated with off-site transport of this waste precludes its treatment in any off-site facility.
- Option studies presented in Sections A2-8 and A2-9, identify size reduction and repackaging in B459 as the preferred option on safety, economic, and environmental grounds. Delaying this work results in increased costs and risk. Completion of the work will allow decommissioning of B459.
- The project estimate for these waste processing operations is £7.05 million, and the work will be undertaken during an approximate 4-year period. It is recommended that £7.7 million is sanctioned to enable the legacy waste processing at Harwell to be undertaken. It is anticipated this work will be complete (at 90% probability) by December 2004.

A2-8. Description of B459

The facility basically consists of two lines of five Hot Cells (High Active and Medium Active Lines) served by a common Activity Maintenance Area (AMA) which accesses the rear faces of both cell lines. Large shielded windows and Master Slave Manipulators (MSMs) are provided on the working faces of the cells, and power manipulators run through both cell lines to aid heavy work tasks. Both cell lines have transfer bays and maintenance areas at each end to support work within the cell lines.

The primary objective of B459 was to support the experimental irradiation programs being performed in the various Harwell Material Testing Reactors (MTRs). This led to additional cells being added to the MA line and the establishment of a metallurgical wing; 459.4. Further extension and refurbishment work was undertaken in the mid 1980s, during which two steel shielded cells, having internal containment boxes, were added to the MA line. In addition, the HA line North Transfer Bay was refurbished and the active ventilation system upgraded requiring the construction of a new plant room; 459.11.

B459 is currently in operational standby with a fully operational safety case in place. For the size reduction operations planned, it will be necessary to construct a Modular Containment System (MCS) in the area adjacent to the high active cell transfer bay.

A2-8-1. *Evaluation of Options*

One of the key elements of the Dounreay Audit (Recommendations R68 and R69) is to complete a detailed inventory of waste and develop a strategic plan for its handling, treatment, and storage. The work to establish an inventory and plans for Intermediate Level Waste (ILW) at Harwell predates the audit. However, the work has been given a fresh focus by the NII recommendations. The summaries below address the specific waste items proposed for early repackaging in B459 as part of the proposed program. The data is a summary of the information in (and supporting) the Harwell ILW database.

A2-8-2. Harwell Miscellaneous Wastes — Including NDS Wastes

Description

The waste is diverse and comprises several thousand items, which include sealed sources, laboratory waste, uranium salts, and solvents. The waste is held in a variety of packages including mild steel drums, source containers, instruments, and cardboard boxes. Approximately 15% (by volume) of this waste will require size reduction in a remote handling facility. Desktop characterisation of this waste has been undertaken [1], but a lengthy sorting process will be necessary to identify many of the items.

The UKAEA undertook disposal of miscellaneous radioactive wastes nationally (National Disposal Service), to ensure wastes arising from medical and educational establishments as well as industry and commerce were disposed in a suitable manner. UKAEA ceased administration of the NDS in 1983. Of the total waste, approximately 75% by volume is NDS waste (99% of items).

Current State

The packaging varies in quality and construction standards and is not subjected to detailed periodic inspection. Concern has been expressed about the possibility of radioactive leakage from such a diverse range of packaging. A number of the larger items requiring remote-handling show advanced states of corrosion and, should a failure of the current primary containment occur, it would:

• Contaminate the unventilated general waste store presenting a significant potential for environmental and, therefore, off-site release; and
• Make waste retrieval and immobilisation a more hazardous and costly task, while producing an increased volume of waste for disposal.

Shortfall Against Modern Standards

The present packaging cannot be considered suitable for further long-term storage of the waste, and the majority of the present packaging does not conform to any QA or transport standards. Its continued use does not meet the requirement to keep the risk of a contamination incident ALARP.

Option Review

A number of options were considered for the treatment for disposal of these wastes, and the advantages and disadvantages are presented in Table A2-8. After detailed internal review, option 1 (process through B459 immediately) was identified as the preferred option on safety, technical, and financial grounds. Options 2 and 3 present the hazard of waste leaking during the delay period and do not accord with the NII preference for treating waste early. Option 4 is discounted on the grounds of excessive transport costs and the delay it will impose to the decommissioning program for A59. Option 5 is not technically feasible for all wastes in this stream, although some smaller sources may be treated through this route in the longer term. Option 6 is discounted since Nirex will not accept large quantities of shielding into the repository.

A2-8-3. FINGAL Vessels

Description

This waste comprises eight stainless steel vessels containing vitrified active liquor. One vessel is held in its original transport flask in B462.20; of the other seven vessels, five are known to be in B462.9. The remaining two are believed to be in B462.9, but the precise storage location is uncertain. The FINGAL process vessel is cylindrical, 150 mm in diameter and 1.49 m long, with a wall thickness of 6 mm. The upper part of the vessel is attached to the vessel head, a spacer, and a shield plug. The whole assembly is 2.36 m long. The depth of glass at the base of the vessels ranges from

Table A2-8. Review of Options for Treating Harwell Miscellaneous Wastes — Including NDS Wastes

Option 1: Process Through B459 Immediately	*Advantages* • B459 has some proven history on this task. • Waste will be size reduced suitably for long-term storage and final disposal. • Facility will be operational for other wastes so small marginal cost. *Disadvantage* • Delays decommissioning.
Option 2: Process Through B459 in 5 Years	*Advantage* • Evokes the benefit in discounting the costs at 6%. *Disadvantages* • Care and maintenance of the facility will be required over the 5 year period. • Some refurbishment will be required over the 5 year period. • Waste may leak out of current packaging during the 5 year period. • Large facility start-up costs will be incurred. • Does not accord with NII preference for early treatment of waste.
Option 3: Process Through New Facility	*Advantage* • Uses a purpose built facility. *Disadvantages* • No new facility currently available. • Waste may leak out of current packaging during design and build period. • NII pressure dictates that wastes are treated sooner rather than later. • Does not accord with NII preference for early treatment of waste.
Option 4: Process Through A59	*Disadvantages* • Negative PR for moving waste to a facility in Dorset when one exists at Harwell. • Type B transport container will be required. • Delays decommissioning of A59.
Option 5: Process Through B220	*Advantages* • B220 has some experience with smaller sources. • Purpose built facility for the smaller sources. *Disadvantages* • B220.29 not suitable for all wastes without significant modification. • AEAT will not provide a firm price for undertaking the work because of high technical risk. • UKAEA do not have direct control waste treatment over operations.
Option 6: Direct Placement into Nirex Container for Disposal	*Advantage* • Negates the need for waste handling. *Disadvantage* • Wastes are shielded. Nirex will not accept significant quantities of shielding into the repository.

about 0.2 m to a maximum estimate of 1.4 m. Desktop characterisation has been undertaken [2] and has identified the main radionuclides of concern as Sr90, Cs 137, Ru106, and Ce144.

Current State

The present arrangement, involving the extended storage of a FINGAL vessel (containing ~200 TBq of bg activity) in an old transport flask on the floor in B462.20, cannot be considered satisfactory. No documentation is currently available to support the use of the flask for waste storage or transport. Whilst there is no evidence of contamination leakage and radiation levels on contact are only a few hundred mSv/h, the hazard presented by the current storage configuration is not considered ALARP, and improvements are required. The preferred storage location is the modern remote-handled ILW store B462.27.

The situation for the other FINGAL vessels stored in B462.9 is more satisfactory, since a higher standard of containment (in a shielded, ventilated building) is provided, although the detail of storage arrangements is not known. However, the need to empty waste from B462.9 and repackage it for long-term storage in B462.27 is recognised, and retrieval of these vessels is, therefore, considered within the scope of this project.

Shortfall Against Modern Standards

The shielded flask in B462.20, containing one of the process vessels, is not an approved transport package under current site requirements and will need to be validated before further use.

Options Review

A number of options were considered for the treatment for disposal of these wastes, and the advantages and disadvantages are presented in Table A2-9. After a detailed internal review, option 1 (process through B459 immediately) was identified as the preferred option on safety, technical, and financial grounds.

Option 2 presents the hazard of waste leaking during the delay period and does not accord with the NII preference for treating waste early. In addition, significant facility start-up costs are likely.

Option 3 is discounted since it compels UKAEA to develop facilities and equipment for handling and storage of boxed waste, the need for which at this stage has not been confirmed.

Option 4 presents the hazard of waste leaking during the delay period and does not accord with the NII preference for treating waste early. In addition, size reduction operations could present significant potential to contaminate a new facility.

Option 5 is discounted, since A59 does not have any particular familiarity with this waste and the public relations and transport costs for transporting vitrified HLW will be high.

Option 6 would involve modification to the facility and poses a contamination hazard, in addition to UKAEA having no direct control over size reduction operations.

A2-8-4. *High βγ Sea Disposal Drums*

Description

The Sea Disposal Drums comprise an inner steel drum containing $\beta\gamma$ wastes with outer concrete shielding inside an outer steel drum with external dimensions of 900 mm diameter by 1150 mm high. Approximately 150 drums have been identified which require dismantling. Desktop characterisation has been undertaken [3] and has identified the presence of Pu and U in a number of drums, in addition to $\beta\gamma$ wastes.

Current State

Although the outer drums are corroded, the overall package is regarded as robust. However, a detailed inventory of the drums is not available to meet Nirex requirements, and the drums must, therefore, be opened for detailed

Table A2-9. Review of Options for Treating FINGAL Vessels

Option 1: Process Through B459 Immediately	*Advantages*
	• Permits early treatment of wastes.
	• Provides a long-term solution acceptable to regulators.
	• Previous experience of handling this waste in B459.
	• Avoids double handling.
	Disadvantage
	• High technical risk associated with retrieving the remaining vessels.
Option 2: Process Through B459 in 5 Years	*Advantage*
	• Evokes the benefit in discounting the costs at 6%.
	Disadvantages
	• Care and maintenance of the facility will be required over the 5 year period.
	• Some refurbishment of B459 will be required over the 5 year period.
	• Waste may leak out of current packaging during the 5 year period.
	• Large facility start-up costs may be incurred.
	• Does not accord with NII preference for early treatment of waste.
Option 3: Dispose Directly into Nirex Box	*Advantage*
	• Avoids cutting through the vessel, therefore no saw set-up costs.
	Disadvantages
	• Commits the UKAEA to box handling equipment and possibly a box store, with immediate storage problems.
	• It is only possible to get one vessel in each box.
	• Box will require shielding.
Option 4: Process Through New Facility	*Advantage*
	• Purpose made facility to size reduce the waste.
	Disadvantages
	• Significant delay before waste is treated which does not accord with regulators preference for early treatment.
Option 5: Process Through A59	*Disadvantages*
	• Container (Flask) costs for transport massive.
	• May attract adverse public reaction.
	• May not get transport authorisation.
	• Delays A59 decommissioning.
Option 6: Process Through B220	*Disadvantages*
	• Facility modification necessary to handle vessel and flask.
	• Significantly increases the decommissioning liability.
	• UKAEA have no direct control over operations.

characterisation and size reduction. There are recorded incidents of the inner drums leaking active liquid into the concrete shielding.

Shortfall Against Modern Standards

Drums were originally constructed to a MAFF specification. However, experience of opening similar drums in A59 at Winfrith shows construction was not always in accordance with the specification. There is no modern standard for this type of package.

Options Review

A number of options were considered for the treatment for disposal of these wastes, and the advantages and disadvantages are presented in Table A2-10. After detailed internal and external [4] review, option 1 (process through B459 immediately) was identified as the preferred option on safety, technical, and financial grounds.

Option 2 presents the hazard of waste leaking during the delay period and does not accord with the NII preference for treating waste early. In addition, significant facility start-up costs are likely for B459.

Option 3 presents the hazard of waste leaking during the delay period and does not accord with the NII preference for treating waste early. In addition, the handling capability of the new facility will need to be enhanced to cope with these drums, which can weigh up to 2 tonnes.

Option 4 is discounted because a type B transport container will be required for the outward journey and a modular flask for the return journey. In addition, obtaining outward transport authorisation may be difficult for this poorly characterised waste, and the presence of alpha nuclides may preclude this option.

Option 5 would involve modification to the facility and poses a contamination hazard, in addition to UKAEA having no direct control over size reduction operations.

Option 6 is currently discounted, since Nycomed Amersham are not prepared at this stage to offer a price for work in their facility, nor are UKAEA able to identify a commercial framework. In addition, the presence of alpha nuclides may preclude this option.

Option 7 was discounted on financial grounds [4].

A2-8-5. Ripple Waste Crates

Description

The waste was produced by UKAEA as a consequence of decommissioning cell 5 in B459, which was used to produce RIPPLE Sr^{90} sources. The decommissioned waste was PVC wrapped and crated with a view to reusing some of the equipment in future productions. There are 12 large crates of Sr^{90} contaminated equipment; nine wooden and three galvanised steel. Desktop characterisation of these wastes was undertaken [5], and this identified the paucity of information available on the levels of total activity in each crate.

Current State

Although the crates appear in good condition, the current plywood packaging is not considered a suitable primary containment for loose Sr^{90} titanate powder (there is no secondary containment). Movement of the crates is regarded a hazardous operation because of the potential for release, however no loose Sr^{90} contamination has been detected in the store, although access is currently limited by other wastes. Should a failure of the current primary containment occur, it would:

- Contaminate the unventilated general waste store presenting a significant potential for environmental and, therefore, off-site release; and
- Make waste retrieval and immobilisation a more hazardous and costly task, while producing an increased volume of waste for disposal.

Table A2-10. Review of Options for Treating High $\beta\gamma$ Sea Disposal Drums

Option 1: Process Through B459 Immediately	*Advantages* • Facility offers containment and remote handling capability. • Facility is currently available and offers early resolution of waste treatment. • No off-site transport requirements. *Disadvantage* • Delays decommissioning.
Option 2: Process Through B459 in 5 Years	*Advantage* • Discounting gives a reduction in cost of doing the work. *Disadvantages* • Continued standby operation of the facility will incur significant costs. • Danger of wastes leaking during the 5 year storage. • Large start-up costs (may need to bring the facility up to modern standards). • Does not accord with the NII preference to treat waste early.
Option 3: Process Through New Facility	*Advantage* • Permits early decommissioning of B459. *Disadvantage* • Does not accord with the NII preference to treat waste early.
Option 4: Process Through A59	*Advantages* • Some tooling already in place to dismantle drums. • Facility staff have previous experience of drum dismantling. *Disadvantages* • Poor characterisation may preclude transport authorisation. • Type B transport container required for outward journey. • Modular flask required for return of ILW to B462.27.
Option 5: Process Through B220	*Advantage* • Permits early decommissioning of B459. *Disadvantages* • Significant facility modifications required. • UKAEA have no direct control over operations.
Option 6: Process Through Nycomed Amersham	*Advantages* • Fixed price. • Technical risks lie with Nycomed Amersham. • Already have the facility in place and operational experience. *Disadvantages* • Not currently available. • NA are not currently prepared to offer a price for this work. • No commercial framework in place. • UKAEA have no direct control over operations.
Option 7: Overpack Directly into Nirex Box	*Advantage* • Eliminates the need to break open the drums. *Disadvantages* • Massive amounts of boxed ILW need to be stored and eventually disposed. • Unlikely to be accepted by Nirex due to poor waste characterisation.

Shortfall Against Modern Standards

The crates do not meet any modern standards.

Options Review

A number of options were considered for the treatment for disposal of these wastes, and the advantages and disadvantages are presented in Table A2-11. After detailed internal and external [6] review, option 1 (process through B459 immediately) was identified as the preferred option on safety, technical, and financial grounds.

Options 2 presents the hazard of waste leaking during the delay period and does not accord with the NII preference for treating waste early. In addition, significant facility start-up costs are likely for B459.

Option 3 presents the hazard of waste leaking during the delay period and does not accord with the NII preference for treating waste early. In addition, the handling capability of the new facility will need to be enhanced to cope with these crates which are highly contaminated, the largest measuring $3 \times 2 \times 1$ meters.

Option 4 is discounted because the type B transport container may be required for the outward journey and a modular flask for the return journey. In addition, obtaining outward transport authorisation may be difficult for this poorly characterised waste.

Option 5 would involve significant modification to the facility and poses a contamination hazard, in addition to UKAEA having no direct control over size reduction operations.

Table A2-11. Review of Options for Treating RIPPLE Waste Crates

Option 1: Process Through B459 Immediately	*Advantages* • Facility staff experienced in handling these wastes. • Facility offers containment and remote handling capability. • Facility is currently available and offers early resolution of waste treatment. • No off-site transport requirements. *Disadvantage* • Delays decommissioning.
Option 2: Process Through B459 in 5 Years	*Advantage* • Discounting at 6% reduces the cost of the task. *Disadvantages* • Care and maintenance of the facility will be required over the 5 year period. • Some refurbishment will be required over the 5 year period. • Waste may leak out of current packaging during the 5 year period. • Does not accord with NII pressure to treat waste sooner rather than later.
Option 3: Process Through New Facility	*Advantage* • Treating waste through a purpose built facility. *Disadvantages* • The requirement to process RIPPLE crates in a new facility will significantly increase the capability requirements. • Radionuclide content may leak out during the design and build program. • Does not accord with NII pressure to treat waste sooner rather than later.
Option 4: Process Through A59	*Disadvantages* • Crate contents insufficiently characterised to obtain transport authorisation. • Significant transport costs to and from Dorset esp. return of ILW to B462.27. • No experience of RIPPLE crates at A59. • May attract adverse public reaction.

Continued

Table A2-11. Continued

Option 5: Process Through B220	*Disadvantages*
	• Significant modifications necessary for the facility.
	• UKAEA do not have direct control over waste treatment operations.
	• Undesirable to add Sr^{90} to the decommissioning waste inventory of the building.
Option 6: Cement Crates and Dispose Directly to Drigg	*Disadvantages*
	• Crate contents not sufficiently characterised for acceptance by BNFL.
	• Unable to get transport authorisation due to poor characterisation.

Option 6 is discounted, since the wastes are not characterised well enough for acceptance by BNFL and no attempt at size reduction conflicts with Cmnd 2919.

A2-9. References

1. Grosset, K. *Characterisation of B462.20 RHILW*, RMC Ltd document reference R99-057(S), Draft issue (March 1999).
2. Smythe, M. J. *Disposal Options For the 'FINGAL' Vessels*, HRS document reference HRS/TN/126 (October 1997).
3. Jenkins, J. A. *Characterisation of Former Sea Disposal Drums in RHILW Category*, HRS document reference HRS/TN/261, AEA Technology reference AEAT-3335 (June 1998).
4. Jenkins, J. A. *Estimated Cost of Dismantling Former Sea Disposal Drums Containing Remote Handled ILW*, AEA Technology document reference AEAT-5859 (July 1999).
5. Goldsmith, A. *Options Study for Disposal of RIPPLE Project Wastes*, HRS document reference HRS/TN/090 (March 1997).
6. Fisher, C. J. *Study of Options for Dealing With the Crates of RIPPLE Waste*, RMC Ltd document reference R97-157(S) (November 1997).

Appendix 3

An Example of a Site Remediation Project — Dounreay Castle Ground Remediation

A. F. MC WHIRTER
S. J. TANDY
Engineering Group, Dounreay, UKAEA, Scotland

A3-1. Background

Dounreay Castle is located approximately 15 km west of Thurso on the north coast of Scotland at Ordnance Survey grid reference NC 98306693. Situated at the mouth of the Mill Lade Stream, the castle is built on a low lying rock plateau outside the northern boundary of the UKAEA Dounreay site. A layout of the site is presented as Figure A3-1. Admittance to the castle is through the UKAEA Dounreay site only, as reactor seawater intakes block access via the coastline, and approach by sea is considered too hazardous.

The castle, dating back to the 16th century, was last occupied in 1863 and is currently in a ruinous state, unroofed, and overgrown. A 19th century single storey cottage abuts the castle's most easterly wall. This is one of the last remaining buildings that once formed part of an extensive postmedieval settlement in the area. The structure of the castle is based on a tower house of "L" shaped plan that is normally associated with the lowlands of Scotland. It is for this reason, coupled with the castle's value in the more immediate local archaeological context, that Historic Scotland has given it Scheduled Monument Status.

As a result of unauthorised discharges from past operations at the UKAEA Dounreay site, the castle environs became affected by radioactive contamination from two sources. The first source was generated during the construction of the UKAEA Dounreay site as a result of effluent dispersion experiments carried out in the mid-1950s. The objective of the experiments was to investigate the dispersion characteristics of radioactive species discharged into the sea, in order to provide data for the design of a sea discharge system for effluent-containing radioactivity. The experiments involved the discharge of a mixed fission product liquor into the Pentland Firth via a temporary tank and pipeline arrangement set-up from the castle courtyard. The tank was only recently removed in 1995, whereas the pipeline was either removed or corroded away prior to this, the date of which is unknown. As a result of leakage and spillage, either during or after these experiments, fission products contaminated the courtyard of the castle.

The second source is believed to have resulted from leakage of the low active drainage system at the site's (D1211) low-level effluent discharge plant during the 1960s and 1970s. The contamination, a mixture of actinides and fission products, migrated along the existing, nonactive drainage system contaminating the castle gate drain (combined sewer) and foreshore.

A3-2. Site Characterisation

In order to establish the magnitude of the problem, the nature of the contamination, its extent, and severity had to be determined as far as reasonably practicable. This was achieved by a combination of desk study and staged site investigation.

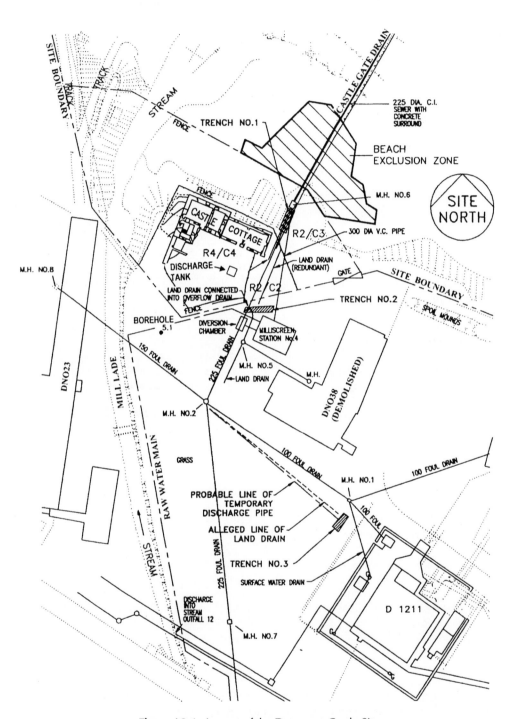

Figure A3-1. Layout of the Dounreay Castle Site.

The desk study involved the review of existing reports, drawings, photographs and maps, discussion with past and present site personnel, and site inspection. The information obtained from the study was used to design and subsequently implement a staged site investigation, which comprised the following aspects:

(a) *Surface survey*: Implemented on a grid system, this involved the measurement of beta and gamma contamination using contamination probes and a sodium iodide crystal detector.
(b) *Intrusive investigation*: Construction of trial pits and boreholes were carried out at selected locations to obtain soil, rock, and groundwater samples that were subsequently analysed for alpha, beta, and gamma emitting radionuclides. As part of the investigation, ground water monitoring points were installed at strategic locations.
(c) *Inspection of the drainage systems*: This involved the physical examination of the manholes and remote CCTV observation of the drain runs. Radiological monitoring was carried out throughout the inspection using contamination probes and sodium iodide crystal detector as appropriate. Samples were taken where available, and analysed as for the intrusive investigation.
(d) *Groundwater monitoring*: This was carried out periodically to identify and monitor any changes of radioactivity in groundwater, upstream of the castle site.

The courtyard was found to be contaminated with beta- and gamma-emitting radionuclides. Activity concentrations of up to 550 Bq/g strontium-90 and 2000 Bq/g caesium-137 were recorded in the upper layers of the soil profile, reducing to 3.8 Bq/g and 0.6 Bq/g, respectively, at a depth of 1.4 m [1,2]. Dose rates were locally in excess of 25 μSv/hr. The contamination was shown to extend into the cottage, but not the tower house of the castle.

The castle gate drain system was shown to contain radioactive contaminated silt up to 41 Bq/g caesium-137 and 0.9 Bq/g americium-241. Unfiltered effluent taken from manhole 6 had an activity concentration of <10 Bq/l caesium-137. Dose rates of 35 μSv/hr were recorded. Ground surrounding the drainage system exhibited a maximum activity concentration of 507 Bq/g Cs-137 and 86 Bq/g americium-241. Sections of the drain were shown to be corroded and in a poor state of repair.

The foreshore area was contaminated with up to 35 Bq/g caesium-137, 18 Bq/g plutonium-239 and 240, 8.6 Bq/g plutonium-238, and 7.4 Bq/g americium-241. The contamination extended to rock head.

Monitoring of the boreholes installed up-stream from the castle environs indicated that migration of contamination was insignificant. The highest activity concentration encountered was 1.9 Bq/g caesium-137, which occurred in borehole 5.1 at the surface (Figure A3-1). Analyses of groundwater were in general below the limit of detection.

The external dose rate produced by the contamination can be compared with natural background. The geological formation associated with this area of Caithness is middle Devonian Old Red Sandstone. This formation produces an absorbed gamma-ray dose rate of 0.052 μGy/hr. The dose equivalent, assuming a weighting factor (W_R) of 1 for gamma, is equal to 0.052 μSv/hr.

Access to the site was restricted by fencing, and areas affected by contamination were designated in accordance with Ionising Radiations Regulations 1985 (Figure A3-1).

In addition to characterising the contamination, a structural survey of the castle was carried out to assess the degree of collapse, the outward lean of the north wall, and any stabilisation works required. The survey identified, among other things, the need to provide stability to the north and east walls, selected windows and repairs to the roof of the single-storey cottage.

A3-3. Option Study

Option studies were commissioned to determine an appropriate remediation solution. After assessing these studies, four main options were identified. These were:

(a) Do nothing.
(b) Full remediation: This would necessitate removal of the castle structure, as contamination is likely to have affected the foundations because of migration.
(c) Target remediation: Achieved either through removal of contaminated material by excavation and replacement with subsequent disposal or treatment of the waste, or by *in situ* treatment.
(d) Encapsulation: Installation of a concrete barrier to reduce the external dose by shielding and prevent airborne contamination.

After considering such factors as cost, health and safety, heritage, regulatory bodies, and technical issues, the preferred option was identified as target remediation. This would be achieved by using an excavation and replacement technique. The reasons for selection of this option were:

(a) Cost-effectiveness;
(b) Waste minimisation;
(c) Proven technology;
(d) Minimal affect on castle structure;
(e) Access to the castle restored; and
(f) ALARP (As Low As Reasonably Practicable).

A3-4. Design

The design proceeded following preparation and approval of the sanction case for this stage of the project by UKAEA. The scope included:

(a) Design for the remediation of the castle courtyard, castle gate drain, foreshore, and mitigation against potential recontamination;
(b) Planning application;
(c) Scheduled ancient monument consent;
(d) Safety case;
(e) Financial case; and
(f) Contract.

A3-4-1. *Remediation Design*

The design target remediation criteria of 1 Bq/g artificial alpha and 4 Bq/g beta/gamma emitters based on ALARP (As Low as Reasonably Practicable) were used. These values, taking account of the discussion on selection of the criteria below, were in line with Basic Safety Standards and National Radiological Protection Board (NRPB) Generalised Derived Limits.

Why set a target of 4 Bq/g for beta and gamma emitting radionuclides? First, both beta and gamma emitting radionuclides are just detectable, separately on common contamination monitors at 4 Bq/g. Secondly, the primary contaminants are the beta and gamma emitters of strontium-90 and caesium-137 (barium-137), which, respectively, have a half-life of approximately 30 years. With the site destined to remain under institutional control for the next 100 years, the contamination would reduce through radioactive decay to negligible levels.

Why set a target of 1 Bq/g for artificial alpha emitting radionuclides? Alpha activity concentration at 1 Bq/g was selected based on the fact that the naturally occurring alpha emitters present within the soil at Dounreay are around 1 Bq/g. As the dose coefficient for thorium-232 (present at approximately 0.1 Bq/g) is higher than that of americium and plutonium, the total inhalation hazard from natural occurring alpha emitters is comparable with that from 1 Bq/g artificial alpha emitters.

As 1 Bq/g artificial alpha is not detectable by normal contamination monitors in the field, an alternative method was used. This was based on ratios of total alpha to caesium-137. Interpretation of the sample analysis from intrusive investigation of the foreshore [3] showed that, from 70 samples, >95% of the results gave a ratio of caesium-137 to total alpha of 4:1. This relationship, with its limitations recognised, was adopted as a practicable way of assessing the alpha activity concentration in the field by using caesium-137 as a finger print, i.e., 4 Bq/g caesium-137 is readily identified by monitors in the field.

The approach of using in-field monitoring as already described was adopted based on the fact that confirmation sampling and radiometric analysis would be carried out on excavated material and at the remediation end point, i.e., the termination depth of the excavation. The samples would also be made available for independent analysis by SEPA, as required.

The actual methodology, based on an excavation and replacement technique, had to be devised to offer the best compromise between archaeological and remediation requirements. It was decided to use a grid system based on a 2 m spacing to allow both control and ease of reference. The advance of the excavation was to be limited to 200 mm

Table A3-1. Waste Categorisation

Activity (Bq/g)	Category
<1 alpha <4 beta/gamma	Material for reuse as backfill
>0.4–<1 alpha >0.4–<40 beta/gamma	Very low high volume waste
>1–<4000 alpha >40–<12,000 beta/gamma	Low level waste (LLW)

depth increments, as calculation showed this to be the maximum penetration depth of monitoring instrumentation, in this case a sodium iodide crystal detector. The excavated material was to be categorised into three categories, as shown in Table A3-1.

Assessment and sentencing of the excavated material would be carried out *in situ* and by monitoring material in the excavator bucket. Material assessed as being above the target level was to be placed in lined polypropylene bags with a safe working load (SWL) of 1 tonne. The material prior to this placement was sampled for radiometric analysis as required.

The bagged material would then be placed in Half Height International Standard Organisation (HHISO) containers of industrial grade for storage of Very Low High Volume Waste and nuclear grade (2910B) for Low Level Waste (LLW). The containers would then be transported to their respective stores on the UKAEA Dounreay site. Material assessed as below the target level would be sampled and stockpiled ready for reuse as backfill.

Traditional archaeological monitoring and recording could not be carried out, primarily on health and safety grounds, because of the radiological hazards involved. A remote monitoring and recording system was, therefore, proposed using a video camera linked to a surveillance center. Additional information would also be gathered by taking photographs and levels at each 200 mm depth increment. The information obtained would then be collated and manipulated using computerised techniques to build a pictorial history of the site.

The castle gate drain outfall from the diversion chamber adjacent to the milliscreen station was also to be replaced and the old outfall removed. The replacement was designed so that the majority of the outfall could be replaced on-line by constructing the new outfall parallel to the existing one. On-line replacement requiring diversion of flows by over pumping was identified for replacement of manhole 6, and in times of high flow when replacing the overflow section from the diversion chamber to the milliscreen station. The outfall diameter was also increased from 225 to 300 mm, making it compatible with the 300 mm outlet pipe from the milliscreen station that was constructed much later than the outfall in the early 1980s.

At the time of the design, the project team was awaiting analysis and interpretation of ground water monitoring, to determine whether measures were required to prevent possible recontamination from existing sources upstream. The decision was made to include in the design, mitigation measures against the potential for recontamination. This comprised a clay cut-off barrier with associated effluent collection and pumping facilities to capture ground water and pipe it to D1211 effluent discharge plant for authorised discharge.

Provision was made for remedial measures and protection to be installed before and during work around the historic structure of the castle. This included removal of loose slates, head stones, and support to window lintels. During the work, protective scaffolding was to be erected when excavating in close proximity to the structure's walls. Precise monitoring was included as an additional requirement for the north wall because of its outward lean. Provision was also made for any necessary underpinning of the cottage that might be required as a result of excavation and removal of the overflow between manhole 6 and the diversion chamber.

Radiological monitoring and controls involving periodicity, limits of detection, area designation, air sampling, contamination monitoring, etc., were determined and specified.

A3-4-2. *Planning Application*

A planning application was prepared which described the work to be carried out, and the hazards and risks associated with its implementation. Following submission to the Local Planning Authority (Highland Council) and clarification on specific points, planning consent was awarded subject to controls stated in the application being implemented.

A3-4-3. *Scheduled Ancient Monument Consent*

A statement was drafted and submitted to Historic Scotland providing the background to the project, the scope of remediation proposed, stabilisation works, and the archaeological monitoring and recording to be undertaken. Historic Scotland, following review, granted Scheduled Monument Consent, subject to conditions in the statement being fulfilled.

A3-4-4. *Safety Case*

A detailed safety case was prepared for the work by independent consultants and involved a Preliminary Safety Report [4] (PSR) at the tender stage of the design to assess the initial safety categorisation for the work. This was followed by a Pre-Commencement Safety Report [5] (PCSR) supported by a Hazard Operability study (HAZOP) [6] during the design to review the safety categorisation and develop the safety case in parallel with the design. The PCSR defined a dose budget for the works of 2 mSv.

A Pretender Health and Safety Plan was developed in accordance with the Construction and Design Management (CDM) regulations 1994, which in part included the PCSR. Additionally, notification of the work was issued to the Health and Safety Executive.

A3-4-5. *Financial Case*

Prior to commencement of the project implementation stage, a detailed sanction paper was prepared for agreement by the Dounreay Board and ratified by the Department of Trade and Industry. The paper included, but was not restricted to, financial and technical appraisal of options, risk analysis and management, safety, project management, and contractual arrangements [7].

A3-4-6. *Contract*

At the time of undertaking this project, UKAEA was introducing the use of the New Engineering Contract (NEC), a form of contract that encouraged the cooperation and openness of all parties.

This project acted as a pilot for the introduction of the NEC and used the New Engineering Construction Contract. Option B, specifically for a priced contract with bill of quantities, was chosen as being the most appropriate because of the uncertainty associated with remediation of contaminated ground.

A3-5. Implementation

Following competitive tender, the implementation contract was awarded to commercial contractors with appropriate experience.

At this stage, UKAEA had obtained sufficient information on the groundwater conditions upstream of the Castle to demonstrate that the potential for recontamination by existing sources was negligible. Hence, the need for a cut-off barrier could be removed from the scope of the project.

After the necessary safety and quality assurance documentation to carry out the work had been prepared and approved, mobilisation of resources and equipment were effected. Set-up of site infrastructure was completed comprising change rooms, security, power, water, fencing, and an archaeological remote monitoring unit.

Following a base-line gamma flux survey using an AEAT Groundhog^{TM} detector, work commenced in the castle courtyard, the area of highest external radiation dose. The reason for this was to remove the contamination that resulted in the elevated dose, so that restrictive working practices and increased dose uptake could be removed in the short-term in accordance with ALARP.

Approximately half-way through remediation of the courtyard, resources were deployed to remediate the foreshore in parallel with the courtyard, finishing with removal of the existing outfall and its replacement. During remediation,

Table A3-2. Analysis of Imported Material from Off-site

	Gross alpha (Bq/g)	Gross beta (Bq/g)	Caesium-137 Bq/g	Americium-241 Bq/g
Range	0.25–0.62	<2.61–5.49	0.0025–0.038	Not analysed
Average	0.39	4.2	0.082	

Number of samples taken and analysed was 10.

Table A3-3. Analysis of Reused Material for Backfill

	Gross alpha (Bq/g)	Gross beta (Bq/g)	Caesium-137 (Bq/g)	Americium-241 (Bq/g)
Range	<0.01–1.65[a]	<1.5–20.45	0.02–5.19	0.02–1.1
Average	0.36	5.0	0.63	0.07

[a] Out of 155 samples analysed, only two samples exceeded 1 Bq/g.

Table A3-4. Analysis of Material at Termination Depth

	Gross alpha (Bq/g)	Gross beta (Bq/g)	Caesium-137 (Bq/g)	Americium-241 (Bq/g)	Strontium-90 (Bq/g)
Range	0.02–9.53[a]	<1.75–93.67	0.01–17.7	<0.008–1.51	0.1–2.5[b]
Average	0.63	11	1.22	0.23	0.8

[a] Out of 160 samples analysed, 11 exceeded 1 Bq/g.
[b] Only five samples analysed for Sr-90.

Table A3-5. Analysis of Swabs Taken at Termination Depth

	Caesium-137 (Bq/g)	Americium-241 (Bq/g)
Range	0.02–214	0.03–8.88
Average	8.0	0.74

Fifty swab samples were taken. The highest results were encountered along the existing position of the outfall.

problems associated with characterisation and assessment of waste arisings, additional contamination finds, and environmental conditions resulted in delays.

Radiological monitoring of operatives was carried out both directly and passively using Personal Integrating Dosimeters (PID), personal air samplers, and Thermoluminescent Dosimeters (TLDs). Appropriate Personal Protective Equipment (PPE) was provided as necessary and included the use of Respiratory Protective Equipment (RPE). Radiological surveys of the site area and buildings were carried out periodically, whilst any presence of airborne contamination was monitored by air samplers strategically located around the site.

As a result of remediation, an area of $900\,m^2$ was excavated down to a maximum depth of 3 m, resulting in a total of 1540 tonnes of LLW, of which 1109 tonnes was assessed as Very Low High Volume Waste. The excavation was backfilled partially by clean, inert imported material, and excavated material that was assessed as being below the target limit. Confirmatory analysis of the backfill was undertaken. Imported material, of which 10 samples were taken, exhibited an activity concentration range, as shown in Table A3-2. Reused material, of which 155 samples were taken, exhibited an activity concentration, as shown in Table A3-3.

A postremediation survey was carried out at the termination depth of the excavation involving direct beta/gamma monitoring by probe, acquisition of swabs and samples, and a gamma flux survey (sodium iodide crystal detector).

The summary of results from sampling is given in Table A3-4 for solid samples and Table A3-5 for swabs.

Selected sample analysis by alpha spectrometry was carried out to confirm the level of plutonium present in samples at or around 1 Bq/g gross alpha. The results are shown in Table A3-6.

The results show that, at termination depth, the contamination remains above the target level in the following areas: one of the castle's southern wall footings, the rock on which the east elevation of the cottage is founded, and an area of rock approximately $140\,m^2$ on the foreshore. An impermeable membrane was placed prior to backfilling to prevent cross-contamination. This approach was not undertaken on the foreshore north of manhole 6 because of the erosion-prone environment.

Table A3-6. Alpha Spectrometry Analysis

Gross alpha (Bq/g)	Plutonium-238 (Bq/g)	Plutonium 239-240 (Bq/g)
1.01	0.004	0.005
1.65	0.175	0.308
1.23	0.054	0.121

Table A3-7. Risk Assessment Exposure Scenarios

Period	Exposure Scenario
Site control 0–50 years	*Recreation*: Person walking on the foreshore for recreation, exposed by inhalation, ingestion, and external exposure for several hours.
	Beachcomber: A beachcomber removes contaminated items. They are also exposed by inhalation, ingestion, and external exposure for a longer period.
Post control >50 years	*Business Park*: The site is restored as a business park, and an employee visits the beach regularly. They are exposed by inhalation, ingestion, and external irradiation. They may also ingest some contaminated marine foodstuff.
	Resident: A person is resident in the remediated area 90% of their time. They are exposed by inhalation, ingestion, and external irradiation. They may also ingest some contaminated marine foodstuffs.

On completion of backfilling operations, a gamma flux walkover survey and an external absorbed gamma dose rate survey at 11 selected locations were carried out. Measurements were taken using an AEAT Groundhog™ detector and Mini Instrument's 6-80 environmental monitor, respectively. Dose rate determined by calculation from the gamma flux survey results indicated that the levels were less than 0.3 μGy/hr across the site. The results of the external absorbed gamma dose rate indicated levels of less than 0.16 μGy/hr.

The personal and environmental monitoring carried out during implementation gave the following results. The total dose uptake for personnel was 52.62 DACh, with the highest individual dose being 17.05 DACh. The total dose measured by PID was 980 μSv, with the highest individual dose being 145 μSv. The result of air sampling showed an average of 0.06 DACh, with the highest recorded at 9.67 DACh. The work was carried out within the dose budget of 2 mSv.

As a result of the archaeological monitoring and recording, five historical periods were defined; Tower House and Barmkin, Tower House & Service Ranges, Tower House & South Range, Farm Cottages, Bothies & Stockyard, and the Nuclear Research Establishment. Several finds were encountered ranging from a quern stone to an enamel mug, most of which were found to be contaminated.

The remediation of the site has, in the main, been completed within the criteria set, i.e., 1 Bq/g artificial alpha and 4 Bq/g beta in accordance with ALARP; archaeological monitoring and recording was carried out successfully using remote techniques to further establish the history of the site; a new outfall was successfully constructed and commissioned; radiological safety of personnel and the environment was controlled within the limits set, despite the harsh environmental conditions and extended project duration. Only minor industrial injuries were sustained during the works and were limited to hand injuries. Despite all these operations around the castle, the building was unaffected. Monitoring of the groundwater upstream of the castle site continues to assess whether the potential for migration of contamination is changing with time.

A3-6. Risk Assessment

A risk assessment was carried out to estimate the future risks associated with the residual contamination. The risk is calculated on the basis of peak risk being realised by the exposure to the residual contamination should it be released into the environment by erosion. Several exposure scenarios were considered as shown in Table A3-7. The exposure scenarios are based on 50 years because of an anticipated reduction to the site's restoration program.

Table A3-8. Peak Risks from Residual Contamination

Exposed group		Peak Risk (y^{-1})	Time of Peak (y)
Site control	Recreation	8.5×10^{-9}	0
0–50 years	Beachcomber	1.7×10^{-8}	0
Post control	Business Park	1.3×10^{-7}	>50
>50 years	Resident	2.7×10^{-6a}	>50^b

[a]This value is reduced to 1.3×10^{-6} y^{-1} using conservative distribution coefficients.
[b]The peak risk for the resident falls below 10^{-6} y^{-1} after 110 years.

A model was derived which took into account marine erosion, hydrology, and hydrogeology. The model splits the area subjected to remediation into three compartments: the courtyard, castle gate drain from the diversion chamber to manhole 6, and the foreshore. The results of the risk assessment are shown in Table A3-8.

The peak risk identified as 2.7×10^{-6} to a possible future resident is in line with current guidance from the Government and its advisors, which advise an annual risk of death of between 10^{-5} and 10^{-6} as being acceptable.

A3-7. References

1. International Atomic Energy Agency. *Classification of Radioactive Waste, A Safety Guide, A publication in the RADWASS Programme*, IAEA Safety Series No.111-G-1.1, IAEA, Vienna (1994).
2. International Atomic Energy Agency. *Establishing a National System for Radioactive Waste Management*, IAEA Safety Series No.111-S-1, IAEA, Vienna (1995).
3. International Commission on Radiological Protection. "Radiological protection and safety in Medicine. ICRP Publication 73," *Annals of the ICRP*, 26 (2) (1996).
4. International Commission on Radiological Protection. " Age dependent doses to members of the public from intakes of radionuclides; Part 5, Compilation of ingestion and inhalation dose coefficients. ICRP Publication 72," *Annals of the ICRP* (1996).
5. International Commission on Radiological Protection. "Protection from potential exposure: a conceptual framework. ICRP Publication 64," *Annals of the ICRP*, 23 (1) (1993).
6. International Commission on Radiological Protection. "Principles for limiting exposure of the public to natural sources of radiation. ICRP Publication 39," *Annals of the ICRP*, 14 (1) (1984).
7. International Commission on Radiological Protection. "Protection of the public in the event of major radiation accidents: principles for planning. ICRP Publication 40," *Annals of the ICRP*, 14 (2) (1984).

Appendix 4

A4-1. Internet Information

Some URLs for Useful Websites for Information and Further Useful Links

A few introductory websites are given below. The web is a rapidly changing source of information and some of the URLs given here may already have moved, changed, or been updated. Using appropriate key words and a good browser (such as Google), you should be able to identify a large number of further pro and antinuclear sources of views and information. The various nuclear companies and power generators also provide related information via their websites.

The British Nuclear Industry Forum (BNIF) website has useful information and many links to industry, government organizations, learned societies, and academic websites with relevant material. http://www.bnif.co.uk/artman/publish/index.shtml

Health & Safety Executive

- HSE policy on decommissioning and radioactive waste management at licensed nuclear sites. http://www.hse.gov.uk/spd/content/spddecom.htm
- HSC research in nuclear energy has links to research in the waste and decommissioning area. http://www.hse.gov.uk/nsd/resindex/index.htm
- Safety assessment principles for nuclear plant: www.hse.gov.uk/nsd/saps.htm

Department for Environment, Food, & Rural Affairs (DEFRA)

- Radioactive waste management. http://www2.defra.gov.uk/environment/radioactivity/waste/index.htm
- Managing Radioactive Waste Safely. http://www2.defra.gov.uk/environment/consult/radwaste/default.htm

Department of Trade and Industry (DTI)

The White Paper "Managing the Nuclear Legacy — A strategy for action", sets out plans for radical changes to current arrangements for the clean-up of the civil nuclear legacy including the creation of a new body — the Liabilities Management Authority (LMA). http://www.dti.gov.uk/energy/nuclear/environment/liabilities/index.shtml

Radioactive Waste Management Advisory Committee (RWMAC)

www.defra.gov.uk/rwmac/index.htm

European Commission

Nuclear safety regulation and radioactive waste management. There are links to nuclear installation safety, radioactive waste management and, decommissioning of nuclear facilities. http://europa.eu.int/comm/energy/en/nuclearsafety/

Nuclear Energy Agency NEA

Radioactive waste management. http://www.nea.fr/html/rwm/

International Nuclear Societies Council

Radioactive waste. http://www2s.biglobe.ne.jp/~INSC/index.html

The US Nuclear Regulatory Commission, NRC

The Nuclear Regulatory Commission, NRC, has put several NUREG technical documents on-line that relate to decommissioning, although not all of them are in final format. These include:

- Radioactive waste. http://www.nrc.gov/NRC/radwaste.html
- Nuclear power plant decommissioning. www.nrc.gov/OPA/reports/dcmmssng.htm
- Overview of decommissioning nuclear plant. www.nrc.gov/OPA/gmo/tip/fsdecommissioning.html
- NUREG-1507: Minimum Detectable Concentrations with Typical Radiation Survey Instruments for Various Contaminants and Field Conditions. http://techconf.llnl.gov/radcri/1507.html
- NUREG-1628: Staff Responses to Frequently Asked Questions Concerning Decommissioning of Nuclear Power Reactors. www.nrc.gov/NRC/NUREGS/SR1628/sr1628.html

The US Environmental Protection Agency, EPA

- MARSSIM: The US Environmental Protection Agency, EPA, provides the Multi-Agency Radiation Site Survey and Investigation Manual (MARSSIM) website. www.epa.gov/radiation/marssim. MARSSIM has information on planning, conducting, evaluating, and documenting environmental radiological surveys of surface soil and building surfaces for demonstrating compliance with regulations. The document, now finalised, is a multiagency consensus document approved by the US Departments of Defense and Energy, the EPA, and the Nuclear Regulatory Commission. Other information found on this website includes the authorising Federal Register Notice, links to associated agencies, ways to obtain printed copies, and a few related tools.
- MARLAP: The Multi-Agency Radiological Laboratory Analytical Protocols Manual (MARLAP) can be found at http://www.eml.doe.gov/marlap/

Argonne National Laboratory of the US Department of Energy

Argonne National Laboratory has developed RESRAD, software that calculates site-specific RESidual RADioactive material guidelines as well as radiation dose and excess lifetime cancer risk to chronically exposed on-site residents. www.ead.anl.gov/project/dsp_fsdetail.cfm?id = 51

Safegrounds

The purpose of the SAFEGROUNDS Learning Network is to deliver a rolling program of best practice guidance about the management of contaminated land on nuclear and defence sites. The Network is initially a collaboration between nuclear liability holders and the regulators, contractors, and consultants to the nuclear industry, but, as it progresses, it will increasingly involve other stakeholders representing public and wider environmental interests. It provides technical papers and background information about remediation of nuclear and defence facilities. www.safegrounds.com

A4-2. Book List

To date, there is no single adequate text to cover the contents of this book. Most conferences provide written information through hand-out material. The danger of a book list is that you might interpret this incorrectly, buy them all, and even worse — try to read them all! You need to be selective. Those included here would be useful general texts for your personal bookshelf. The others would be worth dipping into, or getting your library to buy if they don't have copies already.

Knoll, G. F. *Radiation detection & measurement*, Wiley. ISBN 0471073385, ~£40 (hbk). This is essential for laboratory and lecture sessions.

Spiegel, M. and L. Stephens. *Schaum's outline of statistics*, 3rd edn, ISBN 0071167668. There are a confusingly large number of versions of these books, with slightly different authors (always including Spiegel) and titles.

Living with radiation. NRPB, Chilton, Oxfordshire, UK or the Stationery Office. ISBN 0859514196, £9.95. A general, simple overview of radiation protection.

Wilson, P. D. (ed.). *The nuclear fuel cycle — from ore to waste*, Oxford University Press (1996). ISBN 0198565402, £40.

Lilley, J. *Nuclear physics: Principles & applications*, Wiley. ISBN 0471979368, ~£30 (pbk). A basic general text with applications to medicine and the nuclear industry.

Martin, A. and S. A. Harbison. *An introduction to radiation protection*, 4th edn, Arnold (1996). ISBN 0 412 631105, ~£25 (pbk).

Turner, J. E. *Atoms, radiation and radiation protection*, 2nd edn, Wiley (1995). ISBN 047159581-0, ~£60 (hbk). A wide ranging, more advanced text than Martin and Harbison — with lots of specimen problems, answers, and worked examples.

"Nuclear power in the twenty-first century," Special issue of *Interdisciplinary Science Reviews*, 26 (4) (Winter 2001).

Abelquist, E. W. *Decommissioning Health Physics: A handbook for MARSSIM Users*, Institute of Physics Publishing, Bristol & Philadelphia (2001). ISBN 0 7503 07617.

The tolerability of risk from nuclear power stations, The Health & Safety Executive (1988, revised 1992). ISBN 0-11-886368-1.

Safety assessment principles for nuclear plants, The Health & Safety Executive (1992). ISBN 0-11-882043-5.

Chernobyl 10 years on, OECD (1996).

Nuclear Decom 2001, 16–18 October 2001, Professional Engineering Publishing Ltd, Bury St. Edmunds, UK. *IMechE Conference Proceedings*. ISBN 1 86058 329 6.

"State of the art technology for decontamination and dismantling of nuclear facilities," *Technical report series No 395*, IAEA, Vienna (1999). ISBN 9201024991.

Safety in decommissioning research reactors, IAEA Safety Series No 74, IAEA, Vienna (1986). STI/PUB/713. ISBN 92-0-123086-9.

State of the art technology for decontamination and dismantling of nuclear facilities, IAEA Technical Report Series No 395, IAEA, Vienna (1999). STI/DOC/010/395, ISBN 92-0-102499-1.

"1990 Recommendations of the International Commission on Radiological Protection." ICRP Publication 60, *Annals of the ICRP*, 21(1–3) (1991). See also: http://www.icrp.net/index.asp, http://www.icru.org/ and http://www.ortec-online.com/

Sources and effects of ionizing radiation, Vol I: Sources, Vol II: Effects, United Nations Scientific Committee on the Effects of Atomic Radiation (UNSCEAR), United Nations, New York (2000). ISBN 9211422388 (Vol I), ISBN 9211422396 (Vol II).

The Ionising Radiations Regulations 1999, Statutory Instruments, No 3232, The Stationery Office, London, UK (1999).

Work with ionising radiation. Ionising Radiations Regulations 1999. Approved code of practice and guidance, HSE Books (2000). ISBN 0 7176 1746 7.

The Radman Guide to The (UK) Ionising Radiations Regulation 1999, Collins (July 2000), ISBN 0948237384.

For those readers with little or no basic physics background, a secondary level Physics text should provide a basic introduction to the elements of atomic/nuclear structure and radiation physics and perhaps some applications — including nuclear power and information on biological effects. There are numerous texts at this level; e.g.,

Duncan, T. *Advanced Physics*, John Murray (2000). ISBN 0719576695. This covers all aspects of A-level physics but has a useful section on basic atomic/nuclear physics, including radiation detectors and electronics.

Sang, D. *Bath Science 16–19: Nuclear Physics*, Nelson Thornes (1992). ISBN 0174482086. This is specifically related to nuclear physics and applications.

For those wishing to brush up their physics, a basic undergraduate text would be useful, such as:

Krane, K. S. *Introductory Nuclear Physics*, John Wiley & Sons (1987). ISBN 047180553X, £34. Booklist August 2002.

Appendix 5
Elements and Isotopes

A5-1. Introduction

This appendix describes the nomenclature used in this book for describing radionuclides or isotopes. It gives a brief introduction to a simplistic structure of the atom and radioactive decay.

A5-2. The Nucleus

The nucleus of an atom contains positively charged particles called protons. The number of protons in an atom is known as the atomic number (Z) of the element and determines its chemical properties. Photographs of the tracks of protons in a cloud chamber were first made in 1912.

Atoms are also made up of electrons which have a negative charge (counterbalancing the positive charge of a proton), have almost zero mass, and form a cloud surrounding the nucleus.

Neutrons have similar mass to a proton, are neutral in charge, and were first recognised as discrete particles by James Chadwick in 1932. The mass of an atom is not directly proportional to the number of protons it contains, but rather the number of protons *and* neutrons. Neutrons have no charge and do not affect the main chemical properties of an element. Different forms of the same element can contain different numbers of neutrons. These are known as radionuclides or more commonly as isotopes.

One would normally expect the make-up of an atom's nucleus, containing positively charged protons, to be unstable because of the repulsion forces (like charges repel) between the protons involved. There is, however, a strong force which binds the protons together if they are less than about 10^{-15} m apart. Currently accepted theory suggests that nuclear particles are held together by "exchange forces" in which subatomic particles known as pions (first predicted in 1935 and effects observed in 1947), which are common to both protons and neutrons, are continuously exchanged between them. Neutrons help bind the nucleus together and make it stable. Even so, some nuclei are unstable because they do not have the right ratio of protons-to-neutrons. A large excess of neutrons over protons detracts from the stability of the nucleus. The least stable nuclei contain an odd number of protons and an odd number of neutrons. An unstable nucleus can move to a more stable state by the emission (or capture) of a particle.

Elements that exist naturally on Earth are made up of a combination of radionuclides, as illustrated in Table A5-1 for carbon. In nature, carbon has radionuclides with six neutrons (98.89%), seven neutrons (1.11%), and eight neutrons (a tiny proportion). The average ratio of the radionuclides making up the element gives an atomic weight of 12.011 for carbon, which is dominated by carbon-12.

Elements are therefore expressed as $^A_Z X$, where X is the chemical symbol of the element, A is the mass number being the sum of the protons and neutrons in the nucleus, and Z is the atomic number being the number of protons in

Table A5-1. Carbon Radionuclides

Name	Protons	Neutrons	Mass No. (A)	Atomic No. (Z)	Symbol
Carbon-12	6	6	12	6	$^{12}_6 C$
Carbon-13	6	7	13	6	$^{13}_6 C$
Carbon-14	6	8	14	6	$^{14}_6 C$

the nucleus. Note that $A - Z =$ number of neutrons in the nucleus. Naturally occurring copper, for example, consists of ~69% $^{66}_{29}Cu$ and ~31% $^{65}_{29}Cu$ and has an atomic weight of ~63.55.

A5-3. Radioactivity

Figure A5-1 shows the stable radionuclides of oxygen and nitrogen expressed in terms of the ratio of the number of protons and neutrons in the nucleus. Stable radionuclides lie on the dotted line; those that do not are considered to be unstable. These radionuclides may move back to stability by the emission of α or β particles.

An α (alpha) particle may be regarded as a swiftly moving helium nucleus, $^{4}_{2}He$ (containing two protons and two neutrons) or $^{4}_{2}\alpha$. α radiation has highly ionising properties although requires the least shielding to achieve protection. Its velocity depends upon the nature of the emitting atomic nucleus with values from 1.4 to 2×10^{7} m s^{-1} and a range in air at standard temperature and pressure (STP) of 0.0267–0.0695 m. It is very important not to breathe or ingest α-emitting radionuclides which will then emit their damaging α particles deep inside the body where they could be absorbed by internal organs and lead to cell damage.

By emitting an α particle, a radionuclei can lose mass and move towards the stability line, as shown in Figure A5-1. For example, for polonium:

$$^{212}_{84}Po \rightarrow {}^{4}_{2}\alpha + {}^{A}_{Z}X.$$

To make the equation balance, $A = 208$ and $Z = 82$. The element with atomic number equal to 82 is lead. So, the equation becomes:

$$^{212}_{84}Po \rightarrow {}^{4}_{2}\alpha + {}^{208}_{82}Pb.$$

β (beta) particles are electrons with a negative charge and velocities approaching the speed of light. Relativistic effects have to be taken into account when considering differences between the rest-mass and mass of an electron at such velocities. β particles are ejected from a radionuclei with a continuous spectrum of energies with a specific maximum value depending upon the nature of the atom concerned. The ranges for the travel of these β particles

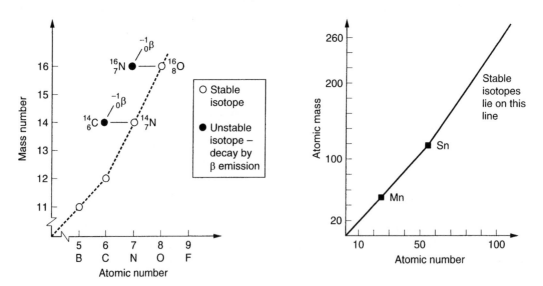

Figure A5-1. Plots of Atomic Mass vs. Atomic Number showing the Stability of Atoms. Note manganese, Mn, has an atomic number (Z) of 55. The element with twice the atomic number is tin, Sn, whose mass is 119. It has twice as many protons, but more than double the mass. Stable radionuclides lie on the dotted line on the diagram. The effect of β decay is to increase the atomic number by one and leave the mass number (A) unchanged. Therefore, any unstable radionuclide which is above the dotted stability line can fall back to stability through β decay.

are some 20–100 times that of an α particle. A β particle with an energy of 0.5 MeV has a range in air at standard temperature and pressure of about 1 m.

A β particle carries a charge (and therefore an atomic number, Z) of -1 and a mass number (A) of 0. One can work out what happens to the nucleus when it throws out a β particle by writing an equation and balancing the atomic number and the atomic mass number on both sides. For the decay of ^{14}C:

$$^{14}_{6}C \rightarrow\ ^{0}_{-1}\beta +\ ^{A}_{Z}X.$$

To make the equation balance, $A = 14$ and $Z = 7$. The element with atomic number equal to 7 is nitrogen. So, the equation becomes:

$$^{14}_{6}C \rightarrow\ ^{0}_{-1}\beta +\ ^{14}_{7}N$$

and nitrogen 14 is stable.

γ (gamma) radiation consists of electromagnetic waves of very short wavelength in the region from approximately 0.01 to 0.1×10^{-10} m. They are identical with very short wavelength X-rays and have approximately 1% of the ionising power of β particles. γ radiation is the most penetrating of radioactive radiation and can, for example, traverse upto 0.3 m of solid steel. Shielding of personnel against the ionising effects of γ radiation is, therefore, most important. γ sources are typically contained in 0.1 m thick lead walled containers. γ radiation may be involved in both beta and alpha emissions during the process of radioactive decay.

There are four radioactive series (the uranium series, the thorium series, the actinium series, and the neptunium series) involving the decay of large mass number parent radionuclides of atomic number (Z) exceeding 90 and very long half-life (except in the case of the neptunium series) which involve ejection of α or β particles (and associated γ radiation) in a move to a stable state. The first seven members of the uranium series are shown below:

$$^{238}_{92}U \xrightarrow{\alpha}\ ^{234}_{90}UX_1 \xrightarrow{\beta}\ ^{234}_{91}UX_2 \xrightarrow{\beta}\ ^{234}_{92}U \xrightarrow{\alpha}\ ^{230}_{90}Th \xrightarrow{\alpha}\ ^{226}_{88}Ra \xrightarrow{\alpha}\ ^{222}_{86}Rn$$

The connecting arrows with an α or β above represent the particle ejected at this stage. The successive atomic numbers given are characteristic of certain elements; thus, 90 is the atomic number (Z) of thorium and 91 is protactinium. So, UX_1 may be written as a radionuclide of thorium with a mass number (A) of 234, instead of the more abundant form of thorium with a mass number of 232. The first seven radionuclides of the radioactive uranium series then become:

$$^{238}_{92}U \xrightarrow{\alpha}\ ^{234}_{90}Th \xrightarrow{\beta}\ ^{234}_{91}Pa \xrightarrow{\beta}\ ^{234}_{92}U \xrightarrow{\alpha}\ ^{230}_{90}Th \xrightarrow{\alpha}\ ^{226}_{88}Ra \xrightarrow{\alpha}\ ^{222}_{86}Rn$$

A5-4. Half-Life

Radioactive nuclides (the term nuclide is preferable to element owing to the frequent occurrence of various species) disintegrate spontaneously to produce fresh radionuclides. The disintegration is due to the ejection by its nucleus of an alpha or beta particle and often accompanying energy as gamma radiation. The number of atoms of a radionuclide which disintegrate per second are directly proportional to the number of unchanged radioactive atoms remaining.

After a time t the number of atoms of the radionuclide may be represented by the number N. The rate of disintegration (i.e., decay) at the time t is, therefore, dN/dt where:

$$dN/dt = -\lambda N$$

and λ is known as the transformation or decay constant. Rearranging this equation and integrating gives:

$$\log_e N = -\lambda t + k$$

where k is a constant. When $t = 0$, $N = N_0$ and substitution in the above equation shows:

$$\log_e N_0 = k.$$

Radioactive decay, therefore, follows the formula:

$$\log_e N = -\lambda t + \log_e N_0$$

and

$$N = N_0 e^{-\lambda t}$$

Showing that the number of radioactive atoms N decreases exponentially with time. The half-life of a radioactive nuclide is defined as the time taken for half of the number of atoms to disintegrate. If no new radioactive material is produced by the disintegration (i.e., if the daughter atoms are not radioactive), then the half-life is also the time taken for the initial activity of the radionuclide to decrease by half. Substituting in the above equation where N becomes $0.5N_0$ in the half-life period $t = T$, then:

$$N = N_0 e^{-\lambda t}$$

$$N/N_0 = 0.5 = e^{-\lambda t}$$

$$\log_e 0.5 = -\lambda T$$

$$\lambda T = \log_e 2 = 2.303 \log_{10} 2 = 0.693$$

$$T = 0.693/\lambda$$

If the half-life of a radionuclide (T) is 10 days, say, then one can calculate how long it will take to reduce the mass of this radionuclide by 90% of the original atoms present as follows:

$$N/N_0 = e^{-\lambda t} = 0.1$$

where t is required to reduce the mass by 90% (10% or 0.1 remaining) of the original number of atoms present. Then, from the equations above:

$$\lambda = 0.693/T = 0.693/10 \text{ per day} \quad \text{and} \quad e^{-0.0693t} = 0.1$$

where t is the required time in days. So:

$$-0.0693t = \log_e 0.1 \quad \text{and} \quad t = -2.303/0.0693 \log_{10} 0.1 = 2.303/0.0693 = 33.24 \text{ days.}$$

It should be noted that the first reactors in the UK used naturally occurring uranium, which only contains some 0.7% of the fissile $^{235}_{92}U$ species. This is the proportion of this radionuclide that has not further decayed through the radioactive series at this point in geological time. It is quite remarkable that the development of sufficient human intuition to utilise the properties of this radionuclide happened to coincide with it still being available in the earth's crust for use and it not having all decayed away!

The radioactive actinium series for fissionable uranium-235 is shown below. Half-lives are shown in parenthesis. The following symbols are used: s for second, min for minute (1 min = 60 s), h for hour (1 h = 3.6 ks), d for day (1 d = 86.4 ks) and a for a year (1 a $\sim$ 31.6 Ms).

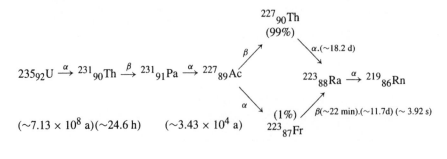

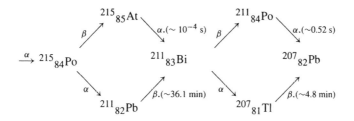

A5-5. Table of Elements

Reproduced from: *Physical Chemistry*, Alberty/Silbey, J. Wiley & Sons (1992) ISBN 0 471 62181 1.

Element	Symbol	Atomic Number (Z)	Atomic Weight	Element	Symbol	Atomic Number (Z)	Atomic Weight
Hydrogen	H	1	1.008	Iodine	I	53	126.904
Helium	He	2	4.002	Xenon	Xe	54	131.29
Lithium	Li	3	6.941	Caesium	Cs	55	132.905
Beryllium	Be	4	9.012	Barium	Ba	56	137.327
Boron	B	5	10.811	Lanthanium	La	57	138.905
Carbon	C	6	12.011	Cerium	Ce	58	140.115
Nitrogen	N	7	14.006	Praseodymium	Pr	59	140.907
Oxygen	O	8	15.999	Neodymium	Nd	60	144.24
Fluorine	F	9	18.998	Promethium	Pm	61	145
Neon	Ne	10	20.179	Samarium	Sm	62	150.36
Sodium	Na	11	22.989	Europium	Eu	63	151.965
Magnesium	Mg	12	24.305	Gadolinium	Gd	64	157.25
Aluminum	Al	13	26.981	Terbium	Tb	65	158.925
Silicon	Si	14	28.085	Dysprosium	Dy	66	162.50
Phosphorus	P	15	30.973	Holmium	Ho	67	164.93
Sulfur	S	16	32.066	Erbium	Er	68	167.26
Chlorine	Cl	17	35.452	Thulium	Tm	69	168.934
Argon	A	18	39.948	Ytterbium	Yb	70	173.04
Potassium	K	19	39.098	Lutetium	Lu	71	174.967
Calcium	Ca	20	40.078	Hafnium	Hf	72	178.49
Scandium	Sc	21	44.955	Tantalum	Ta	73	180.947
Titanium	Ti	22	47.88	Tungsten	W	74	183.85
Vanadium	V	23	50.941	Rhenium	Re	75	186.207
Chromium	Cr	24	51.996	Osmium	Os	76	190.2
Manganese	Mn	25	54.938	Iridium	Ir	77	192.22
Iron	Fe	26	55.847	Platinum	Pt	78	195.08
Cobalt	Co	27	58.933	Gold	Au	79	196.966
Nickel	Ni	28	58.69	Mercury	Hg	80	200.59
Copper	Cu	29	63.546	Thallium	Tl	81	204.383
Zinc	Zn	30	65.39	Lead	Pb	82	207.2
Gallium	Ga	31	69.723	Bismuth	Bi	83	208.980
Germanium	Ge	32	72.61	Polonium	Po	84	210
Arsenic	As	33	74.921	Astatine	At	85	210
Selenium	Se	34	78.96	Radon	Rn	86	222

Continued

Table of Elements (Continued)

Element	Symbol	Atomic Number (Z)	Atomic Weight	Element	Symbol	Atomic Number (Z)	Atomic Weight
Bromine	Br	35	79.904	Francium	Fr	87	223
Krypton	Kr	36	83.80	Radium	Ra	88	226.05
Rubidium	Rb	37	85.467	Actinium	Ac	89	227
Strontium	Sr	38	87.62	Thorium	Th	90	232.038
Yttrium	Y	39	88.905	Protactinium	Pa	91	231
Zirconium	Zr	40	91.224	Uranium	U	92	238.028
Niobium	Nb	41	92.906	Neptunium	Np	93	229-242
Molybdenum	Mo	42	95.94	Plutonium	Pu	94	232-246
Technetium	Tc	43	99	Americium	Am	95	236-248
Ruthenium	Ru	44	101.07	Curium	Cm	96	236-246
Rhodium	Rh	45	102.906	Berkelium	Bk	97	240-250
Palladium	Pd	46	106.42	Californium	Cf	98	242-253
Silver	Ag	47	107.868	Einsteinium	Es	99	244-254
Cadmium	Cd	48	112.411	Fermium	Fm	100	246-255
Indium	In	49	114.82	Mendelevium	Md	101	250-256
Tin	Sn	50	118.710	Nobelium	No	102	
Antimony	Sb	51	121.75	Lawrencium	Lr	103	
Tellurium	Te	52	127.60	Unnilquadium- Unnilseptium	Unq-s	104-7	

A5-6. Reactor Grade Plutonium Decay

Reactor grade plutonium contains a mixture of plutonium isotopes which decay as shown in Table A5-2.

Table A5-2. Plutonium Decay Products

Plutonium isotope	% in Reactor Plutonium	Decay	Half Life (~years)	First Decay Product	Second Decay Product	Decay Chain
Pu-238	~2	alpha	87.74	U-234	Th-230	Uranium series
Pu-239	~55	alpha	24,110	U-235	Th-231	U-235 (Actinium series)
Pu-240	~23	alpha	6,537	U-36	Spontaneous fission	
Pu-241	~15	beta	14.4	Am-241	Np-237	Np-237 (Neptunium series)
Pu-242	~5	alpha	379,000	U-238	Th-234	U-238 (Uranium series)

Pu-241 decays to Am-241, which is a gamma emitter. The ingrowth of americium increases the dose to workers during operations or decommissioning. After some 14.4 years, approximately 7.5% of the total plutonium (reactor grade) will have decayed to americium.

Index

Printed and bound by CPI Group (UK) Ltd, Croydon, CR0 4YY

14/05/2025

01871125-0002